# STRUCTURE AND FUNCTION OF AN ALPINE ECOSYSTEM

LONG-TERM ECOLOGICAL RESEARCH NETWORK SERIES
LTER Publications Committee

Grassland Dynamics: Long-Term Ecological Research
in Tallgrass Prairie
Edited by
Alan K. Knapp, John M. Briggs,
David C. Hartnett, and Scott L. Collins

Standard Soil Methods for Long-Term Ecological Research
Edited by
G. Philip Robertson, David C. Coleman,
Caroline S. Bledsoe, and Phillip Sollins

Structure and Function of an Alpine Ecosystem:
Niwot Ridge, Colorado
Edited by
William D. Bowman and Timothy R. Seastedt

# STRUCTURE AND FUNCTION OF AN ALPINE ECOSYSTEM

## Niwot Ridge, Colorado

*Edited by*

William D. Bowman

Timothy R. Seastedt

2001

# OXFORD
UNIVERSITY PRESS

Oxford   New York
Athens   Auckland   Bangkok   Bogotá   Buenos Aires   Calcutta
Cape Town   Dar es Salaam   Delhi   Florence   Hong Kong   Istanbul
Karachi   Kuala Lumpur   Madras   Madrid   Melbourne   Mexico City
Nairobi   Paris   Singapore   Shanghai   Taipei   Tokyo   Toronto   Warsaw

and associated companies in
Berlin   Ibadan

Copyright © 2001 by Oxford University Press, Inc.

Published by Oxford University Press, Inc.
198 Madison Avenue, New York, New York 10016

Oxford is a registered trademark of Oxford University Press

All rights reserved. No part of this publication may be reproduced,
stored in a retrieval system, or transmitted, in any form or by any means,
electronic, mechanical, photocopying, recording, or otherwise,
without the prior permission of Oxford University Press.

Library of Congress Cataloging-in-Publication Data
Structure and function of an alpine ecosystem : Niwot Ridge, Colorado / edited by
   William D. Bowman, Timothy R. Seastedt.
      p.   cm.—(Long-Term Ecological Research Network series)
   Includes bibliographical references.
   ISBN 0-19-511728-X
   1. Mountain ecology—Colorado—Niwot Ridge.
   I. Bowman, William D.   II. Seastedt,
Timothy R.   III. Series.
QH105.C6 S76   2000
577.5'38'0978863—dc21         00-036746
0-19-511728-X

9 8 7 6 5 4 3 2 1

Printed in the United States of America
on acid free paper

# Acknowledgments

We are grateful to our coauthors for their hard work, patience, and willingness to participate in this effort. Several people provided substantial help by reviewing all or portions of the book manuscript. Diane McKnight and Russ Monson of the University of Colorado provided very helpful comments on an earlier draft of the entire book, and John Blair and Alan Knapp of Kansas State University provided help and encouragement at a time when it was badly needed. Fred Swanson, USDA Forest Service, contributed useful criticisms and comments on the chapters covering the physical environment. Susan Sherrod and Mariah Carbone put in significant effort to edit and collate the text, tables, and figures. Bill Ervin kindly provided several photographs. The research presented in this volume could not have been accomplished without the support of numerous individuals associated with the Mountain Research Station and the Niwot Ridge Long-Term Ecological Research Program, including Jim Halfpenny, Mark Noble, David Inouye, David Yamaguchi, Chuck Aid, Steve Seibold, and Nancy Auerbach. Special thanks to Rick Ingersoll, Mike Hartman, Tim Bardsley, and Chris Seibold for the critical roles they have played over the years in facilitating data collection. Financial support was provided by several sources, most notably the National Science Foundation Long-Term Ecological Research Program, several NSF individual investigator awards, the Andrew W. Mellon Foundation, and the EPA Exploratory Grants Program.

# Contents

Contributors  ix

Foreword  xi

1  Introduction: Historical Perspective and Significance of Alpine Ecosystem Studies  3
   William D. Bowman

## I Physical Environment

2  Climate  15
   David Greenland and Mark Losleben

3  Atmospheric Chemistry and Deposition  32
   Herman Sievering

4  Geomorphic Systems of Green Lakes Valley  45
   Nel Caine

5  Hydrology and Hydrochemistry  75
   Mark W. Williams and Nel Caine

## II Ecosystem Structure

6 The Vegetation: Hierarchical Species-Environment Relationships 99
Marilyn D. Walker, Donald A. Walker, Theresa A. Theodose, and Patrick J. Webber

7 Vertebrates 128
David M. Armstrong, James C. Halfpenny, and Charles H. Southwick

8 Soils 157
Timothy R. Seastedt

## III Ecosystem Function

9 Primary Production 177
William D. Bowman and Melany C. Fisk

10 Plant Nutrient Relations 198
Russell K. Monson, Renée Mullen, and William D. Bowman

11 Controls on Decomposition Processes in Alpine Tundra 222
Timothy R. Seastedt, Marilyn D. Walker, and David M. Bryant

12 Nitrogen Cycling 237
Melany C. Fisk, Paul D. Brooks, and Steven K. Schmidt

13 Soil-Atmosphere Gas Exchange 254
Steven K. Schmidt, Ann E. West, Paul D. Brooks, Lesley K. Smith, Charles H. Jaeger, Melany C. Fisk, and Elisabeth A. Holland

14 Plant-Herbivore Interactions 266
Denise Dearing

## IV Past and Future

15 Paleoecology and Late Quaternary Environments of the Colorado Rockies 285
Scott A. Elias

16 Environmental Change and Future Directions in Alpine Research 304
Jeffrey M. Welker, William D. Bowman, and Timothy R. Seastedt

Index 323

# Contributors

David M. Armstrong
Department of Environmental, Population,
  and Organismic Biology
University of Colorado
Boulder, CO 80309-0334

William D. Bowman
Department of Environmental, Population,
  and Organismic Biology
Mountain Research Station and Institute of
  Arctic and Alpine Research
University of Colorado
Boulder, CO 80309-0334

Paul D. Brooks
Institute of Arctic and Alpine Research
University of Colorado
Boulder, CO 80309-0450

David M. Bryant
Department of Natural Resources
University of New Hampshire
Durham, NH 03824

Nel Caine
Department of Geography and Institute of
  Arctic and Alpine Research
University of Colorado
Boulder, CO 80309-0450

Denise Dearing
Department of Biology
University of Utah
Salt Lake City, UT 84112

Scott A. Elias
Institute of Arctic and Alpine Research
University of Colorado
Boulder, CO 80309-0450

Melany C. Fisk
Department of Natural Resources
Cornell University
Ithaca, NY 14853

David Greenland
Department of Geography
University of North Carolina
Chapel Hill, NC 27599-3220

James C. Halfpenny
A Naturalist's World
Box 989
Gardiner, MT 59030

Elisabeth A. Holland
National Center for Atmospheric Research
Division of Atmospheric Chemistry
P.O. Box 3000
Boulder, CO 80307

x Contributors

Charles H. Jaeger
Institute of Ecology
University of Georgia
Athens, GA 30602-2202

Mark V. Losleben
Mountain Research Station
Institute of Arctic and Alpine Research
University of Colorado
Boulder, CO 80309-0450

Russell K. Monson
Department of Environmental, Population,
  and Organismic Biology
University of Colorado
Boulder, CO 80309-0334

Renée Mullen
The Nature Conservancy
Conservation Planning Programs
2404 Bank Dr., Ste. 314
Boise, ID 83705

Steven K. Schmidt
Department of Environmental, Population,
  and Organismic Biology
University of Colorado
Boulder, CO 80309-0334

Timothy R. Seastedt
Department of Environmental, Population,
  and Organismic Biology
Institute of Arctic and Alpine Research
University of Colorado
Boulder, CO 80309-0450

Herman Sievering
Department of Environmental Science and
  Geography
Global Change and Environmental Quality
  Program
University of Colorado
Denver, CO 80217

Lesley K. Smith
Cooperative Institute for Research in
  Environmental Sciences
University of Colorado
Boulder, CO 80309-0216

Charles H. Southwick
Department of Environmental, Population,
  and Organismic Biology
University of Colorado
Boulder, CO 80309-0334

Theresa A. Theodose
Department of Biology
University of Southern Maine
Portland, ME 04103

Donald A. Walker
Institute of Arctic Biology
University of Alaska
Fairbanks, Alaska 99775-7000

Marilyn D. Walker
Institute of Northern Forestry Cooperative
  Research Unit
School of Agriculture and Land Resources
  Management
University of Alaska
Fairbanks, AK 99775-6780

Patrick J. Webber
Department of Botany and Plant Pathology
Michigan State University
East Lansing, MI 48824

Jeffrey M. Welker
Department of Renewable Resources
University of Wyoming
Laramie, WY 82071

Ann E. West
Robert S. Kerr Environmental Research
  Center
U.S.E.P.A.
P.O. Box 1198
Ada, OK 74820

Mark W. Williams
Department of Geography
Institute of Arctic and Alpine Research
University of Colorado
Boulder, CO 80309-0450

# Foreword

My association with John Marr and the Mountain Research Station began 50 years ago. John and I met at one of the first AAAS national meetings immediately following World War II. We became close friends, with common interests in high mountains and in the Arctic. During the following years, we worked closely together in the field, in teaching, in research, and in writing. During the Korean War, we both worked under Peveril Meigs for the U.S. Army Quartermaster Research and Development Command on mountain and desert environmental problems applicable to central Asia. We did much of this by using analogs from western North America.

It was during this time (1951) that John established the Institute of Arctic and Alpine Research at the University of Colorado in Boulder. The Institute soon became the keystone of terrestrial research and teaching in polar and alpine science, a leadership position that it has enhanced and maintained through the years. At the heart of the Institute has been the Mountain Research Station, providing the interdisciplinary access and logistics to Niwot Ridge for countless mountain researchers in ecology, geology, climatology, and physiology from around the world through all 75 years of the Station's existence.

Among John Marr's early graduate students that worked with him at the Mountain Research Station and on the slopes and summit of Niwot was a remarkable group of excellent and hardy high mountain ecologists: Bill Osburn, Al Johnson, Bill Rickard, Les Viereck, Bettie Willard, and Dexter Hess. It was here that I first met Hal Mooney, who was still an undergraduate at the University of California-Santa Barbara, Martyn Caldwell, and Mary Vetter Sauchyn. The latter three, alpine ecologists all, came to Duke University and worked on their doctorates with me, using Niwot Ridge as their main research site. Following in the Niwot alpine and arctic tradition, there has been a steady stream of outstanding doctoral recipients who were supervised by John Marr, Pat Webber, Mike Grant, Bill Bowman, and others working in association with INSTAAR and the Mountain Research Station. These were Phil Miller, Marilyn Walker, Skip Walker, Vera Komárková, Charles Olmsted, Ann Odasz, David Buckner, and David Cooper, to name a few.

During the late summer of 1952, a group of us (John Marr, Al Johnson, Bill Osburn, Bill Rickard, Dwight Billings, Hal Mooney, and Mark Paddock) were trying to finish the weather station at D1 on the crest of Niwot Ridge at an elevation of 12,400 feet (3,750 m). With our knapsacks of lunch and instruments, we hiked up the Green Lakes Valley to the south slopes of Niwot, and thence to the top. John and I lagged somewhat behind the others who were ascending the south slope with most of the instruments, including a multipronged lightning rod that Al Johnson was carrying in his pack sack. Unfortunately, the lightning, thunder, wind, and rain were a little impatient and couldn't wait until we finished the job. The storm caught us on the slopes above the lakes. John and I crawled under some overhanging rocks (not the ideal or safest place). The others were in an even more dangerous place, huddled around a vertical rock spur about 300 meters farther up the slope—remember that Al had the lightning rod. Well, we all were lucky. After a half hour of this, the storm moved off to the east, the sun came out, and we made it to the top. After lunch, we began completing the weather station at D1, and if you will look at the photograph (Fig. 1), there are Bill Osburn, John Marr, and the lightning rods on the anemometer part of observation D1 just as we finished. Thinking back, that 1952 group of American alpine ecologists might have been reduced in number by being in the wrong place in a long-forgotten storm—and no weather data yet recorded!

The careful and intensive study of Niwot Ridge is a valuable contribution to our understanding of the complex ecology of such a mountain region. It will help immeasurably in planning for the region's ecological and economic integrity in the next decades. Such systems are extremely important to the economy and water resources of the Rocky Mountains with their surrounding agricultural and urban valleys. These valleys are occupied by human populations who are more dependent on the mountain ecosystems than they realize. Only the Swiss and Austrian Alps exist in a comparable ecologically integrated and well-studied mountain-valley set of mesocosms that can serve as models for the future of the southern Rocky Mountains.

Dwight Billings<br>
Department of Botany<br>
Duke University<br>
Durham, North Carolina<br>
June 1996

*Editor's Note:* At the onset of starting work on this book, we contacted Dwight Billings, the foremost alpine ecologist of the twentieth century, to see if he would be amenable to writing a Foreword. Despite his very busy schedule, he agreed, and responded rather quickly, providing us with the preceding text. Sadly, Dwight died a short time after writing this. We dedicate this book to Dwight and to John W. Marr (Fig. 2), whose leadership and excellent work paved the way for the research presented in this volume. We also dedicate this book to our sons, Gordon, Liam, and Tate, who someday will be pioneers in their own fields.

Figure 1. Bill Osburn and John Marr reading the newly installed anemometer with lightning rod on Niwot Ridge (D1), August 1952 (photo by Dwight Billings).

Figure 2. John W. Marr at the Mountain Research Station, June 1957 (photo by Dwight Billings).

# STRUCTURE AND FUNCTION OF
AN ALPINE ECOSYSTEM

# 1

# Introduction
## Historical Perspective and Significance of Alpine Ecosystem Studies

William D. Bowman

Alpine tundra is an intriguing ecosystem—for its beauty as well as for the harsh climate in which it exists. Contrasted against jagged rock precipices and snow and ice and subjected to rapid changes in weather, the tundra, with its proliferation of diminutive flowers, appears deceptively fragile. John Muir, in detailing the alpine of the Sierra Nevada, was at a loss to adequately describe "the exquisite beauty of these mountain carpets as they lie smoothly outspread in the savage wilderness" (Muir 1894). Despite this aesthetic fascination for the alpine, it is one of the least studied ecosystems in the world.

Significant effort has been expended to describe the physiological ecology of alpine organisms (e.g., Bliss 1985; Carey 1993; Körner 1999) and community patterns (Komárková 1979; Billings 1988), but there have been no syntheses detailing alpine ecosystem processes and patterns to the degree that they have been described in the arctic (e.g., Chapin 1992) and forest (e.g., Likens and Bormann 1995) ecosystems. The goal of this book is to provide a description of the Niwot Ridge/Green Lakes Valley alpine ecosystem of the Front Range in the Colorado Rocky Mountains, including the spatial and temporal patterns of animals, plants, and microorganisms and the associated ecosystem processes. The book focuses on the strengths of the research carried out on Niwot Ridge during the past four decades, particularly physical factors influencing alpine ecology (climate and geomorphology), patterns and functions of the vegetation, and N biogeochemistry. While the book focuses on a particular site, the results can be extrapolated to much of the southern and central Rocky Mountains, and thus it pertains to a broader geographic and scientific scope and will be of direct interest to ecologists in general as well as to those interested in ecosystems in extreme environments.

4  Introduction

There are numerous justifications for a synthesis of alpine ecosystem studies. While alpine tundra occupies only about 3% of the global land surface (Körner 1995) and thus has little impact on atmosphere-biosphere exchange, its presence at the extreme climatic tolerance for many organisms and its presence on every continent make it a good "indicator" system for regional environmental change. Most alpine areas have been directly impacted very little by human activities and thus are good systems to study the indirect anthropogenic effects of air pollution and climate. Perturbations in climate, atmospheric chemistry, and increased transmission of solar UVB radiation through the stratosphere may have measurable impact on the alpine prior to other terrestrial ecosystems. In many regions, particularly the western United States, much of the water supply for streams and rivers is derived from snowmelt in the alpine. The quality of the water, as defined by the chemistry and aquatic life, is influenced by ecosystem processes in alpine tundra (chapter 16). Last, in terms of basic research, the alpine is an excellent system to investigate the interaction between the physical and biological environments. Spatial heterogeneity in microclimate and biota is relatively great as a result of high topographic relief (chapter 4) and is manifested at a landscape level by differential spatial accumulation of snow during the winter (Fig. 1.1; Walker et al. 1993; chapter 5). The variation in snow cover determines a biotic gradient of microbial, plant, and animal community composition within the same macroclimate and soil parent material (chapter 6). Within the same microclimatic/community zone, different plant species may dominate adjacent areas, facilitating studies of biotic impacts on ecosystem function (chapters 10, 16). Some plant species may be cosmopolitan in their distribution across the alpine landscape, facilitating the study of the interactions between microclimate and species on ecosystem processes.

## Characteristics of Alpine Tundra

The alpine is defined as the biotic zone on mountains above the natural limit of tree establishment dominated by herbaceous plants (Marr 1961; Billings 1974; see Billings 1988 and Körner 1995, 1999 for description of latitudinal and altitudinal extent of the alpine). The extent of the alpine is correlated with climatic factors (Fig. 1.2a), and although climate variation is no greater than other ecosystems (Fig. 1.2b), the variation incorporates extremes near the adaptational limits of many vascular plant taxa. The lower elevational limit of the alpine may be related to several climatic factors but corresponds relatively well with a 10°C isotherm for the warmest summer month (Wardle 1974). In the southern Rocky Mountains, the limit of tree line, marked by the presence of tree islands, appears to be constrained by an interaction of strong winds, growing season length, and low growing season temperatures (Wardle 1974; Hansen-Bristow 1981; Arno 1984). Unlike other areas (e.g., the British Isles, European Alps, Himalaya), treeline in the southern Rockies is largely unaffected by logging or pastoral activities. The definition of the upper limit of alpine tundra is not well established in the literature, and vascular plants have been found above 6000 m in the Himalaya, although closed canopies occur somewhere

Figure 1.1. Differential snow cover in the alpine, shown at the landscape level, associated with high predominantly westerly winds and topography. The differential snow cover is associated with landscape variation in microclimate (chapter 2), plant community distribution (chapter 6), soil moisture and organic matter (chapter 8), primary production (chapter 9), and nitrogen cycling (chapter 12) (Holy Cross Wilderness, Sawatch Range, central Colorado; photo by William D. Bowman).

around 5000 m (Miehe 1991, as cited in Körner 1999). The upper elevational zone is not reached in the southern Rockies, as the highest peaks (4300+ m) often have substantial cover of vascular plants growing on their summits.

The alpine and arctic are often not differentiated in discussions of tundra biology. They share similar vegetational physiognomies and floristic associations and have low growing season temperatures (Billings 1973; Löve and Löve 1974). Rates of primary production and nutrient cycling are also low in both arctic and alpine tundras (Webber 1978; Marion and Miller 1982; Giblin et al. 1991; chapters 9, 12). However, there are also substantial differences that influence ecosystem structure and function. Relative to the arctic, the alpine of mid-latitudes experiences higher potential solar irradiation (including UVB), shorter day length, lower partial pressures of $CO_2$ and $O_2$, greater vapor pressure gradients, less permafrost and thus often drier soils, and greater wind velocities (Billings 1973), imposing greater environmental stress on alpine biota. Additionally, the alpine has greater spatial heterogeneity than most other ecosystems, resulting from high microtopograpic relief and its interaction with climate, and variation in substrate ages, depths, and textures associated with distribution of glacial ice and periglacial activity (chapter 4).

6    Introduction

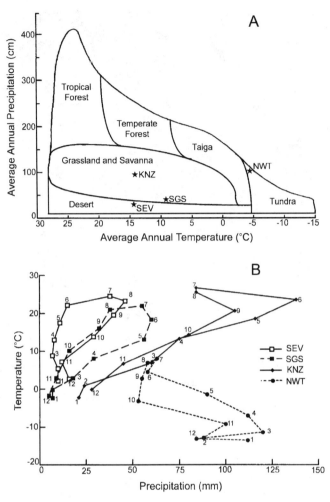

Figure 1.2. Climate diagrams indicating the position of the Niwot Ridge alpine tundra site relative to other regional LTER sites that constitute both a latitudinal and an altitudinal gradient, using (a) annual means in temperature and precipitation and (b) seasonal variation in temperature and precipitation. NWT = Niwot Ridge, KNZ = Konza Prairie, SGS = Short Grass Steppe, and SEV = Sevilleta. Numbers adjacent to symbols in b are months of the year.

Integrating this heterogeneity in ecological studies is one of the challenges (and opportunities) in alpine research.

## Research on the Niwot Ridge Ecosystem

The focus of this book is on one of the best known alpine ecosystems of the world, the Niwot Ridge/Green Lakes Valley region (Fig. 1.3). Niwot Ridge is named for a chief of the Arapaho Indians who lived in the valleys of the Front Range and adjacent high plains during the time of American settlement in the mid-nineteenth century. Near the midpoint of the north-south trending Front Range, Niwot Ridge is part of the southern Rocky Mountain physiographic region (Fig. 1.4; Benedict 1991). The ridge is approximately 8 km in length, running in an east-west orientation, with its highest elevation at 4100 m on Navajo Peak on the Continental Divide (Fig. 1.5). Based on the composition of the landforms (chapter 4), vegetation (chapter 6), and vertebrates (chapter 7) Niwot Ridge is representative of the alpine for much of the central and southern Rocky Mountains (Billings 1988). However, regional variation in biotic composition and ecosystem processes may exist as a function of parent material of the soils, the extent and timing of glaciation of the site, and climatic variation (e.g., amount of wind).

Niwot Ridge has been the focus of extensive alpine research for over four decades, due in large part to the establishment nearby of the University of Colorado's Mountain Research Station (MRS; http://www.colorado.edu/mrs/). Originally known as "Science Lodge," the MRS has provided logistical and scientific support for alpine research since its founding in 1921. While there were no formal publications on alpine ecology prior to 1950, a significant amount of natural history was recorded in conjunction with summer field courses held at the MRS. The start of a nearly continuous long-term ecological research program came in 1950 when John Marr established the Front Range Ecology Project, installing several environmental monitoring stations to quantify climatic and vegetational gradients associated with elevational change (see chapter 2 for details of climate program). Marr's *Ecosystems of the East Slope of the Front Range, Colorado Rocky Mountains* (1961) was a landmark in mountain ecology, providing a framework for future efforts. The importance of Niwot Ridge as a site for alpine research was recognized by its inclusion in the International Biological Programme (1971–1974) as the only U.S. alpine tundra participant. Niwot Ridge was designated a UNESCO Biosphere Reserve in 1979.

In the fall of 1980, Patrick Webber with a team of 25 scientists from five departments at the University of Colorado received support from the National Science Foundation to establish the Long-Term Ecological Research program (LTER) on Niwot Ridge. The initial proposal included 16 projects studying five core areas that were selected for LTER research. These core areas included nutrient cycling, primary productivity, population dynamics of key species, organic matter dynamics, and patterns and frequency of disturbance (Callahan 1984). The pioneering work of Marr (1961), Niwot Ridge's participation in the International Biological Programme (e.g., Caldwell et al. 1974), collaboration in a major Department of Energy (DOE) snow modification study (Steinhoff and Ives 1976), the inclusion of Niwot Ridge in the National Oceanic and Atmospheric Administration's (NOAA) long-term air-quality monitoring, and the selection of Niwot Ridge in 1977 as one of 67 national Experimental Ecological Reserves predisposed Niwot Ridge as an ideal site for the LTER program.

The Niwot Ridge LTER program has evolved during its two decades of existence. The science focus has changed, and scientists have entered and left the program. However, the strength of a viable LTER program is that the data, the science, and

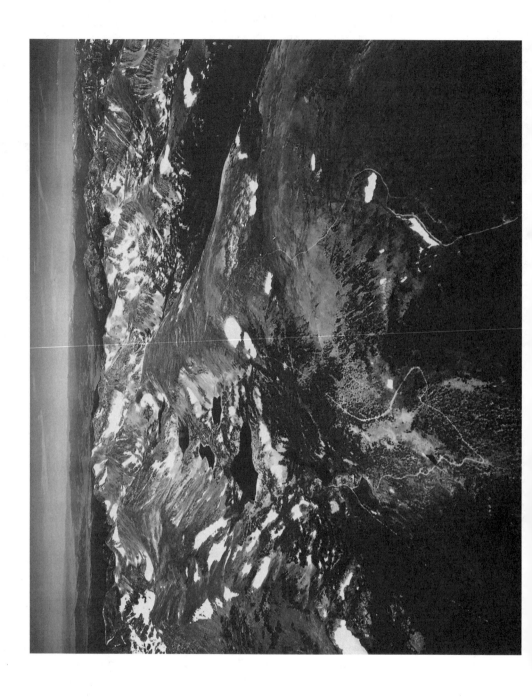

Figure 1.3. Niwot Ridge, Colorado. (left) Aerial view, looking west to the Continental Divide and Middle Park beyond (photo courtesy of Don Cline). (above) Niwot Ridge, looking east from Kiowa Peak. The lake in the background is Lefthand Reservoir, also named for Chief Niwot of the Arapaho Indians (photo by William D. Bowman).

new science opportunities with well-documented data should transcend the scientists. From this perspective, the Niwot LTER program has been remarkably successful. An important aspect of program evolution was the use of the emerging Internet as a focal point for LTER project description and integration (Ingersoll et al. 1997). Electronic communication and information transfer not only increased information flow among scientists both within and among research sites but also created a means of information transfer to the public that previously had not existed. The exposure of science on the World Wide Web has allowed Niwot LTER scientists to promote the significance and relevance of their research to a much broader audience.

The contribution of research on Niwot Ridge to ecology and biogeochemistry continues to expand. The strong climatic and landscape constraints on the biology of the system have been traditionally emphasized in describing alpine tundra structure and functioning, and this is a major theme of this book. Large changes in site soil and air microclimate exist across short topographic gradients, exerting a strong influence on the composition and functioning of alpine organisms. However, interest in biotic responses and feedbacks to changes in atmospheric chemical inputs and climate change has stimulated a number of studies that emphasize that the biotic system responds to these changes in complex ways that alter input-output relationships. Such studies are described in this volume, but many are just beginning and will be

10  Introduction

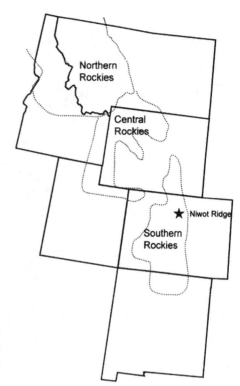

Figure 1.4. Physiographic zones of the Rocky Mountains and the location of the Niwot Ridge research site (modified from Benedict 1991).

the subject of future synthesis efforts. Finally, the current scientific interest in the causes and consequences of biotic and functional diversity is also a major ongoing research effort of Niwot scientists. Those interested in these topics should use the Niwot Ridge LTER homepage to update the science discussions presented here (http://culter.colorado.edu/).

## Organization of the Book

This book is divided into four main sections. The first section provides an overview of the physical factors that are of significant ecological interest, including climate, hydrology, geomorphology, and atmospheric chemistry. To a large degree, the former three physical factors define the alpine as much as the biota. The second section describes the composition, dynamics, and spatial arrangement of biota and soils of Niwot Ridge. The third section focuses on the functioning of alpine tundra, concentrating particularly on energy fluxes and nitrogen cycling. The final section addresses questions of the past and future of tundra biota. How has climate change in the past influenced the present biota of Niwot Ridge, and how might anthropogenically induced environmental change in the future influence the structure and function of alpine tundra?

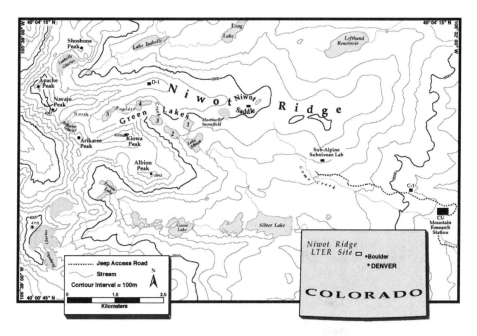

Figure 1.5. Niwot Ridge, Colorado, including the Mountain Research Station, Niwot Saddle research site, climate stations (chapter 2), and the Green Lakes Valley (courtesy of Thomas Davinroy).

References

Arno, S. F. 1984. *Timberline: Mountain and arctic forest frontiers.* Seattle: The Mountaineers.
Benedict, A. D. 1991. *The southern Rockies.* San Francisco: Sierra Club Books.
Billings, W. D. 1973. Arctic and alpine vegetations: Similarities, differences, and susceptibilities to disturbance. *BioScience* 23:697–704.
———. 1974. Arctic and alpine vegetation: Plant adaptations to cold summer climates. In *Arctic and alpine environments,* edited by J. D. Ives and R. G. Barry. London: Methuen.
———. 1988. Alpine vegetation. In *North American terrestrial vegetation,* edited by M. G. Barbour and W. D. Billings. Cambridge: Cambridge University Press.
Bliss, L. C. 1985. Alpine. In *Physiological ecology of North American plant communities,* edited by B. F. Chabot and H. A. Mooney. New York: Chapman and Hall.
Caldwell, M. M., L. L. Tieszen, and M. Fareed. 1974. The canopy structure of tundra plant communities at Barrow, Alaska and Niwot Ridge, Colorado. *Arctic and Alpine Research* 6:151–159.
Callahan, J. T. 1984. Long-term ecological research. *BioScience* 34:363–367.
Carey, C. 1993. *Life in the cold: Ecological, physiological, and molecular mechanisms.* Boulder: Westview Press.
Chapin, F. S., III. 1992. *Arctic ecosystems in a changing climate: An ecophysiological perspective.* New York: Academic Press.
Giblin, A. E., K. J. Nadelhoffer, G. R. Shaver, J. A. Laundre, and A. J. McKerrow. 1991. Biogeochemical diversity along a riverside toposequence in arctic Alaska. *Ecological Monographs* 61:415–435.

Hansen-Bristow, K. J. 1981. Environmental controls influencing the altitude and form of the forest-alpine tundra ecotone, Colorado Front Range. Ph.D. diss. University of Colorado, Boulder.

Ingersoll, R. C., T. R. Seastedt, and M. A. Hartman. 1997. A model information management system for ecological research. *BioScience* 47:310–316.

Komárková, V. 1979. *Alpine vegetation of the Indian Peaks area, Front Range, Colorado Rocky Mountains.* vol. 7, *Flora et Vegetatio Mundi.* Vaduz, Liechtenstein: J Cramer.

Körner, Ch. 1995. Alpine plant diversity: A global survey and functional interpretations. In *Arctic and alpine biodiversity,* edited by F. S. Chapin III and Ch. Körner. Berlin: Springer-Verlag.

———. 1999. *Alpine plant life, functional ecology of high mountain ecosystems.* Berlin: Springer.

Likens, G. E., and F. H. Bormann. 1995. *Biogeochemistry of a forested ecosystem.* 2nd ed. New York: Springer-Verlag.

Löve, A., and D. Löve. 1974. Origin and evolution of the arctic and alpine floras. In *Arctic and alpine environments,* edited by J. D. Ives and R. G. Barry. London: Methuen.

Marion, G. M., and P. C. Miller. 1982. Nitrogen mineralization in a tussock tundra soil. *Arctic and Alpine Research* 14:287–293.

Marr, J. W. 1961. Ecosystems of the East Slope of the Front Range in Colorado. *University of Colorado Studies, Series in Biology* 8:1–134.

Miehe, G. 1991. Der Himalaya, eine multizone Gebirgsregion. In *Ökologie der Erde,* vol. 4, *Spezielle Ökologie der gemäßigten und arktischen Zonen außerhalb Euro-Nordàsiens,* edited by H. Walter and S. W. Breckle. Stuttgart: Fischer.

Muir, J. 1894. *The mountains of California.* New York: Century.

Steinhoff, H. W., and J. D. Ives, eds. 1976. Ecological impacts of snowpack augmentation in the San Juan Mountains of Colorado. Final report to the Division of Atmospheric Water Resources Management, Bureau of Reclamation, Denver, Colorado.

Walker, D. A., J. C. Halfpenny, M. D. Walker, and C. A. Wessman. 1993. Long-term studies of snow-vegetation interactions. *BioScience* 43:287–301.

Wardle, P. 1974. Alpine timberlines. In *Arctic and alpine environments,* edited by J. D. Ives and R. G. Barry. London: Methuen.

Webber, P. J. 1978. Spatial and temporal variation in the vegetation and its productivity, Barrow, Alaska. In *Vegetation and production ecology of an Alaskan arctic tundra,* edited by L. L. Tieszen. New York: Springer-Verlag.

# Part I

# Physical Environment

# 2

# Climate

David Greenland
Mark Losleben

## Introduction

Climate is one of the most important determinants of biotic structure and function in the alpine. High winds and low temperatures are defining elements of this ecosystem, requiring adaptations of the alpine biota. Interaction between topography and snowcover strongly influences spatial heterogeneity of microclimate, which in turn influences and is influenced by the distribution of vegetation. For nearly 50 years investigators have used Niwot Ridge to examine and document the climate and its interaction with the biota of the alpine tundra. This chapter reviews some of the many findings of these ongoing bioclimatic investigations.

Climate studies started on Niwot Ridge in October 1952 when Professor John W. Marr and his students set up a transect of climate stations across the Front Range between Boulder and the Continental Divide (Marr 1961). There were originally 16 stations in groups of four representing different slope exposures in what he defined as the Lower and Upper Montane Forest, the Subalpine Forest, and the Alpine Tundra ecosystems of the Front Range. After 1 year, the network was reduced to four stations, called A1, B1, C1, and D1, which all had ridge-top locations and ranged from lower montane (A1) to high alpine (D1) (Fig. 1.5). From time to time, these stations were supplemented by other stations that supported particular studies. This was especially true during the International Biological Programme years in the early 1970s when focus on work on the Saddle research site of the Ridge began (Fig. 2.1). Following the establishment of Niwot Ridge and Green Lakes Valley as a Long-Term Ecological Research (LTER) site in 1980, even more intensive climatological work has been conducted. The construction of the Tundra Laboratory in August 1990 facilitated intensive winter climatological studies. Geographical lo-

**16** Physical Environment

Figure 2.1. The Saddle research site in June, exhibiting differential snow distribution across the landscape, ranging from very low snow cover in the Fellfield in the foreground to >5 m in the deepest areas near the center of the snowfield. Left to right: Kiowa, Navajo, Apache, and Shoshoni Peaks are in the background (photo by William D. Bowman).

cations and elevational data on most of the stations has been provided by Greenland (1989) and is also found in the LTER electronic database (http://culter.colorado.edu/).

The climate of Niwot Ridge is characterized by large seasonal and annual variability with very windy and cold winters, wet springs, mild summers, and cool, dry autumns. The low thermal lag between 600 and 700 mb, reflecting its high elevation, results in a more rapid change of the Niwot Ridge climate to regional- and possibly global-scale climate change relative to lower elevations. High elevations are likely influenced by different weather patterns than those of lower elevations. For example, even though the locations of the Niwot Ridge and Shortgrass Steppe LTER sites in Colorado are spatially close, an upper-level air mass can override a lower level air mass with minimal mixing. These sites therefore do not show similar long-term trends in annual precipitation or temperature. The proximity of the Denver metropolitan area immediately to the east of Niwot Ridge results in occasional anthropogenic emission influence when upslope easterly wind conditions occur, primarily during non-winter months. The occurrence of easterly air flow is more frequent in summer with spring and fall being transitional months (chapter 3). These upslope events are believed to have important influences on atmospheric deposition of nitrogen (chapter 3) but do not dominate annual patterns in temperature or precipitation. Thus, for the most part, the climate of Niwot Ridge is a well-mixed, free tropospheric, alpine environment with occasional anthropogenic influences.

Niwot Ridge and the Green Lakes Valley have numerous microclimates enmeshed in the macroclimatic regime. These microclimate patterns are generated by wind and precipitation interactions with variation in topography to produce substantial variation in local energy budgets and precipitation deposition. This translates to large differences in the temperature and moisture regimes observed over the alpine landscape. The diversity of microclimates should be regarded as a major feature of the overall climate of Niwot Ridge and is typical of many alpine landscapes.

In this chapter, we first describe the results of the more-standard climatological measurements at the site within the framework of average (synoptic) and mesoscale climatic systems. We relate the standard values to the major biotic communities and ecotones of the Front Range. We then outline what is currently known about temporal trends and variability of the Niwot Ridge climate. The second part of the chapter examines the microclimates of the site, which are affected by the larger scale predominantly cold, dry, westerly wind sweeping across the tundra surface and the heterogeneity of the surface itself. The microclimates are best summarized in terms of the energy and water budgets of the alpine surface. The interaction between the different scale climates of the Ridge and their relationships to the vegetation are a central aspect of the environment. Finally, we summarize our findings and identify some directions for future research.

## Macro- and Mesoscale Climate

### Macroclimate

The climatology of the Niwot Ridge/Green Lakes Valley is strongly determined by its mid-latitude, continental location and by the elevational and topographic situation. The high elevation results in relatively low air temperatures at all times of the year (Fig. 2.2). Air temperatures are further depressed by high wind velocities passing over snow and ice (Greenland 1977). The mid-continental location leads to a large seasonal temperature range, which is more marked than at lower elevations. Storms caught up in the upper westerly airflow in winter and spring bring most of the precipitation. Snow is also derived from more easterly storms at the lower elevations by cyclonic upslope flow developing to the east of the Continental Divide. These cyclonic storms often re-form after the main flow of the air has passed over the southern Rocky Mountains. These upslope storms are responsible for the spring maximum of precipitation (Fig. 2.3). Rainfall in the summer is associated with localized convectional storms that frequently reach their maximum intensity in midafternoon. Fall is the driest season.

Global-scale surface pressure patterns show Niwot Ridge is under relatively high pressure during the winter and relatively low pressure in the summer (Strahler and Strahler 1992). However, the pressure difference is not great either spatially or between the seasons. A more-dominant global-scale influence is the mid-latitude lo-

## 18  Physical Environment

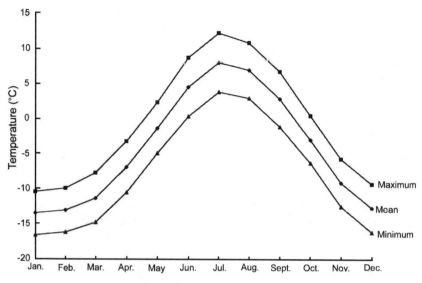

**Figure 2.2.** Average monthly mean, maximum, and minimum temperature at the D1 climate station (3750 m) on Niwot Ridge, 1961–1990.

cation of the westerly winds of the troposphere that are a persistent feature at the site and strongly influence the weather at Niwot Ridge. At a higher level of detail, general weather types with high pressure over the western United States occur primarily in the fall and winter. A synoptic type with a central high about 40° to 45° N off the West Coast and a low-pressure area to the south is the most frequent summer pattern (Barry et al. 1977). Barry (1973) showed that 700-mb height pressure patterns were effective in separating different temperature and precipitation conditions on the Ridge, particularly in winter. Barry also distinguished several frequently occurring general circulation types and the temperature and precipitation departures from average associated with them. For example, he found that warmer-than-average days in January were associated with anticyclonic westerly and northwesterly flow types. Increasing frequency and intensity of precipitation at high altitudes tend to occur under northwesterly and southwesterly circulation. Despite the importance of these large-scale controls, some extreme climatic events at the Ridge, such as the high precipitation of the 1921 year, are not closely related to the larger geographic scale events (Greenland 1995).

These large-scale patterns give rise to the annual average temperature at the uppermost climatic station (D1), located at 3750 m, of −4.1°C and the range from −13.5°C in January to 8.0°C in July (Table 2.1; Fig. 2.2). The mean annual precipitation at the D1 climate station is 993 mm, with seasonal maxima in winter and spring. Monthly extremes of both temperature and precipitation display a considerable range (Figs. 2.2 and 2.3). The climate of D1 result can be classified as "polar tundra due to high altitude" (ET(H)) under the Köppen system, AD′rc′$_1$ under the 1948 Thornthwaite system, which despite the absence of trees at the site is known

Table 2.1. Summary Climatic Statistics for D1 climate station on Niwot Ridge (3750 m) for the Period 1961–1990

|  | Temperature (°C) | | | Precipitation (mm) | | |
|---|---|---|---|---|---|---|
|  | Mean | Max | Min | Mean | Wettest Year (1983) | Driest Year (1962) |
| Jan | −13.5 | −10.5 | −16.6 | 112 | 108 | 79 |
| Feb | −13.1 | −10.0 | −16.2 | 84 | 82 | 61 |
| Mar | −11.4 | −7.8 | −14.8 | 120 | 229 | 38 |
| Apr | −7.0 | −3.3 | −10.6 | 112 | 152 | 61 |
| May | −1.4 | 2.3 | −5.0 | 90 | 220 | 58 |
| Jun | 4.5 | 8.7 | 0.3 | 58 | 73 | 41 |
| Jul | 8.0 | 12.2 | 3.8 | 63 | 106 | 41 |
| Aug | 6.9 | 10.8 | 2.9 | 58 | 76 | 20 |
| Sep | 2.8 | 6.7 | −1.2 | 55 | 34 | 33 |
| Oct | −3.1 | 0.4 | −6.4 | 53 | 102 | 25 |
| Nov | −9.2 | −5.8 | −12.6 | 100 | 249 | 22 |
| Dec | −12.8 | −9.4 | −16.2 | 88 | 150 | 33 |
| Total Annual mean ± SD | −4.1±1.2 | −0.5±1.31 | −7.7±1.14 | 993 | 1581 | 512 |

*Note:* Mean temp, warmest month (*SD*) = 8.0 ± 1.43. Mean max temp, warmest month = 12.2 ± 1.62. Mean temp, coldest month = −13.5 ± 1.95. Mean min temp, coldest month = −16.6 ± 2.08. Annual range, mean monthly temps = 21.5. No months with temp >0 = 4. No months with temp >15 = 0. Total precip in months with temp >0 = 234. Highest monthly mean temp = 10.7 (Jul 1988). Overall max temp = 15.1 (Jul 1988). Lowest monthly mean temp = −19.0 (Feb 1985). Overall min temp = −24.4 (Dec 1978).

descriptively as "Wet, Taiga, little or no water deficit, microthermal" and "Subpolar/Alpine, arid" under the Holdridge classification system (Greenland et al. 1985).

## Mesoscale Climate

Topography has an important effect on wind flow and related factors at a smaller scale. For example, at the Saddle research site, wind often travels around the barrier formed by the higher ground directly to the west so that the wind has a northwesterly or southwesterly component as opposed to being due west. A vertical eddy current forms in the lee of the Continental Divide resulting in high snow accumulation in the higher east-facing cirques.

Other mesoscale circulations occur when the area of the Front Range and adjoining plains are not dominated by strong westerly winds or the flows related to well-developed mid-latitude cyclones. Two mesoscale systems are of particular importance for aerosol transport. First, when the regional flow of air is from the north, topography-induced wind deflection, related to the presence of the topographic feature known as the Cheyenne Ridge in southern Wyoming, may lead to up-slope flows from the east into the Front Range (Young and Johnson 1984; Snook 1994). Second, when the regional wind is from the south, topography-induced cyclonic turning may also give rise to up-slope wind. In both cases, the wind passes over the

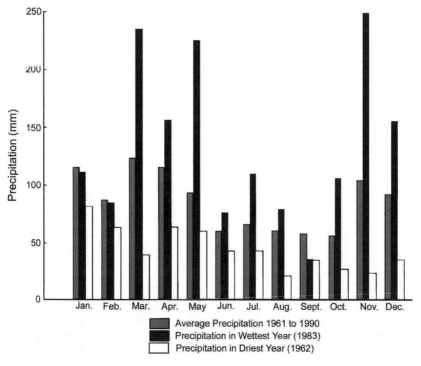

Figure 2.3. Average monthly total precipitation at the D1 climate station (3750 m) on Niwot Ridge, 1961–1990.

Front Range urban corridor and potentially transports nitrates, sulfates, and other aerosols to Niwot Ridge (chapter 3).

## Bioclimatic Zones

Using data from 1952 to 1982, Greenland et al. (1985) identified three bioclimatic variables useful for distinguishing the altitudinal bioclimatic zones of the Front Range. The variables were (1) the ratio of growing season degree days >0°C to growing season precipitation, (2) summer (June to August) mean temperature, and (3) growing season soil moisture deficit, where the growing season is defined by a mean minimum daily temperature ≥0°C. Using these criteria and general climatic characteristics, the climate at the uppermost alpine climate station (D1) is characterized by (1) an annual potential evapotranspiration of 270 mm whereas annual actual evapotranspiration was 261 mm, (2) a growing season moisture deficit of only 9 mm, (3) 710 growing season degree days based on mean temperature and 265 growing season degree days based on minimum temperature, (4) a summer mean temperature was 5.5°C, and (5) a ratio of growing season degree days to growing season precipitation of 1.40°C/mm. The alpine zone was defined in this study as having a ratio of growing season degree days to growing season precipitation of less than 1.48°C/mm, the summer mean temperature is less than 7.4°C, and the growing season moisture deficit is less than 13 mm.

## Temporal Trends

Temporal analysis of four decades (1951–1996) of instrumented climate records at the Alpine (D1, 3750 m) climate station indicate: (1) decreasing autumn temperatures ($-0.043$°C/yr), but no significant annual cooling; (2) a daily average maximum/minimum temperature ratio that is relatively constant on a seasonal basis; (3) no significant trend in the lapse rate between 3749 and 3018 m; (4) a sharp decrease over a very short time (sometimes known as a step-function decrease) in winter pressure variability (19%) occurring in the 1991–1992 winter; (5) a decrease in incident summer solar radiation ($-1.04$ W/m$^2$/yr) between 1965 and 1996 (although increases occurred between 1965–1971 and 1992–1994); and (6) an annual precipitation increase of 11.0 mm/yr. The size of these trends differs somewhat from those found in earlier analyses based on shorter data periods (Barry 1973; Greenland 1989).

We cannot distinguish between any human-caused climate change from natural climatic variability on Niwot Ridge. The increase in annual precipitation and decrease in summer shortwave radiation are consistent with general $CO_2$ doubling modeling scenarios of increased annual precipitation and increased water vapor, but the lack of an annual temperature increase on Niwot Ridge is inconsistent with model predictions of increasing temperatures in the future (Giorgi et al. 1994; Williams et al. 1996). Temperature records for 1953–1991 for three Colorado mountain stations paired with three plains stations (Brown et al. 1992) show similar trends to those shown between Niwot Ridge and the nearby Shortgrass Steppe LTER site. A noteworthy decrease in temperatures on Niwot Ridge occurred during the early 1980s (Greenland 1989). Losleben (1997) suggested that this decrease was associated with the high atmospheric turbidity related to the volcanic eruption of Mount Alaid in 1981, El Chichon in 1982, and the 1982/1983 El Niño event, which was manifested at Niwot Ridge by higher than average precipitation and cloud cover.

The possibility that the Niwot Ridge climate responds to El Niño-Southern Oscillation (ENSO) events is important to its long-term variability and needs further study. As previously mentioned, the low temperatures in the early 1980s may be partly due to the large amount of cloud cover related to the high precipitation values in 1983, which in turn were associated with the very strong El Niño. Even though Niwot Ridge appears to be in a transitional zone between the southern United States and Canada with respect to ENSO signals, Woodhouse (1993) used dendrochronological methods to show weak but detectable tree ring responses to El Niño events. Even more noticeable is a decrease in growth in the years following La Niña events that are characterized by dry springs in the Front Range and nearby plains. Woodhouse (1993) found that tree growth response to fluctuations in the Southern Oscillation Index, which is a measure of ENSO, varied over time, most likely due to variations in strength and timing of El Niño events themselves.

## Microclimate

### Spatial Heterogeneity

The outstanding feature of the microclimate of Niwot Ridge is its spatial heterogeneity. This heterogeneity is due in part to high topographic variability (chapter 4) and to spatial variation in semipermanent snow banks, which in the summer provide sources of soil moisture and result in a variety of different plant communities at the alpine surface (Walker et al. 1994; chapter 6). A gradient exists with different fluxes of energy and moisture to and from the atmosphere. Measurement of these fluxes are further complicated by the advection (horizontal) movement of air across the tundra surface. Snow surfaces in winter cover a larger area but there is still heterogeneity in both snow depth and snow duration due to the clearing of snow from some areas by the strong winds. The spatial heterogeneity presents large problems in microclimatic observation and application of theory. The addition of the spatial to the vertical dimension makes theoretical treatment more complex but not intractable. Studies using remote sensing techniques, such as those of Duguay and others cited later, are overcoming some of these problems.

### Energy and Water Budgets

The most commonly applied method of examining the microclimate has been the energy and moisture budget approach, in which the values of the various fluxes of energy and moisture to and from the tundra surface are observed or estimated over specified time periods. Many investigators have used sophisticated observational energy budget approaches on the Ridge including LeDrew (1975), Isard (1989), Isard and Belding (1989), Greenland (1991, 1993), and Cline (1995, 1997).

### Energy Budgets in Summer

The energy budget values for the fellfield plant community (cf. chapter 6 for full description of plant communities), an area that experiences strong wind scour, has very low snow retention and is one of the most widespread communities found in the alpine of the Front Range, indicate that surface heat energy flows are variable in both time and space (Fig. 2.4). There are measurable differences in the heat budget values between the different plant communities (Table 2.2). Strong temporal variation exists at dry sites with change being driven by the amount of precipitation and the accompanying cloud cover. The differences among communities can be observed on both diurnal and annual time scales. Differences in energy budgets resulting from large changes in precipitation between years may be similar in size to differences in energy budget values in any one year resulting from spatial variability in soil moisture. Spot measurements over six different alpine communities and a summer snow surface show high variability in surface temperature and albedo, which affect the surface heat budget. Two of the six vegetation communities studied, the moist shrub tundra and the snowbed community, appear to have special characteristics, which set them apart from the others. The snowbed surface is notable for its relatively high

Climate 23

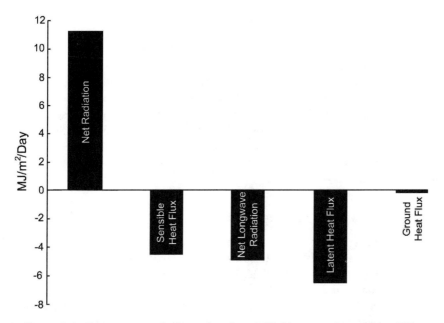

Figure 2.4. Summer energy budget values for a fellfield community on Niwot Ridge.

Table 2.2. Estimated Daily Radiation and Heat Budget for Mid-Summer Tundra Surfaces on Niwot Ridge

| Variable | DM | FF | MST | MM | SB | WM | Snow |
|---|---|---|---|---|---|---|---|
| Incoming shortwave radiation | 19.3 | 19.3 | 19.3 | 19.3 | 19.3 | 19.3 | 19.3 |
| Albedo | 0.18 | 0.17 | 0.18 | 0.18 | 0.17 | 0.18 | 0.45 |
| Outgoing shortwave radiation | −3.5 | −3.2 | −3.5 | −3.5 | −3.2 | −3.5 | −7.9 |
| Surface temperature | 280.4 | 282.2 | 279.9 | 280.1 | 275.4 | 279.1 | 273.0 |
| Outgoing longwave radiation | −30.3 | −31.1 | −30.1 | −30.2 | −28.2 | −29.7 | −27.2 |
| Incoming longwave radiation | 26.2 | 26.2 | 26.2 | 26.2 | 26.2 | 26.2 | 26.2 |
| Net radiation | 11.7 | 11.2 | 11.9 | 11.8 | 14.0 | 12.3 | 10.3 |
| Ground heat flux | −0.2 | −0.2 | −0.7 | −0.7 |  | −0.7 |  |
| Latent heat flux | −6.9 | −6.5 | −6.1 | −7.3 |  | −11.7 |  |
| Sensible heat flux | −4.6 | −4.5 | −5.1 | −3.8 |  | 0.1 |  |
| Bowen ratio (sensible/latent) | −0.67 | 0.70 | 0.84 | 0.51 | −0.01 |  |  |

*Source:* After Greenland 1991.

*Note:* Surfaces are Dry Meadow (DM), Fellfield (FF), Moist Shrub Tundra (MST), Moist Meadow (MM), Snowbed (SB), Wet Meadow (WM), and Snow. Values are in MJ/m$^2$ except for surface temperature which is °K. Heat flow toward the surface is regarded as positive and flow away from the surface is regarded as negative. See original paper for sources, methods, and assumptions.

albedo (reflectivity) while snow is still present, but it is also characterized by the very short time in which it is actually exposed to the atmosphere during the summer season due to its late melt out. This surface becomes relatively dry after snowmelt. The moist shrub tundra surface is distinguished by high cover of shrubs (*Salix planifolia* and *S. glauca*), which, because of their relatively large size, increase shading and decrease the wind velocity. Both the shading and the lower wind speed potentially decrease the actual latent heat loss from the otherwise moist surface, which has potentially higher latent heat loss values. The wet meadow, moist meadow, dry meadow, and fellfield (chapter 6), are characterized by having a distinct gradient in soil moisture availability from wet to dry and consequent expected effects on their heat balance values.

## *Energy Budgets in the Snowmelt Season*

The energy budget of the snowpack during the snowmelt season (ca. April 25–June 6) was studied during 1994 and 1995 (Cline 1995, 1997). Prior to the onset of snowmelt, the combination of a high surface albedo of about 0.9 and constant longwave radiation emittance caused the daytime maximum net radiation to remain below 100 Wm$^{-2}$. Net radiation values began to increase as the albedo of the snow decreased, over the snowmelt season, to a minimum of less than 0.6. Over the entire period, net radiative fluxes provided 75% of the energy for snowmelt, with the rest being provided mostly by downward transfer of sensible heat. The integrated energy flux values over the period (Fig. 2.5) were composed of the sources of net radiation (349.6 MJ) and sensible heat (103.0 MJ), while the sinks were net long-

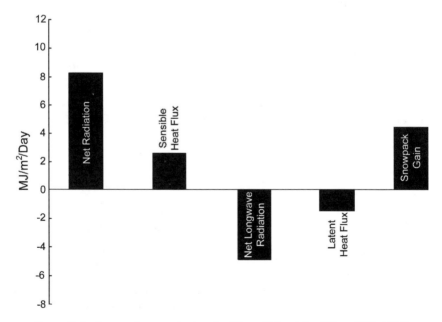

Figure 2.5. Snowmelt energy budget for Niwot Ridge (after Cline, 1995, 1996).

wave radiation (−211.3 MJ) and latent heat flux (−56.7 MJ). The net energy gain to the snowpack was 184.6 MJ. A change in direction from upward to downward net turbulent heat flow coincided with the onset of melt draining from the base of the snowpack. The 1995 snow season had much more snow than 1994, most of it falling in the late spring. The 1995 snow had a melt rate over twice that of 1994, and 51% (compared to 25% in 1994) of the energy for snowmelt came from downward-flowing sensible and latent heat (Cline 1997).

## Advection

The significant downward flux of sensible heat during the snowmelt season indicates the importance of advection in the microclimate of Niwot Ridge since most of this heat came from air that had been transported horizontally from other locations. Cold, dry air from the west of the Continental Divide sweeping across the different tundra surfaces is frequent during the summer. Advection is important to the microclimate of the tundra for two main reasons. First, there is the possibility of enhanced latent heat flow, due to the dryness of the air being transported. Calculations, based on short period data, suggest that this occurs. Further estimations indicate a cooling effect by the advected air (Isard and Belding 1989; Greenland 1991). Second, there is the possibility of small-scale interaction, such as depressed latent heat flow, due to both cold and relatively moist air passing over a drier site. However, the relatively high wind speeds will move air across these surfaces too quickly for a thermal and moisture equilibrium to be reached between the surface and the air.

Additional investigations of the importance of advection (Olyphant and Isard 1987) indicate that advection normally accounts for less than 5% of the total sensible heat flow over vegetated tundra surfaces and up to 20% over snow. These measurements suggest that local-scale advection in the Saddle research site is normally so small as to fall within the error range of the instruments. However, when the same field data are used as boundary conditions for a two-dimensional, turbulent-boundary-layer numerical model, the amount of sensible heat advected to snow patches at the site is an order of magnitude larger than their empirical measurements suggest (Olyphant and Isard 1988). There are several possible reasons for the discrepancy, as Olyphant and Isard (1988) suggested, but the issue has the most chance of being solved by more-accurate independent field observations. Measurements of this kind were made during the 1987 field season by Isard and Belding (1989) using a very accurate lysimeter. Their results strongly support the proposal that there *is* significant advection of cold dry air from the Continental Divide across the Saddle area, and it *does* affect surface energy budget values. Isard and Belding demonstrate that the impact of this advection on the vertical temperature and humidity gradients is approximately the same and that both fluxes of sensible and latent heat were augmented by advection at the expense of ground heat flux. Consequently, the temperature of the tundra surface as well as the temperature and the humidity of the air near the surface are reduced by advection.

The presence of advection has important implications for the interaction of the atmosphere with the vegetation. If advection were suppressed and temperatures were increased, we should expect a biotic response. Portable greenhouse experiments at the Saddle resulted in a marked increase in leaf area index and presumably biomass production (Keigley 1987), and similar experiments have shown increased plant and soil microbial respiration (chapter 16). A further bioclimatological aspect of the advection of cold, dry air is that its presence suppresses dewfall even at times of negative net radiation (LeDrew 1975). Consequently, a potential supply of water to the plants is removed.

## Water Budget

The alpine is a net source of ground and surface water to lower elevations. Water captured as snow is redistributed by wind or as meltwater to subalpine regions. The relatively low actual evapotranspiration of the alpine relative to precipitation values means that only a fraction of atmospheric inputs of water is directly returned to the atmosphere, and as a result, this precipitation subsidizes water use by anthropogenic activities at lower elevations.

A general annual water budget for Niwot Ridge (Table 2.3, Fig. 2.6) indicates that significant evapotranspiration occurs during the growing season and the rapid period during which much of the snow melts. Only a small soil moisture deficit is seen in late summer. However, there is substantial interannual variation, with some years having a more-pronounced deficit. Larger deficits are also likely to be found in areas that have been blown clear of snow, including dry meadows and fellfields.

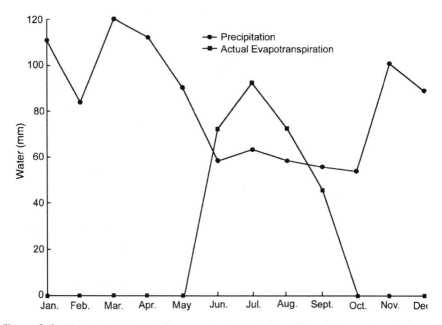

Figure 2.6. Water budget at D1 climate station (3750 m) on Niwot Ridge, estimated from average temperature and precipitation values, 1961–1990.

Table 2.3. Average Annual Water Budget (in mm) for the D1 Climate Station (3750 m) on Niwot Ridge for 1961–1990

| Month | Precipitation | Potential Evaporation | Soil Storage | Actual Evaporation | Deficit | Surplus | Snowmelt | Snowpack Water Equivalent |
|---|---|---|---|---|---|---|---|---|
| Jan | 112 | 0 | 78 | 0 | 0 | 0 | 0 | 354 |
| Feb | 84 | 0 | 79 | 0 | 0 | 0 | 0 | 438 |
| Mar | 120 | 0 | 79 | 0 | 0 | 0 | 0 | 558 |
| Apr | 112 | 0 | 79 | 0 | 0 | 0 | 0 | 669 |
| May | 90 | 0 | 79 | 0 | 0 | 0 | 0 | 759 |
| Jun | 58 | 72 | 79 | 72 | 0 | 434 | 447 | 312 |
| Jul | 63 | 101 | 82 | 92 | 9 | 279 | 312 | 0 |
| Aug | 58 | 86 | 68 | 72 | 14 | 0 | 0 | 0 |
| Sep | 55 | 45 | 78 | 45 | 0 | 0 | 0 | 0 |
| Oct | 53 | 0 | 78 | 0 | 0 | 0 | 0 | 53 |
| Nov | 100 | 0 | 78 | 0 | 0 | 0 | 0 | 153 |
| Dec | 88 | 0 | 78 | 0 | 0 | 0 | 0 | 241 |
| Yearly total | 993 | 304 | | 281 | 23 | 713 | | |

The water budget shown in Fig. 2.6 is representative of one of the more moist locations on Niwot Ridge. As in the case of the energy budgets, there is large interannual and spatial variation in the water budget over the Ridge. Greenland et al. (1984) discussed the relationship of interannual variability of water budgets and the subsequent effect on plant phenology, productivity, soil microbial populations, and stream water chemistry. The retention of snow retards the phenological development of plants, but the presence of increased liquid water may increase flowering and plant productivity. Both soil microbial numbers and soil nitrification rates increase with the greater soil moisture (chapter 12).

The most difficult part of the water budget to measure is evapotranspiration. The most accurate assessment of evapotranspiration was made by Isard and Belding (1989) during the 1987 growing season. They found that evapotranspiration at the Saddle was governed by energy availability for the range of 3–5 days after precipitation events, depending on the depth of the soil. After about 3–5 days, evapotranspiration rates were controlled by the rate at which the gradually drying soil could deliver water to the surface.

Soil moisture availability plays an important role in determining plant species distributions in the different plant communities and in determining annual variation in production (Walker et al. 1994; chapters 6, 9). Taylor and Seastedt (1994) showed a progressive linear decline in soil moisture values during the growing season in dry, moist, and wet meadows, but moist meadows showed the strongest decline and wet meadows the least. Curvilinear fits suggest midsummer minima in moist and dry communities and a midsummer maximum in wet meadows. These investigators suggested a possible increase in soil moisture for the dry meadows since 1953, a finding consistent with the directional increase in precipitation observed during this interval.

## Temporal Change

The principal long-term changes in the values of the energy and water budget components are related to the long-term changes in the solar radiation input and the precipitation. We previously noted increases in solar radiation between 1965–1970 and 1992–1994 and a decrease in incident summer solar radiation of 3 W/m$^2$/yr between 1971 and 1992. These changes will have corresponding direct changes in decreasing and increasing net radiation values during these periods. In addition, because of the increase in annual precipitation of 11.04 mm/yr at 3749 m between 1951 and 1994, latent heat values and actual evapotranspiration values may also be expected to have increased during the times of net radiation increase. Between 1971 and 1992, decreasing summer radiation and increasing annual precipitation, and presumably soil moisture availability, would have had opposite effects on latent heat and actual evapotranspiration values. Further studies would help establish whether the decreasing radiation or the increasing precipitation has the larger effect on evapotranspiration rates at the site.

## Spatial Variability—Remote Sensing

To address the high degree of spatial heterogeneity of the alpine tundra, modeling and remote sensing techniques have been employed to further understand system properties. Duguay and his colleagues pioneered the application of remote sensing technology on Niwot Ridge within the climatic context. In an initial study, Duguay and LeDrew (1991, 1992) used visible, near-infrared, and mid-infrared reflectance from LANDSAT-TM satellite data to estimate surface albedo. The albedo estimates agreed well with field data and the bioclimatic zones of the Front Range. Duguay (1993) extended his estimates to those of the complete radiation budget by using a two-stream atmospheric radiative transfer scheme and again found good agreement with field measurements. Additional results (Duguay 1994, 1995) demonstrated not only the spatial heterogeneity, but also the dynamic nature of this heterogeneity as the snow gradually melts off the tundra. Studies based on LANDSAT-TM satellite data are still limited to the single overpass time every few weeks (Duguay 1995). Duguay and Crevier (1995) have used a relation between the ratio of ground heat flux and net radiation and the Normalized Difference Vegetation Index to obtain values of ground heat flux for Niwot Ridge. The remotely sensed values of ground heat flux were rather large compared to field measurements, but the latter are difficult to make, and the relative spatial variations provided by the remote sensing data are useful. Further investigations have used remote sensing data and a variety of forms of derived topographic and bioclimatic data to increase the accuracy and precision of maximum likelihood landcover classification with respect to the hierarchical Braun-Blanquet vegetation classification system (Duguay and Walker 1995; Peddle and Duguay 1995).

## Summary

Over 40 years ago, Bliss (1956) concluded that the alpine tundra climatic environment was more severe than the arctic tundra environment because of its stronger winds, potentially higher maximum radiation, its higher surface temperatures, and the potential for greater desiccation. To these factors we might add the large interannual climatic variability (Saunders and Bailey 1994).

Overall, we may understand the bioclimatology of the alpine tundra ecosystem in the following terms. *Macro- (synoptic) scale* factors, such as latitude and elevation, and specific location within the general circulation, give rise to the existence of tundra and the general range of values of the heat balance components. *Mesoscale* factors, such as wind and topography, largely determine the summertime distribution of snow banks and soil moisture availability. Factors related to *microscale* climate, particularly soil moisture availability, lead to particular subtypes (plant communities) in tundra vegetation, with the specific plant communities and their soil moisture environment determining the actual values of the heat budget components. At large spatial scales, the presence of alpine tundra is determined mainly by a certain temperature range; however, on the micro- and topo-scales, the actual values of surface heat and water budget components can be strongly influenced by the topography of the surface and the particular plant community. In many cases, soil moisture availability and thermal characteristics, as well as the plants themselves, with their reflective properties and resistance to latent heat loss, tend to determine the actual values of the surface heat budget (Greenland 1993).

Important areas for future research include: (1) an examination of the degree to which Niwot Ridge climate is coupled or decoupled from the lower elevations of the Rockies and adjacent plains and more general global trends in climate, (2) the characterization and determination of the effect of drought periods, (3) the determination of whether the observed decreasing radiation values or increasing precipitation values have an influence on evapotranspiration rates on a decadal time scale; and (4) a more-detailed investigation the role of wind in general on air and soil microclimate. In addition, an investigation of long-term (daily over several years) changes in the values of the energy and water budget components would provide important information for use in the land surface parameterization of larger numeric models. The problems of spatial heterogeneity must be addressed with more sophisticated sensing and modeling techniques. Previous remote sensing studies have been limited by the infrequency of LANDSAT satellite overpasses. Future studies will be able to make use of more-frequent overpasses and multispectral sensors. The remote sensing technology needs to be accompanied by better theoretical treatments of spatial heterogeneity. Partly related to the latter is the intriguing problem of advection. Once again, both observational, possibly using eddy correlation techniques, and theoretical studies of this and its relation to the vegetation would be useful. Such studies would complement and contribute to our understanding of alpine responses to the two environmental variables exhibiting the largest amounts of change, increasing precipitation and increasing atmospheric nitrogen inputs (chapter 16).

References

Barry, R. G. 1973. A climatological transect of the East Slope of the Front Range, Colorado. *Arctic and Alpine Research* 5:89–110.

Barry, R. G., R. S. Bradley, and L. F. Tarleton. 1977. The secular climatic history of the Rocky Mountain area. Final report to the National Science Foundation, GA-40256, March 1977.

Bliss, L. C. 1956. A comparison of plant developments in microenvironments of Arctic and alpine tundras. *Ecological Monographs* 26:303–337.

Brown, T. J., R. G. Barry, and N. J. Doesken. 1992. An exploratory study of temperature trends for paired mountain-high plains stations in Colorado. Proceedings of Sixth Conference on Mountain Meteorology, American Meteorological Society, Boston.

Cline, D. 1995. Snow surface energy exchanges and snowmelt at a continental alpine site. In *Biogeochemistry of seasonally snow covered catchments* (*Proceedings of a Boulder symposium, July 1995*), edited by K. Tonnessen, M. Williams, and M. Tranter. IAHS Publication Number 228.

———. 1997. Effect of seasonality of snow accumulation and melt on snow surface energy exchanges at a continental alpine site. *Journal of Applied Meteorology* 36:32–51.

Duguay, C. R. 1993. Modeling the radiation budget of alpine snowfields with remotely sensed data: Model formulation and validation. *Annals of Glaciology* 17:288–294.

———. 1994. Remote sensing of the radiation balance during the growing-season at the Niwot Ridge Long-Term Ecological Research site, Front Range, Colorado, U.S.A. *Arctic and Alpine Research* 26:393–402.

———. 1995. An approach to the estimation of surface net radiation in mountain areas using remotely sensed and digital terrain data. *Theoretical and Applied Climatology* 52:55–68.

Duguay, C. R., and Y. Crevier. 1995. Use of remote sensing and field measurements to estimate surface energy balance components over alpine tundra surfaces. Vol. 1 of Proceedings of the 17th Canadian Symposium on Remote Sensing, Saskatoon, Saskatchewan, 12–15 June.

Duguay, C. R., and E. LeDrew. 1991. Mapping surface albedo in the east slope of the Colorado Front Range with Landsat 5 Thematic Mapper. *Arctic and Alpine Research* 23:213–223.

———. 1992. Estimating surface reflectance and albedo over rugged terrain from Landsat-5 Thematic Mapper. *Photogrammetric Engineering and Remote Sensing* 58:551–558.

Duguay, C. R., and D. A. Walker. 1995. Environmental modeling and monitoring with GIS at the Niwot Ridge LTER, Colorado, USA. In *Integrating geographic information systems and environmental modeling,* edited by M. Goodchild et al. Fort Collins, CO: GIS World Publishing.

Giorgi, F., C. S. Brodeur, and G. T. Bates. 1994. Regional climate change scenarios over the United States produced with a nested regional climate model. *Journal of Climate* 7:375–399.

Greenland, D. 1977. Living on the 700 mb surface: The mountain research station of the Institute of Arctic and Alpine Research. *Weatherwise* 30:232–238.

———. 1989. The climate of Niwot Ridge, Front Range, Colorado, U.S.A. *Arctic and Alpine Research* 21:380–391.

———. 1991. Surface energy budgets over alpine tundra in summer. *Mountain Research and Development* 11:339–351 and 12: facing p. 1.

———. 1993. Spatial energy budgets in alpine tundra. *Theoretical and Applied Climatology* 46:229–239.

———. 1995. Extreme precipitation during 1921 in the area of the Niwot Ridge Long-Term Ecological Research site, Front Range, Colorado, U.S.A. *Arctic and Alpine Research* 27:19–28.

Greenland, D., J. Burbank, J. Key, L. Klinger, J. Moorhouse, S. Oakes, and D. Shankman. 1985. The bioclimates of the Colorado Front Range. *Mountain Research and Development* 5:352–362.

Greenland, D., N. Caine, and O. Pollak. 1984. The summer water budget and its importance in the alpine tundra of Colorado. *Physical Geography* 5:221–239.

Isard, S. A. 1989. Topographic controls in an alpine fellfield and their ecological significance. *Physical Geography* 10:13–31.

Isard, S. A., and M. J. Belding. 1989. Evapotranspiration from the alpine tundra of Colorado, U.S.A. *Arctic and Alpine Research* 21:71–82.

Keigley, R. 1987. Effect of experimental treatments on *Kobresia myosuroides* with implications for the potential effect of acid deposition. Ph.D. diss. Department of Environmental, Population, and Organismic Biology, University of Colorado, Boulder.

LeDrew, E. F. 1975. The radiation and energy budget of a mid-latitude alpine site during the growing season, 1973. *Arctic and Alpine Research* 7:301–314.

Losleben, M. 1997. An uncommon period of cold and change of lapse rate in the Rocky Mountains of Colorado and Wyoming. Proceedings of the Thirteenth Annual Pacific Climate (PACLIM) Workshop, 15–18 April, 1996, edited by C. M. Isaacs and V. L. Tharp. Interagency Ecological Program, Technical Report 53, California Department of Water Resources.

Marr, J. W. 1961. Ecosystems of the East Slope of the Front Range in Colorado. *University of Colorado, Studies in Biology* 8:1–134.

Olyphant, G. A., and S. A. Isard. 1987. Some characteristics of turbulent transfer over alpine surfaces during the snowmelt-growing season: Niwot Ridge, Colorado, U.S.A. *Arctic and Alpine Research* 19:261–169.

———. 1988. The role of advection in the energy balance of late-lying snowfields: Niwot Ridge, Front Range, Colorado. *Water Resources Research* 24:1962–1968.

Peddle, D., and C. R. Duguay. 1995. Incorporating topographic and climatic GIS data into satellite image analysis of an alpine tundra ecosystem, Front Range, Colorado Rocky Mountains. *GeoCarto International* 10:43–60.

Saunders, I. R., and W. G. Bailey. 1994. Radiation and energy budgets of alpine tundra environments of North America. *Progress in Physical Geography* 18:517–538.

Snook, J. S. 1994. An investigation of Colorado Front Range winter storms using a nonhydrostatic mesoscale numerical model designed for operational use. NOAA Technical Memorandum ERL FSL-10. Forecast Systems Laboratory, Boulder, CO.

Strahler, A. H., and A. N. Strahler. 1992. *Modern physical geography.* 4th ed. New York: Wiley.

Taylor, R. V., and T. R. Seastedt. 1994. Short- and long-term patterns of soil moisture in alpine tundra. *Arctic and Alpine Research* 26:14–20.

Walker, M. D., P. J. Webber, E. H. Arnold and D. Ebert-May. 1994. Effects of interannual climate variation on aboveground phytomass in alpine vegetation. *Ecology* 75:393–408.

Williams, M. W., M. Losleben, N. Caine, and D. Greenland. 1996. Changes in climate and hydrochemical responses in a high-elevation catchment in the Rocky Mountains, USA. *Limnology and Oceanography* 41:939–946.

Woodhouse, C. A. 1993. Tree growth response to ENSO events in the Central Colorado Front Range. *Physical Geography* 14:417–345.

Young, G. S., and R. H. Johnson. 1984. Meso- and microscale features of a Colorado cold front. *Journal of Climate and Applied Meteorology* 23:1135–1325.

# 3

# Atmospheric Chemistry and Deposition

Herman Sievering

## Introduction

The two most significant elements with atmospheric components that influence ecological processes at Niwot Ridge are carbon (C) and nitrogen (N). The enrichment of the atmosphere by carbon dioxide ($CO_2$) is ubiquitous across the globe. Global and regional patterns of the annual increases in atmospheric $CO_2$ as well as the current and anticipated vegetation responses are the subject of ongoing analyses (e.g., Schimel 1995, 1998). Hence, except to emphasize the unique Niwot Ridge contribution to the $CO_2$ database, the material presented in this chapter focuses primarily on N, which has a tremendous potential to influence the structure and function of ecosystems (Vitousek et al. 1997).

Anthropogenic increases in atmospheric deposition can have profound effects on terrestrial and aquatic ecosystems (Ollinger et al. 1993). Of particular concern are increases in the deposition of nitrogen-containing species, including nonprecipitative, dry-deposited gaseous and particulate N, which can be an important component of the N cycle. These species may act as fertilizer N and can be utilized by vegetation and microbes with little or no energy expenditure. Estimation of the magnitude of N dry plus wet deposition to alpine tundra and subalpine forest ecosystems of Niwot Ridge is integral to understanding N cycling within these systems.

This chapter focuses on the estimation of N deposition to the alpine tundra and, to a lesser extent, to the adjacent subalpine forest. The first section presents a brief review of the chemistry of the air environment over Niwot Ridge. The next section discusses the processes of N deposition and exchange with the alpine landscape, source regions for the N in the air over Niwot Ridge, and a procedure for determining atmospheric deposition from ambient air concentrations. Evidence is presented that anthropogenic sources contribute the majority of N in regional air masses and thus to total N deposition. The role that ammonia gas may play in reducing or enhancing N deposition is also described. The contribution that N deposition may make to N accumulation in the subalpine forest ecosystem below Niwot Ridge is briefly discussed in the final section.

## Measurements of Carbon Dioxide of the Colorado Alpine

The ambient air concentration of carbon dioxide has been monitored by the Climate Monitoring and Diagnostics Lab of NOAA at a 3500-m site known at T-van on Niwot Ridge since 1968 (Fig. 3.1; see Fig. 1.5 for location of T-van; Conway et al. 1994). This record, the longest North American record of atmospheric $CO_2$ concentrations, is consistent with other high-altitude, Northern Hemisphere sites. The annual cycle of winter highs and subsequent summer lows due to respiration and photosynthesis is quite distinctive. The atmospheric $CO_2$ concentration growth rate, 0.44% $yr^{-1}$ over the entire record, is similar to that observed elsewhere. A flattening in this trend during 1992 and 1993 may be related to the eruption of Mount Pinatubo and ejection of particulate matter emissions into the stratosphere. Particles (mostly sulfate) with 1–2-year residence times released into the stratosphere by the

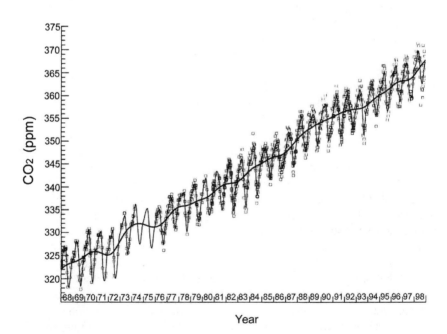

Figure 3.1. $CO_2$ concentrations collected from flasks on Niwot Ridge as part of the NOAA CMDL global monitoring program. Data points are weekly samples, with a smoothed curve fit (third-order polynomial) for the weekly samples and an annual trend fit after filtering out the intra-annual variability (data from T. Conway and P. Tans, NOAA).

Pinatubo eruption increased Earth's albedo during these 2 years. Lower atmospheric temperatures during this period may have increased $CO_2$ uptake by the ocean and/ or altered the ratio of terrestrial photosynthesis/respiration.

Although the lower $CO_2$ partial pressures at high elevation suggest alpine vegetation is potentially more sensitive to changes in $CO_2$ concentrations (Körner and Diemer 1987), a 4-year experiment failed to elicit a growth response of alpine plants under elevated $CO_2$ concentrations (Schäppi and Körner 1996). The effect of N availability was more important than variation in $CO_2$ concentration in determining changes in plant growth in this experiment.

## Air Chemistry Environment

The chemistry of the high altitude environment over Niwot Ridge is dominated by free tropospheric processes. These processes are characterized by low temperatures, low particulate matter, and low relative humidity. This has important implications for the chemical phase and species distribution of most elements in the atmosphere over Niwot Ridge. The free troposphere may be distinguished from the lower portion of the troposphere where physical conditions are largely determined by friction and heat transfer. Heating of Earth's surface by solar radiation imparts energy to the overlying air, enhancing vertical motion and eddy transport. An increase of temperature with height in the atmosphere instead of the normal, adiabatic decrease impedes vertical mixing. This inversion layer causes an accumulation of pollutants in the boundary layer beneath the inversion, making it impossible to determine chemical species working averages that would otherwise be generally applicable. In the Colorado Front Range, these inversions generally keep pollutants below the alpine zone. The free troposphere, above the boundary layer (and "free" of surface influence) is indirectly influenced by local pollution conditions. This part of the troposphere is more homogeneous and concentrations of chemical species are substantially more uniform. The measured concentrations of gases and chemical species have regional and global significance. In the case of the free troposphere over Niwot Ridge, measured concentrations are applicable to the larger Rocky Mountain region and often to the broader western section of the North American continent. This does not mean that local sources of N pollution do not contribute to the free trophosphere N concentrations of the Rocky Mountain region. Rather, such N inputs tend to be homogeneously distributed within the free troposphere.

Free tropospheric N is not only found as $N_2$ but also as gaseous nitric acid ($HNO_3$) and ammonia ($NH_3$) and particulate matter nitrate ($NO_3^-$) and ammonium ($NH_4^+$)—all of which can potentially be used as N sources by terrestrial biota. The addition of anthropogenically derived, atmospheric N has the potential to alter the structure and function of many ecosystems, including alpine tundra (chapter 16).

The N species distribution over Niwot Ridge should reflect the influence of free tropospheric conditions. Thus the majority of nitrate found in air over the alpine zone should be found in the gas phase as nitric acid ($HNO_3$), since little water is present in the liquid phase with low particulate matter air concentrations as well as low relative humidities. These factors preclude the condensation of water vapor on particle surfaces where $HNO_3$ would otherwise be adsorbed and converted to particulate nitrate ($NO_3^-$). The molar concentration ratio of $HNO_3$ to $NO_3^-$ in the free troposphere is generally greater than 2 and, often, greater than 3 (Huebert and Lazrus 1980; Warneck 1988). Further, these concentrations should vary little during any one month or season relative to such concentrations below the inversion layer. Slow increases over several years time may be expected with globally increasing N pollution, since exchange with the free troposphere is slow but significant over the time frame of months to several years. This slowly increasing trend with low variability during any one season/month may also be expected for particulate ammonium ($NH_4^+$) except when volcanic emissions are significant. $NH_4^+$ is predominantly found as ammonium sulfate (($NH_4)_2SO_4$) or bisulfate ($NH_4HSO_4$) in very small particles (Warneck 1988) produced mainly by anthropogenic sulfur emissions. These small particles (<1 μm diameter) can be transported long distances and consequently slowly mix into the free troposphere. On the other hand, gaseous ammonia ($NH_3$) may be emitted or absorbed by alpine plants, and despite the existence of free tropospheric air above the alpine, substantial variability in the concentration of $NH_3$ may prevail during the growing season at Niwot Ridge.

## Deposition of Nitrogen: Rates and Sources

Given the expected high ratio of $HNO_3$ to $NO_3^-$, we may further expect that nitrate as $HNO_3$ will accumulate on and in alpine vegetation. Resistance to transport of $HNO_3$ from the air layer in contact with vegetation into substomatal cavities is effectively zero (Hicks et al. 1987; Huebert and Robert 1985). This is not true for particulate $NO_3^-$.

The net transport of $NH_3$ to tundra plants can be negative; that is, $NH_3$ may be emitted by vegetation. Ammonia can be either deposited to or emitted from plants (Denmead et al. 1976) and soils (Georgii and Lenhard 1978). The direction of the flux depends on the $NH_3$ concentration gradient across the intercellular fluid within the substomatal cavity or soil water and the ambient air concentration. At the compensation point ($C_p$) the concentration of $NH_3$ in the air is in equilibrium with the concentration within plant tissues or soil water, and the net exchange is zero. Changes in the soil environment, especially due to $HNO_3$ deposition to alpine soils affect $C_p$ and the exchange rate of $NH_3$. Typical ranges of air concentrations are 0.5–3 μg N-$NH_3$ m$^{-3}$, with $C_p$ at 20–25°C of <1 μg N-$NH_3$ m$^{-3}$ for forests and >1 μg N-$NH_3$ m$^{-3}$ for low-lying vegetation such as the alpine tundra (Langford et al. 1992). However, $C_p$ has an exponential dependence on temperature, so that lower $C_p$ may be expected for the alpine than for the subalpine forest areas at Niwot Ridge.

As stated previously, free tropospheric concentrations of $HNO_3$ tend to be high relative to particulate $NO_3^-$, with a ratio ranging from 2 to >3. Summertime values of this ratio, obtained by aircraft at 3000–8000 m in ambient, clean air conditions over the continental United States (Huebert and Lazrus 1980), yielded a mean of 2.4 in this ratio. These authors suggested values greater than 3 were due to anthropogenic input. More recent analysis of a number of free tropospheric, largely summertime $HNO_3$ and $NO_3^-$ data sets (Warneck 1988) supports the notion that ratios are affected by anthropogenic activities. Further, the average summertime tropospheric ambient concentration of $HNO_3$ and $NO_3^-$, that is, total nitrate, is less than 100 ng N m$^{-3}$. Compared with these numbers, the growing season molar ratio of $HNO_3$ and $NO_3^-$ at the Niwot Ridge Saddle research site is 4.6, with a total nitrate concentration of 180 ($\pm$23) ng N m$^{-3}$ (Sievering et al. 1996). These data indicate that more than one half, perhaps much more, of the total nitrate in the air over Niwot alpine tundra is anthropogenic in origin.

## Deposition of N

Nitrogen species that contribute to the exchange of N between the tundra and the atmosphere include $HNO_3$, $NO_3^-$, $NH_4^+$, $N_2$, $N_2O$, and $NO_x$. The biogenic fluxes of $N_2$ and $N_2O$ are considered in chapter 13, where it is estimated that the annual emissions of these species together are 0.5 kg N ha$^{-1}$ yr$^{-1}$. This is a fairly small air-tundra exchange flux relative to that of $HNO_3$, $NO_3^-$, $NH_4^+$, and $NH_3$. The data of Parrish et al. (1990), obtained at a lower elevation subalpine forest site (the C1 climate station, 3018 m) on Niwot Ridge where pollution from nearby cities causes higher $NO_x$ concentrations than over the alpine, indicate that a negligible portion (<5%) of N exchange at Niwot Ridge could be due to $NO_x$ (Sievering et al. 1992).

Determining the total magnitude of atmospheric N deposition at Niwot Ridge is difficult, as it may occur via three different mechanisms and several N species. Wet deposition (rain and snow) is the easiest of the three to measure and is extensively monitored as a part of the National Acid Deposition Program (NADP) network, although corrections are required for sampling in the extremely windy conditions at the Niwot Saddle site (Williams et al. 1998). All wet deposition of N is either as $NO_3^-$ or $NH_4^+$. Recent studies have shown that dry deposition—the nonprecipitative, nonhydrometeor (cloud and fog) loading of matter to surfaces—can contribute as much or more to N fluxes than wet deposition can (Geigert et al. 1994; Sievering et al. 1992). Finally, cloud and fog deposition can be dominant at some mountaintop sites such as the White Mountains of New Hampshire or the Great Smoky Mountains in the eastern United States (Lovett and Kinsman 1990) and is difficult to estimate. Fog deposition at Niwot Ridge was found to be a negligible contributor to total N loading in the alpine due both to few low-elevation cloud events and the general absence of intercepting surfaces (Sievering et al. 1989). Cloud deposition has not been addressed.

Dry deposition of $HNO_3$, $NO_3^-$, and $NH_4^+$ may be estimated as the product of the species air concentration and a dry deposition velocity. The dry deposition velocity ($V_d$) is parameterized using three resistance components in series: $R_a$, the aerodynamic resistance; $R_b$, the quasi-laminar sublayer resistance; and $R_c$, the surface resistance (Hicks et al. 1987). The $V_d$ is, then, the inverse of the sum of these resistances. To estimate $V_d$ for fine particulate and gaseous species such as $NH_4^+$ and $HNO_3$, $R_a$ and $R_b$ are independently calculated for stable or unstable wind regimes, and $R_c$ is calculated for specific vegetation type and physiology, N species, and leaf area index (Hicks et al. 1987) or for snow-covered ground conditions. Mountainous, complex terrain will enhance air turbulence over surfaces such as alpine tundra. Thus, $R_a$ (the aerodynamic resistance), which is controlled by turbulence, may be

Table 3.1. Nitrogen Dry (1991–1995) and Wet (1986–1995) Deposition Fluxes for Growing and Snow Seasons as Well as Annual at the 3540 m Saddle Site on Niwot Ridge, Colorado in kg N ha$^{-1}$

|  | Dry | Wet |
|---|---|---|
| Growing Season (June–September) | 1.6 ± 0.6 | 1.2 ± 0.1 |
| Snow Season (October–May) | 1.3 ± 0.3 | 2.0 ± 0.9 |
| Annual total | 2.9 ± 0.8 | 3.2 ± 1.0 |

*Note:* Data are presented in kg N ha$^{-1}$ season$^{-1}$ for each of two seasons at Niwot Ridge, the growing and snow seasons as well as for the entire year. For the alpine of the Colorado Rockies the growing season is generally from June through September and the snow season, the remainder of the year. Corrections for oversampling of winter deposition follow Williams et al. 1998.

reduced, and actual $V_d$ values may be slightly larger than calculated by the resistance approach. The Saddle sampling site is centrally located at a fairly flat, open area with over 200 m of alpine tundra fetch in all directions—more in the prevailing wind direction. Actual $V_d$ values at the Saddle site are not more than 20–30% greater than calculated by the resistance approach (Sievering et al. 1996). Thus, the N dry deposition estimates given in Table 3.1 may be considered to be somewhat conservative.

Identification of $V_d$ values for application at Niwot Ridge are described in Sievering et al. (1992) with the conclusion that $(V_d)_{HNO_3} > (V_d)_{NO_3^-} > (V_d)_{NH_4^+}$, across the wind speed, stability, and turbulence regimes encountered at Niwot Ridge. Applying calculated $V_d$ values to N species, ambient air concentration data provide an estimate of N dry deposition to the tundra (Table 3.1). The dry deposition process must be distinguished under conditions of open vegetation from that of snow cover. The quasi-laminar sublayer resistance ($R_b$) and the surface resistance ($R_c$) may be markedly different under these two regimes.

The wet N deposited in the standard NADP collector at the Saddle site overestimates N wet deposition due to the inclusion of blowing snow (redeposition) in the estimates (Williams et al. 1998). Using an appropriate correction factor, annual wet deposition at the Saddle research site has been estimated at 3.2 ± 1.0 kg N ha$^{-1}$ yr$^{-1}$.

The total growing season N deposition during the first half of the 1990s was 2.5–3 kg N ha$^{-1}$, while the snow season value was approximately 3–3.5 kg N ha$^{-1}$ (Table 3.1), although this input is uneven across the alpine landscape as a result of entrainment of N in snow (Bowman 1992). The annual spatially integrated N deposition is therefore around 6 kg N ha$^{-1}$. A substantial fraction of the N deposited during the snow season is released in May and early June of each year as part of the spring snowmelt (Caine and Thurman 1990). Based on stream water $NO_3^-$ data for two drainage basins on the south side of Niwot Ridge, one vegetated and the other almost devoid of vegetation, roughly half of the spring snowmelt N is assimilated by vegetation, the soil microbial pool, and/or the bulk soil (Sievering et al. 1992). The remaining N is assumed to end up in groundwater and streams, but that assumption has not been rigorously tested.

## Source Regions of Nitrogen

As discussed earlier, at least one half of the annual 6 kg N ha$^{-1}$ deposition is anthropogenic in origin. Analysis of local and regional meteorological data shows that air from the east side of the Colorado Front Range is not influencing the levels of N in the ambient air over the alpine tundra as much as that over the subalpine forest areas (Sievering et al. 1996). However, Parrish et al. (1990) showed Front Range pollutant-N may be mixed into higher level westerly flows that often persist throughout the day with resultant transport of pollutant-N from the boundary layer into the free troposphere.

To better understand what source regions and the degree to which anthropogenic sources may contribute to observed N-species concentrations, meteorological back-trajectories from Niwot Ridge have been constructed (Sievering et al. 1996). These trajectories (Fig. 3.2) show that boundary layer source regions for the air sampled during the growing season at 3300–4000 m on Niwot Ridge are to the west of the Rocky Mountains, often reaching from the Pacific Ocean. The Pacific Ocean marine boundary layer is a contributing source region during one half to two thirds of the growing season of the Colorado alpine. The California-to-northwestern-Mexico area is the dominant source region from one half to one third of this time. The most common mean trajectory, with a 33% probability of occurrence, has only continental input largely from the N-polluted source regions of southern California and northern Mexico. Air concentrations at Niwot Ridge of $HNO_3$, $NH_4^+$, and $NO_3^-$ may be expected to vary, since clean air is brought to Niwot Ridge by the mean trajectories with long transport time over the Pacific Ocean. Snow season back-trajectory analyses show a very similar pattern to that of Fig. 3.2 (Sievering et al. 1996). The back-trajectory analyses along with the local meteorological record show that free tropospheric transport from westerly source regions dominates the alpine on Niwot Ridge. Anthropogenic input of N to the alpine of Niwot Ridge from sources to the east of the Rocky Mountains can be mixed into the higher level westerly flow as Parrish et al. (1990) suggest, but this source will be a part of westerly free tropospheric inputs.

Further support for westerly free tropospheric inputs at the Saddle site is found in Silverstein and Greenland (1991), who concluded that acid precursor sources exist primarily to the southwest of Colorado, but areas to the northwest, west, and south are also likely sources. Losleben (1991), using chlorofluorocarbons as anthropogenic circulation tracers, found that relatively unmixed air masses from the Pacific Northwest reach the subalpine forest below Niwot Ridge during the snow season. Air from the California/Oregon region seems to be well mixed, perhaps through convection, orographic, or frontal lifting. During the growing season there is no indication of any unmixed air masses reaching the subalpine forest. Thus, air masses that originated over the Pacific will be a combination of marine and continental (with anthropogenic contributions) air when they reach Niwot Ridge.

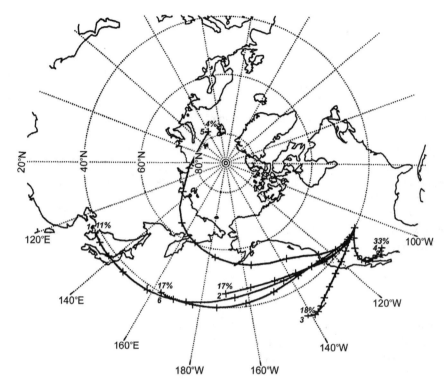

Figure 3.2. Cluster means of back-trajectories for the 3500-m level asl arriving at Niwot Ridge during the growing season. Clustering includes 748 back-trajectories for the June–September 1992–1994 growing season periods. Plus signs denote 1-day intervals. The symbols 10 days back give (top) the percentage of trajectories included in the cluster and (bottom) the arbitrary cluster identifier (Harris and Kahl 1990).

## Deposition and Emission of Ammonia Nitrogen

$NH_3$ may be deposited to or emitted from the alpine vegetation, unlike $HNO_3$, $NO_3^-$, and $NH_4^+$, which can only be deposited. The direction of $NH_3$ flux depends on whether its air concentration is above or below the compensation point ($C_p$). Ammonia concentrations have been successfully obtained at the Saddle research site for only 2 years, unlike $HNO_3$, $NO_3^-$, and $NH_4^+$ which have been measured since 1992. When snow-covered ground conditions prevail, <6 ng m$^{-3}$ of $NH_3$ is present in the air over the tundra. This low value not only supports the expectation that $NH_3$ in the free troposphere over Niwot Ridge should be negligible during the snow season but also that the Ridge and surrounding subalpine forest areas are not a source of $NH_3$ at this time.

$NH_3$ concentrations during the growing season ranging between 20 and 215 ng $NH_3$-N m$^{-3}$ have been measured, with an average of 89 ng $NH_3$-N m$^{-3}$, about tenfold the concentrations in the winter. On the other hand, $HNO_3$ concentrations differ little between summer and winter, indicating that the majority of the $NH_3$ in the air over Niwot Ridge during the growing season is not transported from regional or distant sources. Rather, atmospheric $NH_3$ may be locally derived, probably as emissions from tundra plants.

The net growing season flux of $NH_3$ cannot be estimated using single point $NH_3$ concentrations alone. $NH_3$ emission fluxes measured in cold moorland areas in Scotland are 0.1 kg N ha$^{-1}$ (Sutton et al. 1995) and thus represents a modest flux of gaseous N loss from the system. In comparison, growing season emission flux for $N_2O$ is 0.02 kg N ha$^{-1}$ at Niwot Ridge alpine sites (Neff et al. 1994; chapter 13). The $NH_3$ emissions may account for a modest fraction of the enhanced anthropogenic inputs, thereby reducing the magnitude of net nitrogen enrichment from the atmosphere. At the same time, however, the fate of these emissions need to be considered as well.

During the summer, the compensation point for $NH_3$ in the subalpine forest on Niwot Ridge has been estimated at 300 ng N m$^{-3}$ (Langford and Fehsenfeld 1992). Given the exponential dependence of $C_p$ on temperature, $C_p$ at the alpine tundra may be one half to two thirds that of the subalpine forest, or 150–200 ng N m$^{-3}$. The flux of $NH_3$, $F(NH_3)$, including its direction, may then be estimated as:

$$F(NH_3) = -g(C_a - C_p)/p$$

where $C_a$ is the observed atmospheric concentration of $NH_3$, g is the conductance of $NH_3$ through stomatal openings in tundra plants, and p is atmospheric pressure (Farquhar 1980). Emission will then prevail during the growing season when $C_a < C_p$, that is, $C_a <$ 150–200 ng N m$^{-3}$, and deposition should prevail when $C_a > C_p$ or $C_a >$ 150–200 ng N m$^{-3}$.

The presence of bidirectional $NH_3$ fluxes from alpine tundra on Niwot Ridge is being experimentally assessed using an aerodynamic gradient method, which utilizes the simultaneous measurement of vertical profiles of mean windspeed, temperature, and $NH_3$ concentration over a fairly uniform section of the tundra. Mean $F(NH_3)$ over 1–3-hour periods can then be calculated using two- or three-point vertical profiles of wind speed and $NH_3$ with $C_p$ then estimated using an equation found in Sievering et al. (1994). If observations of $F(NH_3)$ confirm that growing season emission fluxes of $NH_3$ are more prevalent than are deposition fluxes, a portion of the N deposited as $HNO_3$, $NO_3^-$, and $NH_4^+$ would then be emitted back to the atmosphere by tundra plants as $NH_3$.

## Exchange of N in the Subalpine

Due to the substantially warmer temperatures found at the subalpine forest ecosystem below the alpine tundra, $C_p$ should be greater than that for the alpine. Langford and Fehsenfeld (1992) estimated Cp to be 300 ng N m$^{-3}$ in the subalpine forest. Pollution from Front Range cities to the east is brought to subalpine forest sites at Niwot almost daily during the summer by up-slope airflow, with $NH_3$ concentrations in excess of 300 ng N m$^{-3}$ (Roberts et al. 1988; Rusch 1995). As the up-slope flow dissipates in the late afternoon, $NH_3$ concentrations may drop well below $C_p$ for the subalpine forest, resulting in a diurnal bidirectional $NH_3$ flux.

Determination of $NH_3$ concentration changes above and below $C_p$ at the subalpine (C1) site requires high-accuracy, short-term sampling of $NH_3$ and other atmospheric N species in order to measure diurnal changes. $NH_3$ concentrations were measured at 2-hour-long intervals six times per day at the C1 site during the summer of 1996 (Calanni et al. 1999) and showed that $NH_3$ concentrations over the subalpine forest ranged from <50 ng N $m^{-3}$ to >400 ng N $m^{-3}$ during midday (upslope impacted) 2–6-hour periods in June, July, and August. However, during periods when upslope impact is absent, $NH_3$ air concentrations over the C1 subalpine forest are consistently below 160 ng N $m^{-3}$. Thus, short periods of $NH_3$ deposition and longer emission periods may occur diurnally at the C1 site. Using a $C_p$ of 300 ng N $m^{-3}$ along with the concentration data obtained, a net $NH_3$ flux of 0.3–0.8 ng N $m^{-2} s^{-1}$ was estimated for the summer of 1996 (Calanni et al. 1999). Aerodynamic gradient method measurements are ongoing and should provide this tundra $C_p$ and estimation of $NH_3$ flux in the near future.

Upslope events not only bring higher $NH_3$ concentrations to the Colorado Front Range subalpine during the summer but also bring higher concentrations of $HNO_3$, $NH_4^+$, and $NO_3^-$ and may occasionally bring significant amounts of $NO_x$ (Parrish et al. 1990). Given that the net summertime $NH_3$ flux appears to be near zero, concentration data of the remaining N species can be used to estimate N dry deposition to the subalpine forest using species-specific deposition velocities (Peters and Eiden 1992). June–September total N deposition was estimated at between 2.5 and 3 kg N $ha^{-1}$ (Calanni et al. 1999). Assuming subalpine wet N deposition is approximately the same as the alpine, the dry plus wet N deposition to the subalpine forest is about 4 kg N $ha^{-1}$ during the summer, with an annual total of about 7–8 kg N $ha^{-1}$.

Recent construction of two 30-m towers near the subalpine research site for eddy correlation flux measurments, part of the Department of Energy's "AmeriFlux" network, will greatly facilitate future studies of forest-atmosphere gas exchange. In addition to $NH_3$ flux measurements, fluxes of water vapor, $CO_2$, terpenes, isoprene, and ozone are measured.

## Summary and Conclusions

Analysis of ambient air sampling data over the Niwot alpine tundra indicate: (1) The main source region for the ambient air N (non-$N_2$ species) is the greater continental area west of the Rocky Mountains and to a lesser extent the highly polluted metropolitan Front Range area to the east; (2) N dry deposition of $HNO_3$, $NO_3^-$, and $NH_4^+$ species is greater then N wet deposition at the alpine tundra during the June–September growing season; and (3) total annual wet plus dry N deposition is about 6 kg N $ha^{-1}$. Of this, approximately one half occurs during the growing season by the combined deposition of $HNO_3$, $NO_3^-$, and $NH_4^+$. The contribution of $NH_3$, the only other N species that may contribute significantly to N deposition at this site, requires careful scrutiny. $NH_3$ air concentration data indicate that it may be deposited to and emitted from alpine vegetation during the summer months.

Atmosphere-biosphere exchange of N at subalpine forest areas of Niwot Ridge is greater than at the alpine tundra due to upslope flow, which brings substantial quantities of local pollution N to the subalpine almost daily in the summertime but very rarely to the alpine. Total annual N deposition at the C1 subalpine site is estimated to be about 7–8 kg N ha$^{-1}$. This may be compared with conifer forests in the mountainous areas of the eastern United States, where 25–35 kg N ha$^{-1}$ is deposited annually (Johnson and Lindberg 1992). However, the perturbation of the terrestrial N cycle may be reduced at these eastern sites relative to both the tundra and forest areas of Niwot Ridge as a result of higher rates of internal N cycling and greater primary production (Johnson and Lindberg 1992). The ratio of atmospheric deposition to mineralization is about ⅒ to ⅖ at these eastern forest sites and approximately ½ at the Niwot alpine tundra, suggesting a greater potential impact (Sievering et al. 2000, chapter 16).

*Acknowledgments* I thank T. Conway and P. Tans, Carbon Group, NOAA CMDL Lab, Boulder, Colorado, D. Greenland, Professor of Geography, University of North Carolina, and M. Losleben, Climatologist, University of Colorado Mountain Research Station, for their reviews of and data analysis provided for the completion of this chapter. In addition to NSF-LTER funding support, U.S.E.P.A. support under Coop. Ag. #CA 822069-01-3 (D. Mangis, Contract Monitor) is also appreciated.

References

Bowman, W. D. 1992. Inputs and storage of nitrogen in winter snowpack in an alpine ecosystem. *Arctic and Alpine Research* 24:211–215.

Caine, N., and E. Thurman. 1990. Temporal and spatial variations in the solute content of an alpine stream. *Geomorphology* 4:55–72.

Calanni, J., E. Berg, M. Wood, D. Mangis, R. Boyce, W. Weathers, and H. Sievering. 1999. Atmospheric N deposition at a conifer forest: Response of free amino acids in Engelmann spruce needles. *Environmental Pollution* 105:79–89.

Conway, T. J., P. Tans, L. Waterman, K. Thoning, D. Kitzis, K. Mesanie, and N. Zhang. 1994. Evidence for interannual variability of the carbon cycle from the NOAA air sampling network. *Journal of Geophysical Research* 99:22831–22855.

Denmead, O. T., J. R. Freney, and J. R. Simpson. 1976. A closed ammonia cycle within a plant canopy. *Soil Biology and Biochemistry* 8:161–164.

Farquhar, G. D. 1980. On the gaseous exchange of ammonia between leaves and the environment: Determination of ammonia compensation point. *Plant Physiology* 66:710–714.

Geigert, M., N. Nikolaidis, D. Miller, and J. Heitert. 1994. Deposition rates for sulfur and nitrogen to a hardwood forest in northern Connecticut, USA. *Atmospheric Environment* 28:1689–1697.

Georgii, H. W., and U. Lenhard. 1978. Contribution to the atmospheric $NH_3$ budget. *Pure Applied Geophysiology* 116:385–391.

Harris, J. M., and J. Kahl. 1990. A descriptive atmospheric transport climatology for the Mauna Loa Observatory using clustered trajectories. *Journal of Geophysical Research* 95:13651–13667.

Hicks, B., D. Baldocchi, T. Meyers, R. Hosker, and D. Matt. 1987. A preliminary multiple resistance routine for deriving dry deposition velocities from measured quantities. *Water, Air and Soil Pollution* 36:311–330.

Huebert, B. J. and A. L. Lazrus. 1980. Tropospheric gas-phase and particulate nitrate measurements. *Journal of Geophysical Research* 85:7322–7328.

Huebert, B. J., and C. H. Robert. 1985. The dry deposition of nitric acid to grass. *Journal of Geophysical Research* 90:2085–2090.

Johnson, D. W., and S. Lindberg. 1992. *Atmospheric deposition and forest nutrient cycling. Ecological Studies* 91. New York: Springer-Verlag.

Körner, Ch., and M. Diemer. 1987. In situ photosynthetic responses to light, temperature, and carbon dioxide in herbaceous plants from low and high altitude. *Functional Ecology* 1:179–194.

Langford, A., and F. Fehsenfeld. 1992. The role of natural vegetation as a source or sink for atmospheric ammonia: A case study. *Science* 255:581–583.

Langford, A., F. Fehsenfeld, J. Zachariassen, and D. Schimel. 1992. Gaseous ammonia fluxes and background concentrations in US terrestrial ecosystems. *Global Biogeochemical Cycles* 6:459–483.

Losleben, M. 1991. Temporal variations of climatically important anthropogenic trace gas substances at a sub-alpine site. Master's thesis, Department of Geology, University of Colorado, Boulder.

Lovett, G., and J. Kinsman. 1990. Atmospheric pollutant deposition to high-elevation ecosystems. *Atmospheric Environment* 24A:2767–2786.

Neff, J. C., W. D. Bowman, E. Holland, M. Fisk, and S. Schmidt. 1994. Fluxes of $N_2O$ and $CH_4$ from N amended soils in a Colorado alpine system. *Biogeochemistry* 27:23–33.

Ollinger, S., J. Aber, G. Lovett, S. Millham, R. Lathrop, and J. Ellis. 1993. A spatial model of atmospheric deposition for the Northeastern U.S. *Ecological Applications* 3:459–472.

Parrish, D., C. Hahn, D. Fahey, E. Williams, M. Boninger, G. Hubler, M. Buhr, P. Murphy, M. Trainer, E. Hsie, S. Liu, and F. Fehsenfeld. 1990. Systemic variation in the concentration of $NO_x$ (NO plus $NO_2$) at Niwot Ridge, Colorado. *Journal of Geophysical Research* 95:1817–1836.

Peters, K., and R. Eiden. 1992. Modeling dry deposition to a spruce forest. *Atmospheric Environment* 26:2555–2564.

Roberts, J. M., A. Langford, P. D. Goldan, and F. Fehsenfeld. 1988. Ammonia measurements at Niwot Ridge, CO using the tungsten denuder technique. *Journal of Atmospheric Chemistry* 7:137–152.

Rusch, D. 1995. Ammonia and size-distributed ammonium and nitrate measurements at the Niwot Ridge subalpine forest C1 site. MSES Project paper #44.

Schäppi, B., and Ch. Körner. 1996. Growth responses of an alpine grassland to elevated $CO_2$. *Oecologia* 105:43–52.

Schimel, D. S. 1995. Terrestrial ecosystems and the carbon-cycle. *Global Change Biology* 1:77–95.

———. 1998. Climate change—The carbon equation. *Nature* 393:208–209.

Sievering, H., J. Braus, and J. Caine. 1989. Dry deposition of nitrate and sulfate to coniferous canopies in the Rocky Mountains. In *Transactions on effects of air pollution on western forests,* edited by R. K. Olson and A. S. Lefohn. Pittsburgh, PA: Air and Waste Management Assoc.

Sievering, H., D. Burton, and N. Caine. 1992. Atmospheric loading of nitrogen to alpine tundra in the Colorado Front Range. *Global Biogeochemical Cycles* 6:339–346.

Sievering, H., J. Kins, K. Ruoss, G. Roiter, L. Anderson, and R. Dlugi. 1994. Nitrate, nitric acid, and ammonium profiles at the Bayerischer Wald. *Atmospheric Environment* 28A:311–315.

Sievering, H., L. Marquez, and D. Rusch. 1996. Nitric acid, particulate nitrate and ammonium in the continental free troposphere: nitrogen deposition to an alpine tundra ecosystem. *Atmospheric Environment* 30:2527–2537.

Sievering, H., I. Fernandez, and L. Rustad. 2000. Forest canopy uptake of atmospheric nitrogen deposition at eastern U.S. conifer sites: Carbon storage implications? *Global Biogeochemical Cycles*. 14:762–776.

Silverstein, M. C., and D. Greenland. 1991. Wet acid deposition in Colorado. *Physical Geography* 12:55–71.

Sutton, M. A., J. Moncrieff, D. Guerin, and D. Fowler. 1995. Measurement and modelling of ammonia exchange over arable croplands. In *Acid rain research,* edited by G. Hey and J. Erisman. London: Elsevier.

Vitousek, P. M., J. D. Aber, R. W. Howarth, G. E. Likens, P. A. Matson, D. W. Schindler, W. H. Schlesinger, and D. G. Tilman. 1997. Human alteration of the global nitrogen cycle: Sources and consequences. *Ecological Applications* 7:737–750.

Warneck, P. 1988. *Chemistry of the natural atmosphere*. Vol. 41 of *International Geophysics Series*. San Diego: Academic Press.

Williams, M. W., T. Bardsley, and M. Rikkers. 1998. Overestimation of snow depth and inorganic nitrogen wetfall using NADP data, Niwot Ridge, CO. *Atmospheric Environment* 32:3827–3833.

# 4

# Geomorphic Systems of Green Lakes Valley

Nel Caine

## Introduction

There are at least three justifications for the examination of the geomorphology of the area in which ecosystem studies are conducted. First, the present landscape and the materials that make it up provide the substrate on which ecosystem development occurs and may impose constraints, such as where soil resources are limited, on that development. Second, the nature of the landscape and the geomorphic processes acting on it often define a large part of the disturbance regime within which ecosystem processes occur (Swanson et al. 1988). Third, the processes of weathering, erosion, sediment transport, and deposition that define geomorphic dynamics within the landscape are themselves ecosystem processes, for example, involving the supply of resources to organisms. In this last context, it is noteworthy that drainage basins (also called watersheds or catchments) were recognized as units of scientific study during a similar time period in both geomorphology and ecology (Slaymaker and Chorley 1964; Bormann and Likens 1967; Chorley 1969). The drainage basin concept, the contention that lakes and streams act to integrate ecological and geomorphic processes, remains important in both sciences and underlies the studies in Green Lakes Valley reviewed here.

Over the past 30 years, Niwot Ridge and the adjacent catchment of Green Lakes Valley have been the subject of much research in geomorphology. Building on the studies of Outcalt and MacPhail (1965), White (1968), and Benedict (1970), work has emphasized the study of present-day processes and dynamics, especially of mass wasting in alpine areas. These topics have been reviewed by Caine (1974, 1986), Ives (1980), and Thorn and Loewenherz (1987). Studies of geomorphic processes have been conducted in parallel with work on Pleistocene (3 million to 10,000 yr

## 46 Physical Environment

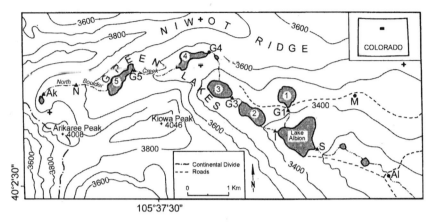

Figure 4.1. Location Maps of the Indian Peaks Areas and Green Lakes Valley. Sites within Green Lakes Valley are identified: Ak, Arikaree; N, Navajo; G5, Green Lake 5; G4, Green Lake 4; G3, Green Lake 3; G1, Green Lake 1; I, Lake Albion Inlet; S, Lake Albion Spillway; M, Martinelli Snowpatch; Al, Albion. The Green Lakes are numbered 1–5.

BP) and Holocene (10,000 yr BP to present) environments in the Colorado Front Range (Madole 1972; Benedict 1973) that have been reviewed by White (1982). This chapter is intended to update those reviews in terms that complement the presentation of ecological phenomena such as nitrogen saturation in the alpine (chapter 5) as well as to refine observations and conclusions of earlier geomorphic studies. In sequence it offers: (1) a brief description of the landforms of Green Lakes Valley and Niwot Ridge, (2) a summary of the historical development of that landscape, (3) a review of contemporary landforming processes in the valley, and (4) an evaluation of climatic influences on landscape development (Barsch and Caine 1984).

The Indian Peaks sector of the Colorado Front Range lies immediately south of Rocky Mountain National Park and straddles the Continental Divide at its most easterly point. This part of the range forms a prominent north-south barrier that contrasts the High Plains—30 km to the east and 2300 m lower in elevation. On the Eastern Slope of the range, environmental research has been concentrated on Niwot Ridge and Green Lakes Valley and in the Loch Vale Watershed (Baron 1992). The Green Lakes Valley is generally oriented west-east, with its western boundary defined by the Continental Divide between Navajo and Arikaree Peaks and its northern boundary by the crest of Niwot Ridge (Fig. 4.1). It is representative of the geomorphic processes of alpine catchments of the Southern Rockies in terms of its size, bedrock, relief, and landforms.

### The Alpine Landscape

The Colorado Rockies provided the type areas from which Barsch and Caine (1984) defined a "Rocky Mountain Relief" type in a global classification of mountain ge-

Table 4.1. Basin Morphometry of the Green Lakes Valley

| | Elevation (m) | Area (ha) | Relief (m) | Relative Relief (m/ha) | Mean Slope (°) | Channel Length (km) | Stream Order | Drainage Density (m/ha) | Stream Gradient (°) | Lake Area (ha) |
|---|---|---|---|---|---|---|---|---|---|---|
| Arikaree | 3790 | 9.2 | 220 | 32.6 | 36.0 | 0.35 | 1 | 38 | 6.4 | 0.2 |
| Navajo | 3730 | 41.8 | 360 | 8.6 | 29.1 | 1.14 | 2 | 27 | 4.4 | 0.2 |
| Green Lake 5 | 3620 | 135.2 | 470 | 3.5 | 22.0 | 4.65 | 3 | 34 | 7.6 | 4.3 |
| Green Lake 4 | 3550 | 220.9 | 540 | 2.4 | 20.0 | 9.33 | 3 | 42 | 3.6 | 8.9 |
| Green Lake 3 | 3450 | 306.0 | 640 | 2.1 | 20.1 | 10.43 | 3 | 34 | 6.9 | 16.4 |
| Albion Inlet | 3345 | 354.7 | 745 | 2.1 | 21.6 | 11.65 | 3 | 33 | 7.6 | 23.2 |
| Spillway | 3345 | 546.2 | 745 | 1.4 | 17.7 | 16.18 | 3 | 30 | 0.0 | 44.8 |
| Albion | 3250 | 709.7 | 840 | 1.2 | 17.5 | 25.67 | 3 | 36 | 2.5 | 55.0 |
| Green Lake 1 | 3415 | 47.3 | 235 | 5.0 | 23.6 | 0.55 | 2 | 12 | 13.2 | 4.1 |
| Martinelli | 3410 | 8.0 | 155 | 19.4 | 18.0 | 0.35 | 1 | 44 | 8.1 | 0.0 |

*Note*: Characteristics are for the catchment area draining to each site. Elevation is that of the lowest point in the basin, stream gradients are estimated for the channel between successive sites, exclusive of lakes, and lake and pond areas are estimated cumulatively and so include all the open water area within the drainage basin.

## 48 Physical Environment

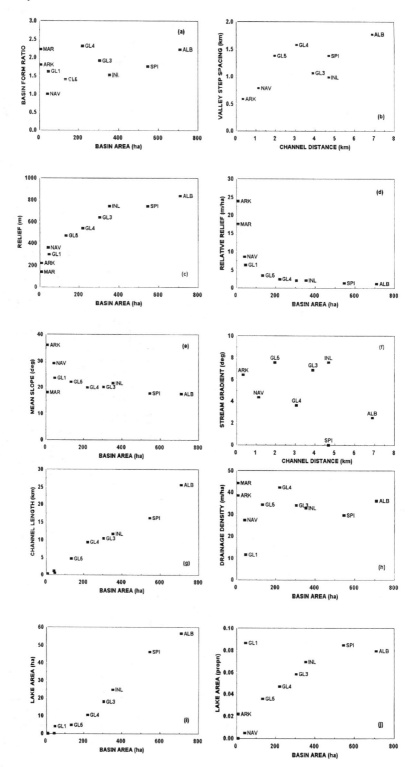

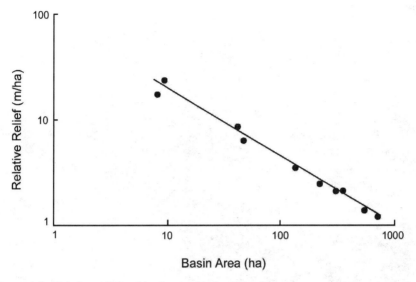

Figure 4.3. Relative relief and basin area for the Green Lakes system. The power relation fitted by least squares has the form: $RR = 82.49\, A^{-0.646}$ with $R^2 = 0.981$.

omorphology typology. This is a landscape that has lower total relief, more-rounded ridges and summits, and a less-pervasive effect of Pleistocene glaciation than many other high mountain systems, such as those of the Alps, Scandinavia, and the Himalaya. It includes a wide variety of finer scaled structural features that have been characterized by O'Neill and Mark (1990) and demonstrates how the slope angles of these mountains are distinctly different from those found in unglaciated areas such as the Appalachians and Oregon.

Studies in Green Lakes Valley have been conducted in a set of 10 nested catchments with areas that vary from less than 10 hectares to the 710 hectares of the entire drainage above the Albion townsite (Fig. 4.1). The morphometric characteristics of these basins are summarized in Table 4.1 and Fig. 4.2. Catchment shape, estimated as a form ratio (i.e., from the equidimensional basin above Navajo to the trapezoidal form of the entire catchment above Albion) varies from 1.0 to 2.5. Within sections of the valley (e.g., between Navajo and Green Lake 4 or between the Lake Albion Inlet and Albion), there is a tendency for the basin shape to elongate with downvalley distance as valley width remains approximately constant (Fig. 4.2a). Relief in the valley increases in predictable fashion with catchment area, but at a lower rate than the increase in basin area (Figs. 4.2c, 4.3). The floor of Green Lakes Valley has the stepped form of a glaciated mountain valley (e.g., Embleton

Figure 4.2. Catchment characteristics in the Green Lakes system. (a) Basin form (L/W) and basin area; (b) valley step spacing and downvalley distance; (c) basin relief and basin area; (d) relative relief and basin area; (e) mean slope and basin area; (f) stream gradient and downvalley distance; (g) channel length and basin area; (h) drainage density and basin area; (i) total lake area and basin area; and (j) relative lake area and basin area.

Figure 4.4. Aerial oblique view of Niwot Ridge from the northeast. The valley of South St. Vrain Creek is in the foreground with the Green Lakes Valley beyond the broad interfluve of Niwot Ridge. Note the correspondence in elevation between Niwot Ridge and the saddle between Kiowa Peak and Mount Albion and the common elevation of the summits along the Continental Divide and along the Gore Range (background) (photo by Nel Caine).

and King 1975) and the spacing of the bedrock steps tends to increase with downvalley distance in the upper valley, to Green Lake 4, and again along the lower valley (Fig. 4.2b). The discontinuity in these trends occurs between Green Lakes 3 and 4 and corresponds to a large step that separates the two different sections of the valley (Caine and Thurman 1990).

The ridges defining Green Lakes Valley have two contrasting forms. Around the upper valley, especially west of Green Lake 4, they consist of bedrock or blockfield-covered, narrow ridges (aretes) between glacial cirques (Fig. 4.4) (White 1976, 1982). Farther east, and most obviously on Niwot Ridge, they consist of broadly convex ridges (Benedict 1970). The surfaces of these ridges are extensively patterned by relict sorted polygons, with active solifluction lobes (wedges of stones created by ice), terraces, and patterned ground, where a high groundwater table is maintained in autumn and winter (Figs. 4.5, 4.6) (Benedict 1970, Fahey 1975).

Below these ridges, the alpine slopes of Green Lakes Valley fit the cliff-talus-sub-talus model that describes a rock-dominated system, the coarse debris system (Caine 1974; Thorn and Loewenherz 1987). These slopes are dominated by rockwalls with steep, though not vertical, profiles that are broken by structurally defined ledges. The alpine cliffs are frequently indented by couloirs (steep avalanche gullies), which channel snow, water, and rockfall debris and so influence the form of the subjacent talus (Rapp 1960; Davinroy 1994). The proportion of the landscape in

Geomorphic Systems of Green Lakes Valley 51

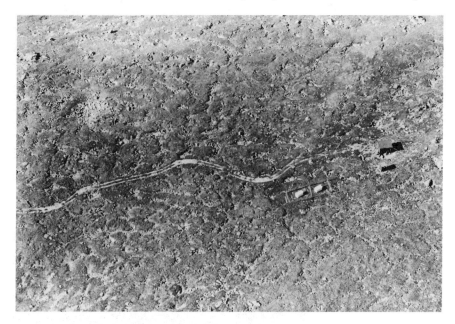

Figure 4.5. Relict patterned ground on the crest of Niwot Ridge at the D1 station. Buildings (right) and jeep road give scale of polygons (photo by Nel Caine).

Figure 4.6. Relict sorted patterns on Niwot Ridge 200 m west of the Saddle research site (photo by Nel Caine).

Figure 4.7. Talus beneath the north side of Kiowa Peak (photo by Nel Caine).

the cliff, talus and sub-talus units (almost 100% above Green Lake 5) is reflected in basin relative relief and mean slope (Fig. 4.2d, e).

The talus slopes of the upper valley are predominantly straight and steep, with angles in excess of 32° (Fig. 4.7) and a planar form (Rapp 1960). Those of the lower valley, especially on its south wall below Kiowa Peak, are more varied. Here, talus development has been more influenced by snow avalanche and debris flow activity, and so the slopes have a more-concave profile form (White 1968). Talus cones (Fig. 4.8), reflecting the gullied topography of the overlying cliff, are also more common and better developed here than in the upper valley.

The lower talus slope and the valley floor beneath it carry a distinctive suite of features, particularly on the north-facing side of the valley. This is where the three valley-side rock glaciers of Green Lakes Valley occur (White 1976). Elsewhere, the toe of the talus slopes may be modified by debris flow deposits or small, lobate terraces, suggesting creep or flow in the blocky debris. A fringe of large blocks is also common where the valley floor is open beyond the foot of the talus.

Figure 4.8. Talus cone below couloir on the south face of Niwot Ridge. Green Lake 5 is in the foreground (photo by Nel Caine).

Most of the streams in Green Lakes Valley are small and carry flows only during the May–September period of snow melt (chapter 5). The channels are generally less than 2 m wide with a stable but rough bouldery floor and occasionally disappear beneath the surface materials in the upper valley. A great variability in flow depths, widths, and gradients (e.g., Fig. 4.2f) contribute to channel hydraulics that are highly variable in both space and time (Furbish 1993). In places such as the inlet to Green Lake 4, a greater organization is forced on the channel form by the influence of deep winter snow cover that forms sub-nival pavements along the drainage line (White 1972; Hara and Thorn 1982).

Channel length increases in simple predictable fashion with increased catchment area (Fig. 4.2g) but drainage density (channel length/area) remains relatively constant, or at least not varying consistently, throughout the nested basins of the system (Fig. 4.2h). Average channel gradients tend to decrease downstream, as basin area increases (Fig. 4.2f) and is clearly conditioned by the stepped form of the valley floor.

## 54 Physical Environment

Table 4.2. Characteristics of the Green Lakes

| | Elevation (m) | Area (ha) | $D^a$ (m) | $D^b_{max}$ (m) | Volume ($*10^6$ m$^3$) | Length (m) | Width (m) | $D_V{}^c$ (m) |
|---|---|---|---|---|---|---|---|---|
| Green Lake 5 | 3620 | 40 | 4.0 | 8.0 | 1.21 | 900 | 250 | 1.50 |
| Green Lake 4 | 3550 | 53 | 4.1 | 13.0 | 2.15 | 850 | 240 | 0.93 |
| Green Lake 3 | 3450 | 75 | 8.6 | 16.0 | 6.47 | 600 | 350 | 1.62 |
| Green Lake 2 | 3400 | 68 | 7.7 | 16.0 | 5.23 | 700 | 240 | 1.44 |
| Green Lake 1 | 3425 | 41 | 3.7 | 7.0 | 1.50 | 500 | 440 | 1.59 |
| Lake Albion | 3345 | 216 | 6.0 | 15.0 | 13.00 | 1000 | 850 | 1.20 |

*Sources:* Data derived from surveys of McNeely (1983) and Harbor (personal communication).

$^a$D = mean depth.
$^b$D$_{max}$ = maximum depth.
$^c$D$_V$ = volume development (D$_V$ = 3 * (D / D$_{max}$).

The six lakes of Green Lakes Valley all occupy natural bedrock basins but have been modified during the last century. Those of the lower valley have been impounded and their levels raised by up to 3.5 m. In the upper valley, Green Lakes 4 and 5 now have levels about 1 m below their natural level because of channelization of their outlet streams.

The morphometry of the lake basins can be estimated from the surveys of McNeely (1983) and Harbor (personal communication). Despite dams, their volumes are relatively slight and amount to no more than 10 days of summer inflows to them (Table 4.2). The total area of lakes and their importance in terms of dominating biogeochemical processes within the landscape tend to increase downvalley (Fig. 4.2i, j). All of the lakes occupy bedrock basins that become less linear with downvalley distance: The length/width ratio changes from more than 3.5 at Green Lakes 4 and 5 to 1.2 at Lake Albion (Table 4.2). Volume development (D$_v$), as estimated by the ratio: 3D/D$_{max}$ (where D = mean depth and Dmax = is maximum depth) (Hutchinson 1957), reflects the origin of the lakes as glacial rock basins and the pattern of valley steps (Table 4.2). D$_v$ is greatest at Green Lake 1, Green Lake 3, and Green Lake 5, all situations where glacial overdeepening of the valley is most marked.

## The Development of the Landscape

### Pre-Quaternary History

The geologic history of Green Lakes Valley is that of the Colorado Front Range, which originated in the regional uplift of the Laramide Orogeny (70–50 Ma) (e.g., Madole et al. 1987). Erosion during uplift subsequently removed between 3.0 and 3.5 km of Paleozoic and Mesozoic rocks from the crest of the range to expose the granites and gneisses of its core (Gregory and Chase 1992, 1994).

The east slope of the Colorado Front Range descends from the Continental Divide (at about 4000 m elevation) to the High Plains (below 2300 m) as a series of summits and ridges. The nature and origin of the erosion surface(s) has been the subject of debate for more than 100 years (Bradley, 1987). It is now usual to interpret this landscape as derived from a single surface of late-Eocene age (Epis and Chapin 1975; Gregory and Chase 1992), which may have had considerable relief and may also have been warped and faulted during late-Tertiary uplift (Bradley 1987).

The history of the late-Eocene surface after its formation is also the subject of continuing debate. For a long time, there was broad agreement that it had been produced at low elevation, perhaps close to sea level (Davis 1911), and then uplifted 2000–3000 m during the Pliocene. This post-Laramide uplift initiated fluvial dissection during an episode of valley incision to define the broad outline of the present drainage pattern (Wahlstrom 1947; Scott 1975; Bradley 1987). Such uplift may have been partially driven by isostatic responses to valley erosion (e.g., Small and Anderson 1995; Wernicke et al. 1996). This would also be consistent with the suggestion that erosion surfaces like that of the Front Range may develop at relatively high elevation with later dissection and erosion initiated by climatic change rather than tectonic uplift (Molnar and England 1990; Gregory and Chase 1992, 1994).

Locally, remnants of the late-Eocene erosion surface on Niwot Ridge and in Green Lakes Valley decline eastward from 3700 m at the alpine climate station (D1) and the saddle between Kiowa Peak and Mt. Albion to 3500 m at Niwot Mountain. It is also represented by the summit elevations of the Indian Peaks themselves (generally between 4000 and 4100 m). Together, these remnants suggest an eastward gradient of ca. 0.06, which is consistent with the elevation of ridges running east from the Continental Divide to the Great Plains.

The erosion surface remnants of the Indian Peaks are covered by coarse deposits that have been extensively reworked by mass wasting and frost processes during the Pleistocene. Initially, these deposits were interpreted as glacial tills from a Pliocene or early Pleistocene icesheet that developed on the range before the present valley system was established (Wahlstrom 1947). They have since been reexamined by Madole (1982), who found no evidence to support a glacial origin.

## Pleistocene Glaciation

The glacial record on the East Slope of the Front Range has been comprehensively reviewed by Madole (1972, 1976, 1986) and Meierding and Birkeland (1980). Glacial deposits and features are generally confined to the valleys above 2500 m elevation and related to three glaciations that are differentiated by moraine morphology, soil development, and surface weathering (Madole 1976). All three glaciations were limited to existing valley systems that they postdate. With the exception of the Lake Devlin deposits (Madole 1976), absolute dating of glacial events is usually restricted to minimum estimates of deglaciation (Fig. 4.9).

Bonnett (1970, quoted in White 1982) reports till of pre-Bull Lake age in the drainage of North Boulder Creek at 2485 m elevation, east of the more obvious moraines between North Boulder Creek and Como Creek. Bull Lake glacial deposits occur more extensively in the North Boulder Creek drainage. These deposits derive from at least two major glacial intervals (stades) that have long been thought to have ended before 70 ka (70,000 years ago) (Richmond 1986; Madole 1976). Recent dat-

| Isotope Stage | Date (ka) | | | |
|---|---|---|---|---|
| 1 | | Arapaho Peak Moraines<br>Audubon Moraines<br>Early Neoglacial (?)<br>Altithermal | Neoglaciation | HOLOCENE |
| | —10— | | | |
| 2 | | "Triple Lakes" Moraines<br>Monarch Moraines<br>Up-valley end moraines<br>Outer end moraines<br>Early Pinedale Moraines (?) | Pinedale Glaciation | LATE PLEISTOCENE |
| | —35— | | | |
| 3 | | | | |
| | —65— | | | |
| 4 | | | | |
| | —79— | | | |
| 5 | | | | |
| | —132— | | | |
| 6 | | Inner Moraine<br>Outer Moraine | Bull Lake Glaciation | |
| | —198— | | | |
| 7 | | | | MIDDLE PLEISTOCENE |
| | —252— | | | |
| 8 | | Outer Moraine (?) | | |
| | —302— | | | |
| 9-11 | | | | |
| | —428— | | | |
| 12 | | Pre-Bull Lake Till | | |
| | —480— | | | |

Figure 4.9. Pleistocene glaciation in the North Boulder Creek Basin. Sources: Benedict (1973); Madole (1976); Richmond (1986); Davis (1988).

ing by cosmogenic beryllium-10 isotope ($^{10}$Be) of the Bull Lake moraines near Fremont Lake, Wind River Range (138 ± 4 ka, $^{10}$Be years) supports this age estimate (Gosse et al. 1995a), leaving a hiatus of more than 50 ka before the subsequent Pinedale glaciation.

The evidence of Pinedale glaciation is a dominant element in the contemporary scenery of the Front Range. Here, as elsewhere in the Rockies, the earliest evidence of Pinedale glaciation dates from about 30 ka (Nelson et al. 1979; Madole 1986; Sturchio et al. 1994). Pinedale glaciation reached its maximum extent by 23 ka, when glacial Lake Devlin was impounded in the valley of a right-bank tributary to North Boulder Creek. The 10,000-yr sequence of proglacial lake sediments that followed this was terminated by catastrophic drainage during deglaciation at 13 ka (Madole 1986). Thereafter, deglaciation of the North Boulder Creek valley, including Green Lakes Valley, seems to have occurred quickly: Harbor (1984) reports a 12-ka basal date from the sediments in Green Lake 5, which matches the age of deglaciation elsewhere in the Rocky Mountains (Davis 1988; Fall et al. 1995; Gosse et al. 1995a).

Within Green Lakes Valley, Pinedale glacial ice accounts for the roches moutonnees (glacially abraded and streamlined bedrock knob) and smoothed bedrock surfaces on the valley floor above Lake Albion. On a larger scale, the stepped long profile of the valley, including the bedrock hollows now occupied by lakes, record the imprint of glacial erosion, though these forms may be largely inherited from earlier glacials (Caine 1986). In the valley above Albion, at 3300 m, the approximate location of the equilibrium line during Pinedale glaciation, there is little evidence of glacial deposition from the main Pinedale glaciation. At the foot of the present Arikaree Glacier (Fig. 4.10), the 5-m-high moraine of "Triple Lakes" age probably represents the latest Pinedale glacial expansion, which may be correlated with the Younger Dryas (around 11,000 years BP) event around the North Atlantic (Zielinski and Davis 1987; Davis 1988; Menounos and Reasoner 1997).

## *Holocene Landform Development*

Early work on the Holocene history of Arikaree cirque, at the head of Green Lakes Valley, suggested three stades (intervals) of Neoglaciation (Ives 1953; Benedict 1968; Mahaney 1971; Carroll 1974). These were widely recognized in the Rocky Mountains but have only recently been well dated (e.g., Zielinski and Davis 1987; Gosse et al. 1995b). Davis (1988) provides a comprehensive review of Neoglacial chronologies in the Rocky Mountains, including the Colorado Front Range.

During the Altithermal, from about 9000 to 5000 BP, fossil evidence from Lake Isabelle suggests milder conditions with a higher tree line than at present (Elias 1985; chapter 15), which supports the suggestion that there were no permanent snowpatches in the Front Range during this interval (Benedict 1973). This would have allowed soil development in the hollows now occupied by snowpatches and explain the existence of paleosols at these sites (Burns 1979, 1980).

Cooler and wetter conditions, with persistent snowpatches and a lower tree line, followed the Altithermal (Elias 1985). This is associated with increased rates of geomorphic activity in the Arapaho Pass area between 5 and 3 ka (Benedict 1985) and

Figure 4.10. Arikaree Glacier: aerial oblique from the northeast. The moraines at the foot of the southern half of the glacier are of "Triple Lakes" age, with late-Holocene glacial deposits on their proximal face. The relict rock glacier (left) distal to these moraines is of late-Pinedale age (photo by Nel Caine).

the development of rock glaciers (lobate body of angular debris with a steep front at the angle of repose) in the Front Range (Madole 1976; Benedict 1981; Dowdeswell 1982; Harbor 1985).

The Audubon stade of Neoglaciation (2.4–0.95 ka) involved further rock glacier development and extensive snow cover above 3700-m elevation (Benedict 1973, 1993). In Green Lakes Valley, the lichen-kill zones around and north of Arikaree Glacier are evidence of a more-extensive snow cover at this time (Fig. 4.10; Mahaney 1971; Carroll 1974).

The Little Ice Age is represented in the Indian Peaks by deposits of the Arapaho Peak stade (300–100 BP) (Benedict 1973), including small moraines on the proximal face of earlier moraines at Arikaree Glacier. Glacier recession in the twentieth century coincides with documentary records (Lee 1900; Fenneman 1902; Henderson 1904; Ives 1940, 1953), which show a marked loss of ice mass from Arapaho Glacier between 1900 and the midcentury (Waldrop 1964) and a great reduction in the intensity of glacial processes (Reheis 1975). In Green Lakes Valley, Arikaree Glacier experienced a similar loss of mass in the first half of this century (Ives 1953) but has since maintained a quasi-balanced mass budget (Johnson 1980).

Holocene hillslope development in the alpine of the Front Range has undergone marked changes in intensity during the past 10,000 years (Benedict 1985). Neoglacial intervals of extensive snow cover and glacier expansion correlate with periods of greater activity in the cliff-talus and rock glacier systems (Madole 1976; White 1976; Dowdeswell 1982; Harbor 1985). On other alpine slopes, the same intervals of wetter, cooler conditions coincide with intervals of more rapid movement in stone-banked terraces and solifluction lobes (Benedict 1967, 1970, 1973, 1976, 1993).

In general, Holocene lake sediments reflect geomorphic activity in the terrestrial environment only poorly (Caine 1974; Andrews et al. 1985; Harbor 1984, 1985). Variations in the organic content of the sediments in the small lake below Arapaho rock glacier show intervals of increased geomorphic activity during the late-Holocene (Davis 1988). However, most of the Holocene sediments in the larger alpine lakes are homogeneous with few textural changes that would reflect changes in hillslope activity (Caine 1986).

## Alpine Geomorphic Processes

Contemporary geomorphic activity in Green Lakes Valley has been described in terms of three overlapping systems that involve different materials and dynamics (Caine 1974, 1986; Thorn and Loewenherz 1987). These are the coarse debris system, the fine sediment system, and the geochemical system, which are treated here separately.

### The Coarse Debris System

As in many mountain areas, the movement and storage of coarse debris (>8-mm size) dominates the sediment budget of Green Lakes Valley (Table 4.3). This sys-

Table 4.3. Sediment Budgets for the Upper Part of the Green Lakes Valley Above Green Lake 4

|  | Volume ($m^3 \, yr^{-1}$) | Work ($10^6 \, J \, yr^{-1}$) |
|---|---|---|
| Coarse debris system | | |
| Rockfall | 10.0 | 44.6 |
| Talus accumulation | 1.5 | 2.87 |
| Talus shift | 206,500 | 14.43 |
| Debris flow | 9.4 | 15.32 |
| Rock glacier flow | 373,000 | 3.59 |
| Subtotal | 579,521 | 80.81 |
| Fine sediment system | | |
| Soil creep and solifluction | 95,440 | 0.90 |
| Surficial wasting | 240 | 0.03 |
| Lake sedimentation | 13.5 | 1.00 |
| Fluvial sediment export | 0.38 | 0.05 |
| Eolian transport | 20 | ? |
| Subtotal | 95,714 | 2.25 |
| Geochemical system | | |
| Solute transport | 1.9 | 13.36 |

*Source:* Caine (1986) with selected updates and additions.

*Note:* Solute transport, corrected for solutes in precipitation, is based upon the sum of major cations and $SiO_2$ concentrations (Table 4.3) and an assumed bedrock density of 2600 kg $m^{-3}$.

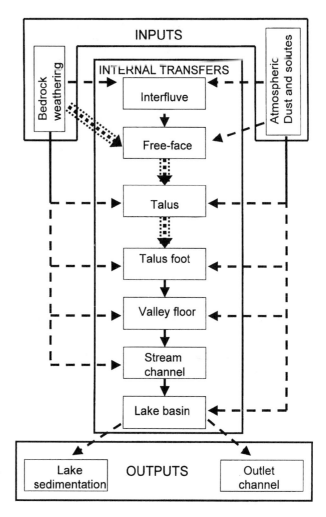

Figure 4.11. Alpine geomorphic systems. Source: Caine (1974).

tem involves the bedrock cliffs, their associated talus, and the accumulated blocky debris in the talus slopes and rock glaciers below them (Fig. 4.11). This system has been effectively closed during the Holocene but was presumably breached during glacial periods when the accumulated debris in it would be incorporated into the tills deposited at lower elevations (Figs. 4.7, 4.8).

A 25-year record in upper Green Lakes suggests that rockfall from the alpine cliffs yields about 12 m$^3$ yr$^{-1}$ of debris, corresponding to a rate of cliff retreat of only 0.02 mm yr$^{-1}$ (Caine 1986). By comparison to the existing volume of talus in the valley, this mass is negligible and has little effect on the landscape. Much of the rockfall material is finer than the existing debris mantle into which it is incorporated, without modifying the slope form. Even large rockfalls (involving up to 50 m$^3$ each), which occur with a decadal frequency, have little effect on the talus for the large blocks in them usually move over the talus to its toe.

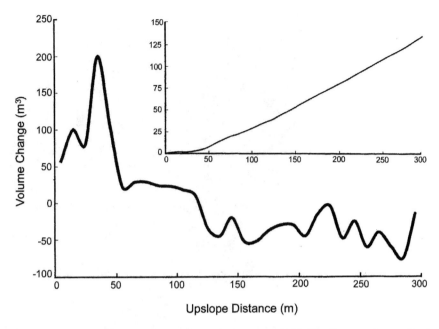

Figure 4.12. Mass budget of the debris flow of June 26, 1988. The flow, on the south valley wall between Green Lake 4 and Green Lake 5, occurred in response to a 52-mm rainstorm (recurrence interval of about 100 years in the Green Lakes area). Inset is longitudinal profile of the flow (true scale).

Movement of the Green Lakes talus was noted by Ives (1940) and has since been measured by White (1981), Caine (1986), and Davinroy (1994). On average, it amounts to less than 3 cm yr$^{-1}$ at the surface, tends to decrease downslope from the head of the talus, and is highly variable (e.g., Gardner 1979). Where flows of water, snow, and debris are concentrated through couloirs, rates of deposition and movement may be two orders of magnitude greater than those on open talus slopes (Gardner 1982, 1986; Davinroy 1994). White (1968) reported the influence of debris flows (Figs. 4.12, 4.13) and wet snow avalanches in forming concave talus profiles. The influence on sediment transport of avalanches on Arikaree Peak has been reported in Caine (1986, 1995a). Dry snow avalanches from Kiowa Peak occasionally transport coarse debris onto the ice cover of Green Lake 3, contributing to the sediments of that lake.

In the talus foot zone, the valley-side rock glacier at Green Lake 5 is the only obviously active one in Green Lakes Valley (White 1981). It has a surface velocity of less than 2.0 cm yr$^{-1}$, an order of magnitude less than the flow rates of valley-floor rock glaciers in the Front Range (White 1981; Benedict et al. 1986). The other two valley-side rock glaciers, at Green Lake 2 and in the Arikaree cirque, appear to have been inactive for millenia. Lichens and rock weathering on the former suggest that it stabilized more than 2000 years ago. The latter appears even older and was ascribed a late-Pinedale age by Carroll (1974).

Figure 4.13. Debris flow deposits on the south valley wall between Green Lake 4 and Green Lake 5. This flow occurred on June 26, 1988 as a response to a rainstorm that produced 52 mm of precipitation in 8 hours. Previous debris flow activity on this slope occurred on July 15, 1965 in response to 21 mm of rain in 4 hours (photo by Nel Caine).

## *The Fine Sediment System*

The fine sediment cascade involves the slow processes of soil creep, frost creep, gelifluction (downward movement of soil due to freeze-thaw), and surficial erosion as well as the more catastrophic ones of landsliding and debris flow. Unlike the coarse debris of talus and blockfields, the finer material of this system may be entrained by water and air flows and incorporated into more-extensive exchanges, potentially extending beyond the alpine environment (Thorn and Darmody 1980, 1985; Burns and Tonkin 1982; Caine 1986; Litaor 1987). However, in the upper Green Lakes Valley, the fine sediment system involves only 15–20% of the mass in the coarse debris system (Table 4.3).

On Niwot Ridge, rates of solifluction and frost creep in the debris mantle are maximized in solifluction lobes and terraces that remain wet in autumn and early winter (Benedict 1970; Table 4.4). In better drained sites, rates of solifluction and creep are an order of magnitude lower (less than $0.1$ cm yr$^{-1}$), and these rates are probably more typical of the alpine tundra (Caine 1986).

Table 4.4. Landforms in Mountain Areas

| Elevation | Zone | | Source |
|---|---|---|---|
| Highest | Glacier | Active glaciers, perennial snow, ice | 1, 2 |
| | Blockfield zone | Bedrock, blockfield, talus | 3, 4 |
| | | Sorted patterned ground, "unbound" solifluction | 3, 4 |
| | Solifluction zone | Hummocks | 3 |
| | | Vegetated solifluction forms, micro-terracettes | 5 |
| Lowest | Forest zone | Treeline, krummholz | |

*Sources:* 1. Rapp and Rudberg (1960); 2. Caine (1978); 3. Rudberg (1972); 4. Hollerman (1967); and 5. Hastenrath (1974).

Estimates of soil erodibility suggest that surface erosion by rainsplash, surface runoff, frost action, and wind may be high where the tundra vegetation is disturbed (Summer 1982). Empirical studies of surface erosion support this and show that rates of soil loss increase by orders of magnitude where the vegetative cover is reduced to less than 25% in snowpatch sites (Caine 1976; Bovis 1978; Bovis and Thorn 1981). These sites become sources of sediments when they are exposed to summer rainstorms, although they may be protected from erosion by a long-lasting snow cover (Caine 1992; Caine and Swanson 1989). Changes in the winter snow cover during the late-Holocene should have modulated these influences over the alpine tundra (Benedict 1993).

Further sites of surficial erosion by wind and rain are the soil mounds and tunnel casts of pocket gophers (Thorn 1982). In fact, Thorn (1978) suggests that this is the most effective mechanism of soil erosion in the tundra of the Colorado Rockies; turning over up to 1500 kg ha$^{-1}$ of soil each year (Stoecker 1976; chapters 7, 8).

Eolian deposition of dust into Green Lakes Valley may exceed 1 g m$^{-2}$ yr$^{-1}$ of silt and clay and appears to be an important source of buffering in the geochemical system (Litaor 1987). The source of the eolian deposits remains unstudied but is likely from the southwestern United States (chapter 3). On tundra surfaces, these fine particles are probably retained to become involved in pedogenesis (Thorn and Darmody 1980, 1985; Burns and Tonkin 1982); on bedrock and boulder surfaces, they may be transported through the fluvial system into lake sediments (Caine 1986).

Little of this activity involving fine sediment is reflected in suspended sediment transport through the stream systems of Green Lakes Valley (Table 4.3). Suspended sediment concentrations in the streams are generally low and yields only exceed 1.0 g m$^{-2}$ yr$^{-1}$ in years of high stream discharge (Caine 1995a). Snowpatch sites appear to be important sources of this sediment, but even there, sediment concentration-discharge relationships are not simple but reflect complex influences of the snow cover (Caine 1989, 1992).

The stream channels of Green Lakes Valley are effectively covered by coarse sediments and remarkably stable. The only visible changes effected by channelized flows occur where high meltwater flows released by the failure of snow and ice dams are superimposed from the snowcover onto the valley floor, usually outside the stream channels (Caine 1995a). The lake sediments are invariably fine grained, except where high-energy slopes impinge directly on the lake shore (as at Green Lake 3), and are predominantly of silt and clay (Caine 1986). Over the last 5000 years, sedimentation into Green Lakes 4 and 5 has averaged only 0.15 mm yr$^{-1}$ (Harbor 1984), which is consistent with the low suspended sediment concentrations of the streamflows.

## The Geochemical System

Geochemical weathering rates in Green Lakes Valley were first estimated for the Martinelli catchment by Thorn (1975, 1976) and have been expanded during LTER studies of the entire valley. Solute concentrations and water discharges have been monitored in nine basins of varying size for 15 years. Solute concentrations in the stream waters are generally low but with regular temporal and spatial patterns (Caine and Thurman 1990).

Caine and Thurman (1990) found no consistent temporal trends in major ion concentrations and yields but more recent analyses of a longer record suggest a tendency toward acidification (Williams et al. 1996). This is most marked in the higher parts of Green Lakes Valley (Caine 1995b), corroborating the forecasts of Lewis (1982) and Kling and Grant (1984). The pattern of acidification has not yet been detected in increased base cation concentrations, which would reflect changing rates of geochemical denudation.

On an annual time scale, patterns of solute removal are dominated by the seasonal cycle, which explains about 65% of the variability in long-term records of specific conductance at Albion and Green Lake 4 (Caine and Thurman 1990). This pattern reflects the influence of snowmelt (Williams et al. 1993; Campbell et al. 1995). In May and June early meltwater flows produce relatively high concentrations in solutes at all sites in Green Lakes Valley (Caine and Thurman 1990; chapter 5). Concentrations then decline to low levels during the summer but rise gradually from late summer into winter as the relative proportions of ground and soil water contributions to streamflow increase. In small headwater basins, an equivalent out-of-phase periodicity in solute concentrations and water flows occurs on a daily basis during the summer (Caine 1989).

Solute concentrations increase consistently downstream (Table 4.5), reflecting a greater mean residence time of water; greater contributions to streamflow from groundwater, and increased biological activity. These influences are also evident in a comparison of the Green Lake 1 and Martinelli catchment records with those from basins of similar area at the head of the main valley (Caine and Thurman 1990).

Total solute yields from Green Lakes Valley average 10 g m$^{-2}$ yr$^{-1}$ (Table 4.5). Corrected for precipitation loading, this represents a bedrock lowering rate of less than 0.003 mm yr$^{-1}$, appreciably lower than rates in mountain catchments elsewhere (e.g., Dethier 1986; Souchez and Lemmens 1987; Pape et al. 1995) but still greater than the rate of sediment yield (Table 4.5).

Spatial patterns of solute yields and concentrations reflect patterns of snow accumulation and meltwater drainage, modulated by soil- or groundwater flows (Dixon 1986; Caine 1992; Litaor 1993). These patterns are also reflected in varia-

Table 4.5. Geochemical Denudation in the Green Lakes Valley

|  | T.D.S. | Cations | $SiO_2$ | Cl | Water (mm) | Sediment |
|---|---|---|---|---|---|---|
| Precipitation | 5.63 | 0.58 | 0.01 | 0.23 | 1218 | 1.00[a] |
| Arikaree | 15.76 | 2.51 | 1.58 | 0.32 | 3845 |  |
| Navajo | 8.89 | 1.71 | 1.27 | 0.25 | 1545 | 0.27 |
| Green Lake 5 | 7.22 | 1.51 | 0.82 | 0.19 | 984 |  |
| Green Lake 4 | 8.46 | 1.76 | 1.07 | 0.20 | 868 | 0.44 |
| Albion Inlet | 6.93 | 1.66 | 0.62 | 0.12 | 592 |  |
| Spillway | 7.84 | 1.87 | 0.60 | 0.11 | 601 |  |
| Albion | 10.85 | 2.45 | 1.24 | 0.25 | 640 | 0.59 |
| Green Lake 1 | 4.13 | 0.97 | 0.38 | 0.03 | 91 |  |
| Martinelli | 9.97 | 1.91 | 1.59 | 0.22 | 1206 | 0.54 |

*Note:* Except for water, all values are g m$^{-2}$ yr$^{-1}$ and are based on 10–14 years of record (1981–1994).

[a] The dustfall estimate includes only the silt-clay fraction (Litaor 1987).

tions in the thickness of the chemically altered surface layers of rocks (Thorn 1975), although these may be affected by differential removal rates (Ballantyne et al. 1989).

## *Materials Budgets*

Sediment budgets for Green Lakes Valley above Green Lake 4 were summarized by Caine (1986) and have been updated with more recent data (Table 4.3). The coarse debris system accounts for over 95% of all geomorphic work accomplished at the present time in the valley. Within that system, rockfall dominates because of the relatively great (vertical) distance of transport that it entails, especially in the larger falls, which often traverse the entire talus slope. The geomorphic work done in talus shift is less than this by a factor of three or four, despite the large mass of material involved. By comparison, other processes of coarse debris movement are relatively slight contributors to work done in the basin, although they may be important on individual slopes.

As in other coarse debris systems (e.g., Rapp 1960), contemporary rates of talus accumulation and shift in Green Lakes Valley are incapable of accounting for the volume of coarse debris accumulated on the valley walls since deglaciation (Caine 1986). About $0.8 \times 10^6$ m$^3$ of talus and a further $1.0 \times 10^6$ m$^3$ of other debris have accumulated in the upper valley over 12 ka, at an average rate of almost 150 m$^3$ yr$^{-1}$, an order of magnitude greater than the rate shown by contemporary measurement (Table 4.3). This disparity argues for coarse debris production at rates up to two orders of magnitude greater than the present during the late Pleistocene and in Neoglacial stades (Harbor 1985). As Wahrhaftig (1987) points out with respect to rock glacier systems, this coarse material is likely to remain stored within the alpine until redistributed or removed by another major glaciation.

The fine sediment system is dominated by the large mass of material involved in soil creep and solifluction, which accounts for almost half of the work done in the fine sediment system. The fluvial transport of silt and clay to lake basins involves a smaller mass but a greater vertical transport, which allows it to contribute more than 50% to the work done in the system. The major uncertainty in this budget concerns eolian transport, for which the volume of sediment involved and distances of transport are still only poorly estimated.

In contrast to the systems involving sediment and debris, the geochemical system is one in which the coupling of slope and stream systems is the closest. For that reason, it is estimated with greater precision as an average for entire drainage basins. In terms of total mass and work estimates, it has relatively low magnitude, accounting for only 16% of the total geomorphic work done in the basin and a much smaller proportion of the mass involved in geomorphic exchanges (Table 4.3). Nevertheless, it dominates the estimate of basin denudation, accounting for more than 92% of the work done in sediment and solute export and lake sedimentation.

## Morphoclimatic Landforms

In mountain environments, the concept of a climatic influence on landscape development has usually been defined in terms of altitudinal limits and elevational zones of landforms and processes (e.g., Budel 1954; Bik 1967; Rudberg 1972).

Above treeline, the periglacial landforms of Niwot Ridge and Green Lakes Valley may be classified into the blockfield and solifluction zones of Rapp and Rudberg (1960) and Rudberg (1972) (Table 4.4). Within these zones, landform patterns like those of other mid-latitude mountains (Chardon 1984; Kotarba 1984) are found. However, they are not defined by elevation alone, requiring an orthogonal criterion based on topographic location (e.g., Caine 1978). This separates the landforms of ridges and valleys and so reflects a historical influence, distinguishing landforms in areas of late-Pleistocene glaciation from those on terrain that has remained extraglacial at that time (Table 4.6).

In the solifluction zone, mapping of small-scale forms on Niwot Ridge by Benedict (1970) and surveys of forms in Green Lakes Valley allow the limits of active periglacial forms to be defined (Table 4.6). The variety of periglacial forms found in the zone suggest elevation limits that vary between 3350 m and 3520 m with some difference between the valley floor environment and that of the interfluve. Active sorted patterned ground suggests a 70-m difference in elevational limits between Niwot Ridge and the neighboring valley (Warburton 1991).

The features of the blockfield zone are separated spatially from those of the solifluction zone, with the single exception of talus that is active (if not intensely so) along the south valley wall to the elevation of Lake Albion (Table 4.6). However, this is in response to bedrock instability and high slope angles rather than to a direct climatic influence. In contrast, blockfield and rock glacier forms are found only above 3650 m, and the latter are restricted to the south wall of the valley, as elsewhere in the Front Range (White 1971).

Above the blockfield zone in the Indian Peaks is a glacial zone, at 3800 m and above, represented in Green Lakes Valley by Arikaree Glacier (Table 4.6). Glacial action here is presently minimal.

Table 4.6. Elevational Limits of Active Periglacial Forms in the Green Lakes Valley and on Niwot Ridge

| | Interfluves | Limit (m) | Valley Locations | Limit (m) |
|---|---|---|---|---|
| Glacier zone | | | Arikaree glacier | 3800 |
| Blockfield zone | Autochthonous blockfields | 3700 | Tongue-shaped rock glacier | ?[a] |
| | | | Allochthonous (?) blockfields | 3700 |
| | | | Lobate rock glacier | 3650 |
| | | | Rockfall talus | 3200 |
| Solifluction zone | Sorted Patterned ground | 3625 | Sorted patterned ground | 3550 |
| | Non-sorted circles | 3500 | Non-sorted circles | 3450 |
| | Stone-banked terraces | 3475 | Sub-nival pavements | 3425 |
| | Hummocks | 3450 | Hummocks | 3350 |
| | Turf-banked lobes and terraces | 3400 | | |

Sources: Outcalt and MacPhail (1965), Benedict (1970), White (1971), Madole (1972), Fahey (1975), and Warburton (1991).

[a] The only tongue-shaped rock glacier in Green Lakes Valley is the inactive one immediately below Arikaree Glacier at 3800 m.

Attempts to define morphoclimatic zones on the basis of contemporary erosion on small plots and denudation rates in stream basins have not usually been successful. For example, Caine (1984) found only slight differences between the "high alpine" and the "alpine tundra" zones. These minor differences are particularly related to (1) the presence of weak glacial effects in the higher zone, (2) the greater activity in the talus slopes of the coarse debris system, and (3) slight increases in geochemical concentrations at lower elevations (Caine 1984).

On the broader scale of the entire Front Range, general contrasts of geomorphic levels are more evident. Although it includes areas of high activity, the subalpine forest is the least active in geomorphic terms today, with the exception, perhaps, of geochemical processing (Bovis 1978; Caine 1984). However, even this relative stability may be catastrophically interrupted by fire (Morris 1986; Morris and Moses 1987), wind storms, or anthropogenic disturbance of the ground cover (Bovis 1978). The stream channels of the subalpine belt often suggest higher rates of sediment transport and channel modification than those of the alpine, but this, too, could be a reflection of more than a century of channel and catchment disturbance.

## Conclusion

The geomorphic systems of the Green Lakes Valley and Niwot Ridge are representative of the alpine areas of the southern Rocky Mountains, including the variety of landforms found in that broader geographic context. The effects of Pleistocene environments remain dominant in the present alpine landscape: in the glacial landforms of the valleys and on remnants of the Eocene erosion surface where periglacial modification of the surface materials occurred at the time of valley glaciation. Indirectly, glacial erosion has greatly affected the valley walls by steepening them and so allowing the Holocene development of talus and cliff systems.

In contrast to the lasting imprint of late Pleistocene and Neoglacial conditions, present-day processes of landscape change appear to be relatively modest. Present-day rates of geomorphic activity are slight when compared with those of other high mountain areas, as they are when compared with those of the past. This is especially true if a denudation rate, that is, the export of sediment and solutes, is used to estimate geomorphic activity for the entire system or for parts of it. Mass loss through the stream channels represents no more than 15% of the geomorphic work done in the basin. The remaining work is done in a closed system from which the accumulating coarse debris will not be discharged to lower elevations until the advent of renewed glacial conditions. This relative calm in geomorphic activities allows for the expression of biotic activities to assume a relatively large role in the biogeochemistry of the alpine.

All of this contributes to the high variability in alpine landforms, which is an important influence on other environmental processes and on biotic patterns. The resulting variety also accounts for much of the visual attractiveness of this mountain landscape.

References

Andrews, J. T., P. W. Birkeland, J. Harbor, N. Dellamonte, M. Litaor, and R. Kihl. 1985. Holocene sediment record, Blue Lake, Colorado Front Range. *Zeitschrift fur Gletscherkunde und Glazialgeologie* 21:25–34.

Ballantyne, C. K., N. M. Black, and D. P. Finlay. 1989. Enhanced boulder weathering under late-lying snowpatches. *Earth Surface Processes and Landforms* 14:745–750.

Baron, J. S. 1992. Biogeochemistry of a subalpine ecosystem: Loch Vale watershed. *Ecological Studies* 90. New York: Springer-Verlag.

Barsch, D., and N. Caine. 1984. The nature of mountain geomorphology. *Mountain Research and Development* 4:287–298.

Benedict, J. B. 1967. Recent glacial history of an alpine area in the Colorado Front Range, U.S.A. I. Establishing a lichen-growth curve. *Journal of Glaciology* 6:817–832.

———. 1968. Recent glacial history of an alpine area in the Colorado Front Range, U.S.A. II. Dating the glacial deposits. *Journal of Glaciology* 7:77–87.

———. 1970. Downslope soil movement in a Colorado alpine region: rates, processes and climatic significance. *Arctic and Alpine Research* 2:165–226.

———. 1973. Chronology of cirque glaciation, Colorado Front Range, U.S.A. *Quaternary Research* 3:584–599.

———. 1976. Frost creep and gelifluction features: A review. *Quaternary Research* 6:55–76.

———. 1981. The Fourth of July Valley: Glacial geology and archeology of the timberline ecotone. Ward, CO: Center for Mountain Archeology Research Report 2: 139 pp.

———. 1985. Arapaho Pass: Glacial geology and archeology at the crest of the Colorado Front Range. Ward, CO: Center for Mountain Archeology Research Report 3: 197 pp.

———. 1993. A 2000-year lichen-snowkill chronology for the Colorado Front Range, U.S.A. *The Holocene* 3:27–33.

Benedict, J. B., R. J. Benedict, and D. Sanville. 1986. Arapaho Rock Glacier, Front Range, Colorado, USA: a 25-year resurvey. *Arctic and Alpine Research*, 18:349–352.

Bik, M. J. J. 1967. Structural gomorphology and morphoclimatic zonation in the Central Highlands, Australian New Guinea. In *Landform studies from Australia and New Guinea,* edited by J. N. Jennings and J. A. Mabbutt. Canberra, Australia: A.N.U. Press.

Bonnett, R. B. 1970. The glacial sequence of upper Boulder Creek drainage basin in the Colorado Front Range. Ph.D. diss., Ohio State University.

Borman, F. H., and G. E. Likens. 1967. Nutrient cycling. *Science* 155:424–429.

Bovis, M. J. 1978. Soil loss in the Colorado Front Range: Sampling design and areal variation. *Zeitschrift für Geomorphologie Supplementband* 29:10–21.

Bovis, M. J., and C. E. Thorn. 1981. Soil loss variation within a Colorado alpine area. *Earth Surface Processes and Landforms* 6:151–163.

Bradley, W. C. 1987. Erosion surfaces of the Colorado Front Range: A review. In *Geomorphic systems of North America,* edited by W. L. Graf. Boulder, CO: Geological Society of America Centennial Special Vol. 2.

Budel, J. 1954. Klima-morphologische Arbeiten in Athiopien im Fruhjahr 1953. *Erdkunde* 8:139–156.

Burns, S. F. 1979. Buried soils beneath alpine snowbanks may date the end of the Altithermal. Geological Society of America, *Abstracts with Programs* 11:267–268.

———. 1980. Alpine soil distribution and development, Indian Peaks, Colorado Front Range. Ph.D. diss., University of Colorado, Boulder.

Burns, S. F., and P. J. Tonkin. 1982. Soil-geomorphic models and the spatial distribution and development of alpine soils. In *Space and time in geomorphology,* edited by C. E. Thorn. London: Allen and Unwin.

Caine, N. 1974. Geomorphic processes of the alpine environment. In *Arctic and alpine environments,* edited by J. D. Ives and R. G. Barry. London: Methuen.

———. 1976. The influence of snow and increased snowfall on contemporary geomorphic processes in alpine areas. In *Ecological impacts of snowpack augmentation in the San Juan Mountains, Colorado.* edited by H. W. Steinhoff and J. D. Ives. CSU-DWS 7052-4, Fort Collins, CO: Colorado State University.

———. 1978. Climatic geomorphology in mid-latitude mountains. In *Landform evolution in Australasia,* edited by J. L. Davies and M. A. J. Williams. Canberra, Australia: A.N.U Press.

———. 1984. Elevational contrasts in contemporary geomorphic activity in the Colorado Front Range. *Studia Geomorphologica Carpatho-Balcanica* 18:5–31.

———. 1986. Sediment movement and storage on alpine hillslopes in the Colorado Rocky Mountains. In *Hillslope processes,* edited by A. D. Abrahams. Winchester, MA: Allen and Unwin.

———. 1989. Diurnal variations in the inorganic solute content of water draining from an alpine snowpatch. *Catena* 16:153–162.

———. 1992. Sediment transfer on the floor of the Martinelli Snowpatch, Colorado Front Range, U.S.A. *Geografiska Annaler* 74A:133–144.

———. 1995a. Snowpack influences on geomorphic processes in Green Lakes Valley, Colorado Front Range. *Geographical Journal* 161:55–68.

———. 1995b. Temporal trends in the quality of streamwater in an alpine environment: Green Lakes Valley, Colorado Front Range, U.S.A. *Geografiska Annaler* A: 207–220.

Caine, N., and F. J. Swanson. 1989. Geomorphic coupling of hillslope and channel systems in two small mountain basins. *Zeitschrift für Geomorphologie* 33:189–203.

Caine, N., and E. M. Thurman. 1990. Temporal and spatial variations in the solute content of an alpine stream, Colorado Front Range. *Geomorphology* 4:55–72.

Campbell, D. H., D. W. Clow, G. P. Ingersoll, M. A. Mast, N. E. Spahr, and J. T. Turk. 1995. Processes controlling the chemistry of two snowmelt-dominated streams in the Rocky Mountains. *Water Resources Research* 3:2811–2821.

Carroll, T. 1974. Relative age-dating techniques and a late Quaternary chronology, Arikaree Cirque, Colorado. *Geology* 2:321–325.

Chardon, M. 1984: L'etagement des paysages et les processus geomorphologiques actuels dans les Alpes occidentales. *Studia Geomorphologica Carpatho-Balcanica* 18:33–43.

Chorley, R. J. 1969. The drainage basin as the fundamental geomorphic unti. In *Water, earth and man,* edited by R. J. Chorley. London: Methuen.

Davinroy, T. C. 1994. Rates and controls of rock movement through alpine couloirs, Colorado Front Range. Master's thesis, University of Colorado, Boulder.

Davis, W. M. 1911. The Colorado Front Range. *Annals Association of American Geographers* 1:21–83

Davis, P. T. 1988. Holocene glacier fluctuations in the America Cordillera. *Quaternary Science Reviews* 7:129–157.

Dethier, D. P. 1986. Weathering rates and the geochemical flux from catchments in the Pacific Northwest, U.S.A. In *Rates of chemical weathering of rocks and minerals,* edited by S. M. Colman and D. P. Dethier. Orlando, FL: Academic Press.

Dixon, J. C. 1986. Solute movement on hillslopes in the alpine environment of the Colorado Front Range. In *Hillslope processes,* edited by A. D. Abrahams. Winchester, MA: Allen and Unwin.

Dowdeswell, J. A. 1982. Relative dating of late Quaternary deposits using cluster and discriminant analysis, Audubon Cirque, Mount Audubon, Colorado Front Range. *Boreas* 11:151–161.

Elias, S. E. 1985. Paleoenvironmental interpretation of Holocene insect fossil assemblages from four high altitude sites in the Front Range, Colorado. *Arctic and Alpine Research* 17:31–48.

Embleton, C. and C. A. M. King. 1975. *Glacial and periglacial geomorphology.* Vol. 1, *Glacial geomorphology.* London: Edward Arnold.

Epis, R. C., and C. E. Chapin. 1975. Geomorphic and tectonic implications of the post-Laramide, late Eocene erosion surface in the southern Rocky Mountains. *Geol. Society of America Memoir* 144: 45–74.

Fahey, B. D. 1975. Non-sorted circle development in a Colorado alpine location. *Geografiska Annaler* 57A:153–164.

Fall, P. L., P. T. Davis, and G. A. Zielinski. 1995. Late Quaternary vegetation and climate of the Wind River range, Wyoming. *Quaternary Research* 43:393–404.

Fenneman, N. N. 1902. The Arapahoe Glacier in 1902. *Journal of Geology* 10:839–851.

Furbish, D. J. 1993. Flow structure in a bouldery mountain stream with complex bed topography. *Water Resources Research* 29:2249–2263.

Gardner, J. S. 1979. The movement of material on debris slopes in the Canadian Rockies. *Zeitschrift für Geomorphologie Supplementband* 23:45–57.

———. 1982. Alpine mass wasting in contemporary time: Some examples from the Canadian Rocky Mountains. In *Space and time in geomorphology,* edited by C. E. Thorn. London: Allen and Unwin.

———. 1986. Sediment movement in ephemeral streams on mountain slopes, Canadian Rocky Mountains. In *Hillslope processes,* edited by A. D. Abrahams. Winchester, MA: Allen and Unwin.

Gosse, J. C., J. Klein, E. B. Evenson, B. Lawson, and R. Middleton. 1995a. Beryllium-10 dating of the duration and retreat of the last Pinedale glacial sequence. *Science* 268:1329–1333.

———. 1995b. Precise cosmogenic $^{10}$Be measurements in western North America: Support for a global Younger Dryas cooling event. *Geology* 23:877–880.

Gregory, K. M., and C. G. Chase. 1992. Tectonic significance of paleobotanically estimated climate and altitude of the late Eocene erosion surface, Colorado. *Geology* 20:581–585.

———. 1994. Tectonic and climatic significance of a late Eocene low-relief, high level geomorphic surface, Colorado. *Journal of Geophysical Research* 99B:20141–20160.

Hara, Y. and C. E. Thorn. 1982. Preliminary quantitative study of alpine subnival boulder pavements, Colorado Front Range, U.S.A. *Arctic and Alpine Research* 14:361–367.

Harbor, J. 1984. Terrestrial and lacustrine evidence for Holocene climatic/geomorphic change in the Blue Lake and Green Lakes Valleys of the Colorado Front Range. Master's thesis, University of Colorado, Boulder.

———. 1985. Problems with the interpretation and comparison of Holocene terrestrial and lacustrine deposits: An example from the Colorado Front Range, U.S.A. *Zeitschrift für Gletscherkunde und Glazialgeologie* 21:17–24.

Hastenrath, S. 1974. Glaziale und periglaziale Formbildung in Hoch-Semyen, Nord-Athiopien. *Erdkunde* 28:176–185.

Henderson, J. 1904. Arapaho Glacier in 1903. *Journal of Geology* 12:30–33.

Hollerman, P. W. 1967. Zur Verbreitung rezenter periglazialer Kleinformen in den Pyrenaen und Ostalpen. *Gottinger Geographische Abhandlungen* 40: 198 pp.

Hutchinson, J. E. 1957. *A treatise on limnology.* Vol. 1: *Geography, physics and chemistry.* New York: Wiley.

Ives, R. L. 1940. Rock glaciers in the Colorado Front Range. *Geological Society of America Bulletin* 51:1271–1294.

———. 1953. Later Pleistocene glaciation in the Silver Lake Valley. *Geographical Review* 43:229–253.
Ives, J. D. 1980. Geomorphology overview. In *Geoecology of the Colorado Front Range: A study of alpine and subalpine environments,* edited by J. D. Ives. Boulder, CO: Westview Press.
Johnson, J. B. 1980. Mass balance studies on the Arikaree Glacier. In *Geoecology of the Colorado Front Range: A study of alpine and subalpine environments,* edited by J. D. Ives. Boulder, CO: Westview Press.
Kling, G. W., and M. C. Grant. 1984. Acid precipitation in the Colorado Front Range: An overview with time predictions for significant effects. *Arctic and Alpine Research* 16:321–329.
Kotarba, A. 1984. Elevational differentiation of slope geomorphic processes in the Polish tatra Mountains. *Studia Geomorphologica Carpatho-Balcanica* 18:117–133.
Lee, W. T. 1900. The glacier of Mt Arapahoe, Colorado. *Journal of Geology* 8:647–654.
Lewis, W. M., Jr. 1982. Changes in pH and buffering capacity of lakes in the Colorado Rockies. *Limnology and Oceanography* 27:167–172.
Litaor, M. I. 1987. The influence of eolian dust on the genesis of alpine soils in the Front Range, Colorado. *Soil Science Society of America Journal* 51:142–147.
———. 1993. The influence of soil interstitial waters on the physicochemistry of major, minor and trace metalsin stream waters of the Green Lakes Valley, Front range, Colorado. *Earth Surface Processes and Landforms* 18:489–504.
Madole, R. F. 1972. Neoglacial facies in the Colorado Front Range. *Arctic and Alpine Research* 4:119–130.
———. 1976. Glacial geology of the Colorado Front Range. In *Quaternary stratigraphy of North America,* edited by W. C. Mahaney. Stroudsburg, PA: Dowden, Hutchinson and Ross.
———. 1982. Possible origins of till-like deposits near the summits of the Front Range in north-central Colorado. *U.S. Geological Survey Professional Paper* 1243: 31 pp.
———. 1986. Lake Devlin and the Pinedale glacial history of the Colorado Front Range. *Quaternary Research* 25:43–54.
Madole, R. F., W. C. Bradley, D. S. Loewenherz, D. F. Ritter, N. W. Rutter, and C. E. Thorn. 1987. In *Geomorphic systems of North America,* edited by W. L. Graf. Boulder, CO: Geological Society of America, Centennial Special Vol. 2.
Mahaney, W. C. 1971. Note on the "Arikaree Stade" of the Rocky Mountains neoglacial. *Journal of Glaciology* 10:143–144.
McNeely, R. 1983. Preliminary physical data on Green Lakes 1–5, Front Range, Colorado. University of Colorado Long-Term Ecological Research Working Paper 83/4.
Meierding, T. C., and P. W. Birkeland. 1980. Quaternary glaciation of Colorado. In *Colorado geology,* edited by H. C. Kent and K. W. Porter. Denver, CO: Rocky Mountain Association of Geologists.
Menounos, B., and M. Reasoner. 1997. Evidence for cirque glaciation in the Colorado Front Range during the Younger Dryas chronozone. *Quaternary Research* 48:38–47.
Molnar, P., and P. England. 1990. Late Cenozoic uplift of mountain ranges and global climate change: Chicken or egg. *Nature* 346:29–34.
Morris, S. E. 1986. The significance of rainsplash in the surficial debris cascade of the Colorado Front Range foothills. *Earth Surfaces Processes and Landforms* 11:11–22.
Morris, S. E., and T. A. Moses. 1987. Forest fire and the natural soil erosion regime in the Colorado Front Range. *Annals of the Association American Geographers* 77:245–254.

Nelson, A. R., A. C. Millington, J. T. Andrews, and H. Nichols. 1979. Radiocarbon-dated upper Pleistocene glacial sequence, Fraser Valley, Colorado Front Range. *Geology* 7:410–414.
O'Neill, M. P., and D. M. Mark. 1990. On the frequency distribution of land slope. *Earth Surface Processes and Landforms* 12:127–136.
Outcalt, S. I., and D. D. MacPhail. 1965. A survey of neoglaciation in the Front Range of Colorado. *University of Colorado Studies, Series in Earth Science* 4: 124 pp.
Pape, G. A., R. I. Dorn, and J. C. Dixon. 1995. A new conceptual model for understanding geographical variations in weathering. *Annals of the Association of American Geographers* 85:38–64.
Rapp, A. 1960. Talus slopes and mountain walls at Tempelfjorden, Spitsbergen. *Norsk Polarinstitut Skrifter* 119: 96 pp.
Rapp, A., and S. Rudberg. 1960. Recent periglacial phenomena in Sweden. *Biuletyn Periglacjalny* 8:143–154.
Reheis, M. J. 1975. Source, transportation and deposition of debris on Arapaho Glacier, Front Range, Colorado. *Journal of Glaciology* 14:407–420.
Richmond, G. M. 1986. Stratigraphy and correlation of glacial deposits of the Rocky Mountains, the Colorado Plateau and the ranges of the Great Basin. *Quaternary Science Reviews* 5:99–127.
Rudberg, S. 1972. Periglacial zonation—a discussion. *Gottinger Geographische. Abhandlung* 60:221–233.
Scott, G. R. 1975. Cenozoic surfaces and deposits in the Southern Rocky Mountains. *Geological Society of America Memoir* 144:227–248.
Slaymaker, O., and R. J. Chorley. 1964. The Vigil network system. *Journal of Hydrology* 2:19–24.
Small, E. E., and R. S. Anderson. 1995. Geomorphically driven late Cenozoic uplift in the Sierra Nevada, California. *Science* 270:277–280.
Souchez, R. A., and M. M. Lemmens. 1987. Solutes. In *Glacio-fluvial sediment transfer*, edited by A. M. Gurnall and M. J. Clark. Chichester, England: Wiley.
Stoecker, R. 1976. Pocket gopher distribution in relation to snow in the alpine tundra. In *Ecological impacts of snowpack augmentation in the San Juan Mountains, Colorado*, edited by H. W. Steinhoff and J. D. Ives. Fort Collins: Colorado State University, CSU-DWS 7052-4.
Sturchio, N. C., K. L. Pierce, M. T. Murrell, and M. L. Sorey. 1994. Uranium-series ages of travertines and timing of the last glaciation in the northern Yellowstone area, Wyoming-Montana. *Quaternary Research* 41:265–277.
Summer, R. M. 1982. Field and laboratory studies on soil erodibility, southern Rocky Mountains, Colorado. *Earth Surfaces Processes and Landforms* 7:253–266.
Swanson, F. J., T. K. Kratz, N. Caine, and R. G. Woodmansee. 1988. Landform effects on ecosystem patterns and processes. *Bioscience* 38:92–98.
Thorn, C. E. 1975. Influence of late-lying snow on rock-weathering rinds. *Arctic and Alpine Research* 7:373–378.
———. 1976. Quantitative evaluation of nivation in the Colorado Front Range. *Geological Society of America Bulletin* 87:1169–1178.
———. 1978. A preliminary assessment of the geomorphic role of pocket gophers in the alpine zone of the Colorado Front Range. *Geografiska Annaler* 60A:181–187.
———. 1982. Gopher disturbance: Its variability by Braun-Blanquet vegetation units in the Niwot Ridge alpine tundra zone, Colorado Front Range, U.S.A. *Arctic and Alpine Research* 14:45–51.

Thorn, C. E., and R. G. Darmody. 1980. Contemporary eolian sediments in the alpine zone, Colorado Front Range. *Physical Geography* 1:162–171.

———. 1985. Grain-size distribution of the insoluble component of contemporary eolian deposits in the alpine zone, Front Range, Colorado, U.S.A. *Arctic and Alpine Research* 17:433–442.

Thorn, C. E., and D. S. Loewenherz. 1987. Alpine mass wasting in the Indian Peaks area, Front Range, Colorado. In *Geomorphic systems of North America,* edited by W. L. Graf. Boulder, CO: Geological Society of America, Centennial Special Vol. 2.

Wahlstrom, E. E. 1947. Cenozoic physiographic history of the Front Range, Colorado. *Geological Society of America Bulletin* 58:551–572.

Wahrhaftig, C. 1987. Foreword to *Rock glaciers,* edited by J. R. Giardino, J. F. Shroder, and J. D. Vitek. Winchester, MA: Allen and Unwin.

Waldrop, H. A. 1964. *The Arapaho Glacier: A sixty-year record.* University of Colorado Studies. Series in Geology 3.

Warburton, J. 1991. Absence of frost sorting at an experimental site, Green Lakes Valley, Colorado Front Range, U.S.A. *Permafrost and Periglacial Processes* 2:113–122.

Wernicke, B., R. Clayton, M. Ducea, C. H. Jones, S. Park, S. Ruppert, J. Saleeby, J. K. Snow, L. Squires, M. Fliedner, G. Jiracek, R. Keller, S. Klemperer, J. Luetgert, P. Malin, K. Miller, W. Mooney, H. Oliver, and R. Phinney. 1996. Origin of high mountains in the continents: The southern Sierra Nevada. *Science* 271:190–193.

White, S. E. 1968. Rockfall, alluvial and avalanche talus in the Colorado Front Range. *Geological Society of America Special Papers* 115:237.

———. 1971. Rock glacier studies in the Colorado Front Range, 1961 to 1968. *Arctic and Alpine Research* 3:43–64.

———. 1972. Alpine subnival boulder pavements in Colorado Front Range. *Geological Society of America Bulletin* 83:195–200.

———. 1976. Rock glaciers and block fields: Review and new data. *Quaternary Research* 6:77–97.

———. 1981. Alpine mass movement forms (noncatastrophic): Classification, description and significance. *Arctic and Alpine Research* 13:127–137.

———. 1982. Physical and geological nature of the Indian Peaks, Colorado Front Range. In *Ecological studies in the Colorado alpine: A festschrift for John W. Marr,* edited by J. C. Halfpenny. Occasional Paper 37, Institute of Arctic and Alpine Research, University of Colorado, Boulder.

Williams, M. W., A. Brown, and J. M. Melack. 1993. Geochemical and hydrologic controls on the composition of surface water in a high-elevation basin, Sierra Nevada. *Limnology and Oceanography* 38:775–797.

Williams, M. W., J. S. Baron, N. Caine, R. Sommerfeld, and R. Sanford. 1996. Nitrogen saturation in the Rocky Mountains. *Environmental Science and Technology* 30:640–646.

Zielinski, G. A., and P. T. Davis. 1987. Late Pleistocene age for the type Temple Lake moraine, Wind River Range, Wyoming, U.S.A. *Geographie Physique et Quaternaire* 41:397–401.

# 5

# Hydrology and Hydrochemistry

Mark W. Williams
Nel Caine

## Introduction

Seasonally snow-covered areas of Earth's mountain ranges are important components of the global hydrologic cycle. Although their area is limited, the snowpacks of these areas are a major source of the water supply for runoff and ground water recharge over wide areas of the mid-latitudes. They are also sensitive indicators of climatic change. The release of ions from the snowpack is an important component in the biogeochemistry of alpine areas and may also function as a sensitive indicator of changes in atmospheric chemistry.

The demand for water in the semiarid areas of the western United States is reflected in extensive systems of reservoirs, canals, and flow diversions that have been constructed over the past century. Most of the water resources tapped by these systems derives from the mountain environments of the Rocky Mountains, where contributions of the alpine have long been recognized (Martinelli 1975). In Colorado, 9000 km$^2$ of alpine terrain, less than 4% of the state's area, provide more than 20% of the state's streamflow and is especially important in maintaining late-summer flows (Martinelli 1975).

Lakes in the Rocky Mountains are relatively uncontaminated compared with many other high-elevation lakes in the world, with the median value of $NO_3^-$ concentrations less than 1 $\mu$eq L$^{-1}$ (Psenner 1989). However, in comparison with downstream ecosystems, these high-elevation ecosystems are relatively sensitive to changes in the flux of energy, chemicals, and water because of extensive areas of exposed and unreactive bedrock, rapid hydrologic flushing rates during snowmelt, limited extent of vegetation and soils, and short growing seasons (Williams 1993). Hence, even small changes in atmospheric deposition have the potential to result in large changes in ecosystem dynamics and water quality (Williams et al. 1996a). Furthermore, these ecosystem changes may occur in alpine areas before they occur in downstream ecosystems (Williams et al. 1996b). Apart from its use in municipal supply, agriculture, recreation, and power generation, this water also mediates transfers of geomorphic and biological materials. For this reason, the drainage basin, or catchment, has long been recognized as a basic geomorphic unit in environmental research (e.g., Chorley 1967; Bormann and Likens 1969). Thus, the definition of local water and chemical budgets underlies much work on erosion and denudation and on the implications of environmental change.

This chapter reports on streamflow in a single alpine catchment, Green Lakes Valley, which appears to be representative of the alpine environments of the Southern Rockies and of the Front Range in particular. The discharge of water through the Green Lakes Valley is dominated by snowmelt during the summer months (May–September), which accounts for more than 90% of the annual streamflow (Caine 1995a). These flows conform to predictable patterns that reflect the seasonal and diurnal energy inputs to snowmelt and, in the small basins that are typical of the alpine environment, are not greatly affected by travel times. Hydrologic and hydrochemical studies in this catchment have been conducted on a variety of spatial scales (from 0.08 to 7 km$^2$) over the past 15 years and those empirical records form the basis of this chapter.

## Study Area

Snow conditions in the Green Lakes Valley reflect its midcontinental, high-elevation location. Eighty percent of the annual precipitation of around 1000 mm occurs as snow that accumulates without significant melting from October to June in most years. For most of this period the snowpack is "cold," that is, with a temperature below 0°C, apart from a shallow surface layer in which the temperature fluctuates diurnally and occasionally reaches melting temperatures even in midwinter. The characteristics of a "cold" snowpack have two important hydrologic implications: (1) It is especially prone to redistribution by the wind during and after snowfall, and (2) the release of meltwater from it to the soil and streams is delayed until the snowpack temperature is raised to 0°C.

The relocation of snow on the alpine landscape by winter winds with a strong westerly component gives a patchy accumulation pattern that tends to be reproduced each winter (Cline 1993; Walker et al. 1993). Areas that are exposed to the westerly winds usually retain little snow, whereas accumulations of more than 15 m depth are common in sheltered, lee situations (Caine 1995a). Berg (1986) recorded 69% of mean hourly wind speeds in winter at 7.5 m s$^{-1}$ or more and suggested that wind drifting of snow occurs on 75% of winter days. This, combined with generally low vapor pressures, suggests a high potential for water loss by sublimation (Schmidt 1982) as well as the potential for redistributing the water mass within the system. On the basis of two winters of record, Berg (1986) suggested that sublimation losses could exceed 250 mm yr$^{-1}$, about 30% of the annual snowfall.

Streamflows and chemical contents in Green Lakes Valley have been studied at a cascade of sites along the main drainage channel and in two tributaries draining from Niwot Ridge (Fig. 4.1). Continuous records of water levels at four sites, Albion (ALB), Green Lake 4 (GL4), Navajo (NAV), and Martinelli (MAR) form the basis of this study. At intermediate sites, the Lake Albion Spillway (SPI), the Lake Albion Inlet (INL), Green Lake 5 (GL5), and Green Lake 1 (GL1), flow estimates are based on intermittent measurements only and so are less satisfactory. Discharge estimates for Arikaree Glacier (ARK), where the flow occurs through many small and subsurface channels, are extrapolated from the NAV record and so are even more prone to error. Grab samples are collected for chemical analysis at these sites at weekly intervals from the initiation of snowmelt through late autumn and on a monthly basis during the winter at GL4 and ALB. Water samples are analyzed for pH, conductance, acid neutralizing capacity (ANC), and major solutes. At all of these sites, and others in the valley, hydrologic work over the past 15 years has been combined with water quality studies to address the potential impacts of atmospheric deposition and subsequent acidification (Caine 1995b; Williams et al. 1996b).

Storage and release of solutes from the seasonal snowpack is the main pathway for fluxes of wet and dry deposition in the catchment. Previous research has shown that atmospheric deposition in the Rocky Mountains has become more acidic over time (Lewis and Grant 1980). Of particular concern, precipitation chemistry measured by the National Atmospheric Deposition Program (NADP) shows that there has been about a 200% increase in $NO_3^-$ loading from wet deposition at Niwot Ridge over the last decade, increasing from about 8 kg ha$^{-1}$ yr$^{-1}$ on average for 1985–1987 to 16.5 kg ha$^{-1}$ yr$^{-1}$ for 1990–1992 (Williams et al. 1996b). Earlier and comparable measurements of annual $NO_3^-$ deposition extend the record back to 1982 and indicate an even larger increase in $NO_3^-$ loading, with a mean $NO_3^-$ loading of 5.7 kg ha$^{-1}$ yr$^{-1}$ for 1982–1986 (Reddy and Caine 1988). A simple linear regression analysis shows that increases in precipitation accounts for about one half of the increase in annual $NO_3^-$ loading ($r^2 = 0.56, p < 0.01$) and about one half comes from increases in the annual volume-weighted mean concentration of $NO_3^-$ in precipitation ($r^2 = 0.59, p < 0.01$).

## The Annual Hydrograph

Although total streamflow in the Green Lakes Valley increases with drainage area in the downvalley direction, specific discharges (mm) decline by a factor of two or more in the same direction (Fig. 5.1a). This pattern is consistent with a regional pattern in the Front Range during the 1965–1995 period. Flows at the same site in the Green Lakes Valley vary interannually by a factor of two or three and are more variable in the smaller subbasins than for the entire drainage system (Table 5.1). Streamflow amounts show no time trend over the last 15 years (Fig. 5.1b) although inclusion of the records from the 1968–1974 period at GL4 suggests an increase in annual water yield of about 7.5 mm yr$^{-1}$ over the last 30 years at that site. This increase would be consistent with an average increase in precipitation of 14 mm yr$^{-1}$

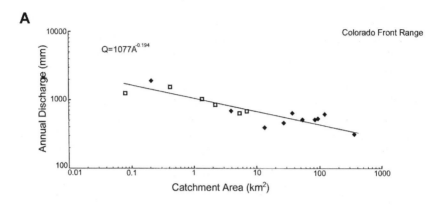

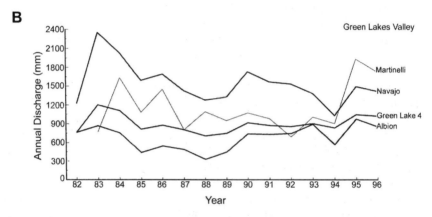

Figure 5.1. (a) Specific discharge (mm) versus drainage area for Front Range catchments. Sites in the Green Lakes Valley are based on records for 1982–1996. The U.S. Geological Survey sites shown for the Front Range are those with more than 8 years of record from 1966 to 1995 (only those sites that are not affected by flow diversions). The best fit relationship is shown (Q = 1077A$^{-0.194}$, $r^2$ = 0.769, $n$ = 16). (b) Temporal variations in specific annual discharge at four sites in Green Lakes Valley, 1982–1996.

observed at the alpine climate station (D1, 3750 m) between 1967 and 1994 (Williams et al. 1996a). The year-to-year variations in this record are not nearly as great as those found at lower elevations where summer precipitation is more likely to account for high stream discharges (Jarret 1989). This reflects the modulating influence of snow and ice in the alpine hydrological cycle maintaining a long, relatively constant amount of flow on the seasonal hydrograph and allowing storage to be carried over from one water year to the next (Martinelli 1975).

The seasonal pattern of flows in Green Lakes Valley (Fig. 5.2) is typical of marginally glacierized alpine catchments. Flows are only maintained through the winter at the lowest elevations (ALB) where baseflow discharges decline to 20 L s$^{-1}$ before the start of spring snowmelt in May. In the higher, smaller basins, streamflow usually ceases by late November and does not start again until the following May.

Table 5.1. Annual Peak Flows in Green Lakes Valley

| | Albion | | Green Lake 4 | | Martinelli | |
|---|---|---|---|---|---|---|
| | Flow (m³/d) | Date | Flow (m³/d) | Date | Flow (m³/d) | Date |
| 1982 | 132,008 | June 28 | 50,606 | July 29 | 1115 | July 10 |
| 1983 | 326,171 | July 10 | 74,525 | June 21 | 4102 | July 10 |
| 1984 | 102,269 | June 30 | 51,918 | June 16 | 2785 | June 30 |
| 1985 | 144,480 | June 8 | 63,651 | June 6 | 5017 | June 8 |
| 1986 | 99,871 | July 6 | 40,135 | July 5 | 3575 | July 4 |
| 1987 | 77,514 | June 9 | 31,582 | June 9 | 1135 | June 10 |
| 1988 | 84,750 | June 26 | 67,700 | June 26 | 2652 | June 21 |
| 1989 | 77,820 | July 30 | 23,709 | June 16 | 2213 | June 16 |
| 1990 | 303,225 | June 10 | 28,223 | July 8 | 2370 | June 11 |
| 1991 | 158,583 | June 15 | 25,198 | June 13 | 3120 | June 15 |
| 1992 | 179,006 | June 25 | 28,748 | June 20 | 1662 | June 23 |
| 1993 | 166,705 | July 14 | 46,492 | June 21 | 3440 | June 29 |
| 1994 | 69,429 | June 21 | 42,553 | June 20 | 2759 | June 1 |
| 1995 | 157,153 | July 11 | 41,175 | July 11 | 6630 | July 11 |
| 1996 | 101,111 | June 21 | 35,018 | June 21 | 4500 | July 17 |
| Mean (m³/d) | 145,340 | June 25 | 43,416 | June 24 | 3138 | June 21 |
| (mm/d) | 20.5 | | 20.7 | | 39.2 | |

80  Physical Environment

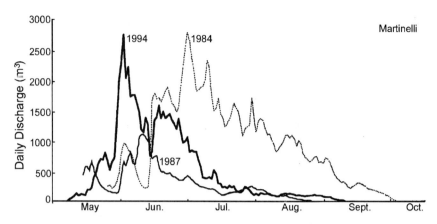

Figure 5.2. Seasonal hydrograph at Martinelli sampling site, Green Lakes Valley. The 1994 hydrograph approximates the average seasonal discharge for the period 1982–1994; flows in 1984 were above the average throughout the valley and in 1987 lower than the average (Table 5.1).

The seasonal hydrograph usually starts to rise in mid- to late May, when areas of relatively shallow snow become isothermal at 0°C and so are able to transmit meltwater to the underlying rock and soil (chapter 2). Prior to this time, meltwater produced close to the snow surface is translocated and refrozen within the snow cover, often as ice lenses that influence later meltwater flows through it (Furbish 1988). In any year, the initiation of increasing flow is usually synchronous at all sites in the valley (Table 5.1). In contrast, there is much greater variability in the timing of early season flows in different years at the same site. This reflects weather conditions during spring and the depths of snow that must become isothermal at 0°C before meltwater release can occur. Thus, there is often a correlation between the annual flow volume (indicative of the volume of snow in the basin at the end of winter) and the date of peak flow ($r = 0.59$ with $n = 12$ at MAR). In 1987, following a winter of low snow accumulation, the hydrograph started to rise on May 3 at MAR and on May 7 at ALB and peaked throughout the valley on June 9 (Table 5.1). In 1984 and 1995, years with much greater than average snow accumulations, the start of the rising limb was delayed until May 15 at MAR and May 20 at ALB.

Peak flows usually occur simultaneously along the main drainage of the valley (June 24 on average), with the south-facing sub-basins reaching the seasonal maximum a few days earlier (Table 5.1). Exceptions to this simultaneity occur in years when reservoir storage at Lake Albion must be recharged before high flows can occur at ALB, as in 1989 and 1993. On only one occasion (1982 at GL4) was the seasonal high flow associated with rainfall.

Not only is the timing of peak flows conservative, their volumes also tend to be relatively invariant. At any site, 15 years of record show them to vary by a factor of only three (with a slightly greater range in small basins such as MAR, Table 5.1). Years of high total seasonal discharge are not necessarily those with a high peak flow; rather they are associated with sustained high discharges after the peak (Fig. 5.2).

Table 5.2. Daily Recession Coefficients in the Green Lakes Valley

|  | Albion | Green Lake 4 | Martinelli |
|---|---|---|---|
| 1981 | 0.90 | 0.95 | — |
| 1982 | 0.90 | 0.93 | 0.94 |
| 1983 | 0.87 | 0.93 | 0.90 |
| 1984 | 0.80 | 0.91 | 0.91 |
| 1985 | 0.87 | 0.94 | 0.90 |
| 1986 | 0.90 | 0.94 | 0.89 |
| 1987 | 0.85 | 0.94 | 0.88 |
| 1988 | 0.87 | 0.97 | 0.87 |
| 1989 | 0.90 | 0.96 | 0.90 |
| 1990 | 0.89 | 0.93 | 0.87 |
| 1991 | 0.88 | 0.94 | 0.87 |
| 1992 | 0.85 | 0.94 | 0.84 |
| 1993 | 0.84 | 0.94 | 0.83 |
| 1994 | 0.91 | 0.95 | 0.87 |
| 1995 | 0.83 | 0.94 | 0.84 |
| 1996 | 0.90 | 0.95 | 0.84 |
| Mean | 0.87 | 0.94 | 0.88 |
| S.D. | 0.03 | 0.01 | 0.03 |

*Note:* Coefficients are estimated for daily discharges on the recession limb of the seasonal hydrograph.

At all sites, the seasonal decline in flow usually extends from June into October and is less steep than the increase in flow observed in spring (Fig. 5.2). It is readily modeled by a decay function, and the daily recession coefficients are quite similar (approx. 0.877) at the three sites in the basin for which continuous records are available for more than a few years (ALB, GL4, and MAR; Table 5.2).

This seasonal pattern has been empirically fitted to a third-order polynomial that has similar form each year (Caine 1992a). This annual pattern accounts for between 30 and 80% of the variance in hourly discharges at all spatial scales within Green Lakes Valley. These results could be used to remove the seasonal contribution to flows to examine patterns and trends in short-term phenomena contributing to the flow regime.

## Biogeochemistry

Solutes produced by geochemical weathering generally increase in concentration in surface waters with an increase in basin area (Caine and Thurman 1990; chapter 4). In the Green Lakes Valley, $Ca^{2+}$ and $Na^+$ concentrations increase rapidly with basin area (Fig. 5.3a). This increase of base cations (and ANC) with basin area reflects two complementary processes: (1) an increase in hydrologic residence time with basin area and (2) an increase in subsurface water temperature with basin area. Identification of causal mechanisms for this phenomenon is complicated by the fact

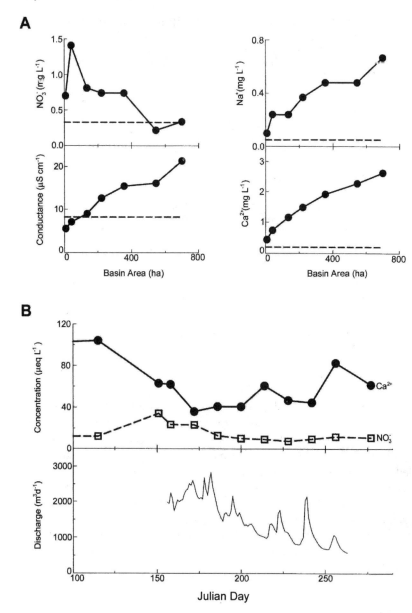

Figure 5.3. Seasonal hydrograph and chemographs for $NO_3^-$ and $Ca^{2+}$, Green Lake 4, 1990.

that both physical retention and biotic processes affecting these concentrations change with basin size. Both the increase in residence time and average temperatures increase the rate of geochemical weathering and the production of buffering solutes in surface waters. Concentrations of $HCO_3^-$ and $Ca^{2+}$ are much higher in surface waters than in precipitation, reflecting the geochemical origin of these solutes.

In contrast, concentrations of $NO_3^-$ in surface waters decrease with increasing basin area (Fig. 5.3a). Assimilation of N from atmospheric deposition by terrestrial and aquatic organisms immobilizes N, resulting in the decrease in $NO_3^-$ concentrations with basin area.

Solute concentrations also show a repeating pattern as a function of the seasonal hydrograph. Solutes with a predominantly geochemical source, such as silica and base cations, decrease during increasing flow because of dilution by snowmelt, then increase later in the season when subsurface discharge begins to contribute significantly to surface waters (Fig. 5.3b). In contrast, the strong acid anions generally increase in concentration at the start of increasing flow (Fig. 5.3b) then decrease rapidly with time. This initial increase in concentration of strong acid anions is generally attributed to the storage and release of strong acid anions from the seasonal snowpack in snowmelt runoff (Williams and Melack 1991).

The chemical content of snow meltwater before contact with the soil shows both a 24-hr (diel) periodicity and seasonal changes. Once precipitation is deposited as snow in high-elevation catchments, solutes are eluted in the form of an ionic pulse. The ionic pulse is caused by physico-chemical processes at the scale of individual snow particles. These processes cause a redistribution of solutes to the outside of the particles and to boundaries between particles (e.g., Bales 1992). The first fraction of meltwater percolating through the snowpack readily leaches these solutes. Environmental factors that are associated with an increase in the magnitude of the ionic pulse include an increase in snow depth, a decrease in the rate of snowmelt, and melt-freeze cycles (Williams and Melack 1991). To illustrate, the ratio of maximum concentrations of $NH_4^+$, $NO_3^-$, and $SO_4^{2-}$ in snowpack meltwater before contact with the soil to bulk concentrations from a colocated snowpit were sampled concurrently in 1994 and 1995. Release of solutes from storage in the seasonal snowpack in the form of an ionic pulse caused meltwater concentrations in 1994 to be 3–4 times those of bulk snowpack concentrations and in 1995 about 20 times those of bulk snowpack concentrations; other solutes showed a similar pattern (Fig. 5.4). For similar reasons, solute concentrations in snow meltwater are inversely related to daily and seasonal discharges, with the highest solute concentrations at minimum discharge and the lowest solute concentrations at maximum discharge (Caine 1989; Bales et al. 1993).

## Diurnal Hydrographs

The 24-hour periodicity in snowmelt-generated streamflows, which is a clear response to diurnal patterns of energy input to snowmelt (Cline 1995, 1997), has been widely reported from small stream basins in arctic and alpine environments (e.g., Slaymaker 1974; Marsh and Woo, 1984). Changes in the amplitude, timing and form of the diurnal flow pattern during the flow season have also been reported and explained as a response to progressive changes in the snow-covered area and snow depths (Jordan 1983; Caine 1992a). Within Green Lakes Valley, the 24-hour cycle is most marked in small catchments (e.g., NAV and MAR) and is usually less pronounced in basins of more than 1.0 $km^2$ area (e.g., GL4 and ALB; Fig. 5.5).

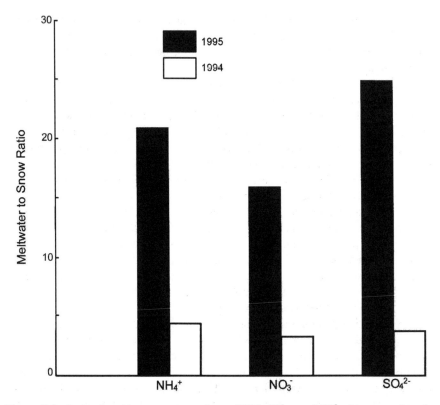

Figure 5.4. Ratio of maximum concentrations of $NH_4^+$, $NO_3^-$, and $SO_4^{2-}$ in snowpack meltwater (before contact with the ground) to bulk concentrations from a colocated snowpit sampled concurrently, 1994 and 1995. 1994 was characterized by an average snowpack and relatively rapid snowmelt, and 1995 was characterized by a snowpack about 150% of normal with snowmelt starting much later than normal. Release of solutes from storage in the seasonal snowpack in the form of an ionic pulse caused meltwater concentrations in 1994 to be 3 to 4 times those of bulk snowpack concentrations and in 1995 about 20 times those of bulk snowpack concentrations; other solutes showed a similar pattern (redrawn from Williams et al. 1996a).

Caine (1992a) partitioned the flow season at the Martinelli watershed into three segments according to the nature of the 24-hour signal. Prior to mid-June, approximately the time of peak flow, the signal is weak and not maintained throughout the record. For the two months following peak flow, from mid-June to mid-August, the diurnal periodicity in the flows is well marked and consistent (Fig. 5.5c). During this interval, it explains up to 50% of the residual variance in the series, after removal of the seasonal trend, and shows a forward phase shift of as much as 10 hours. Concurrently, its amplitude declines. By mid-August, when the surface flowpath between the residual snow cover in the basin and its outflow channel is lost, the 24-hour periodicity is also lost. At this time, the streamflow responds only to a declining groundwater contribution and to ephemeral inputs from rainstorms (Caine 1992a).

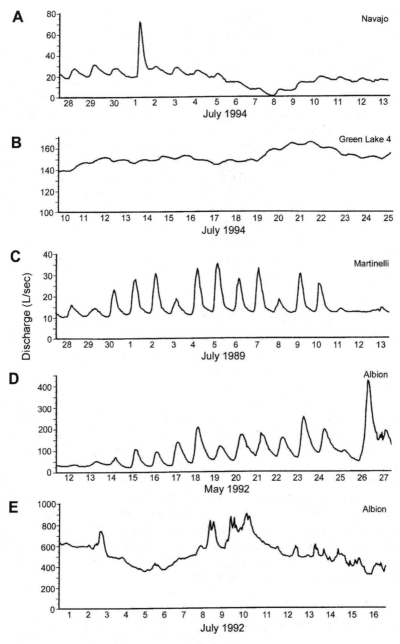

Figure 5.5. The diurnal flow pattern at four sites in Green Lakes Valley. (a) Navajo, catchment area 41.8 ha, in July 1994 (after peak flow for the year). (b) Green Lake 4, catchment area 2.2 km², in July 1994. (c) Martinelli, catchment area 8.0 ha, in June 1992 (including the seasonal peak flow). (d) Albion in May 1992. At this time, no flow from the basin above the Lake Albion spillway was reaching the gauge, which, therefore, had a catchment area of 1.6 km² at the lowest elevations in the basin. (e) Albion in July 1992, when the record includes flows from the entire 7.1 km² area of Green Lakes Valley.

Such variations in the 24-hour periodicity were interpreted as reflecting the hydrologic conditions of the basin. During the early part of the season, large parts of the basin snowpack have subfreezing temperatures and so any 24-hour periodicity in the flows is generated on areas of shallow, isothermal snow (often near the basin margins). These source areas change rapidly, some of them clearing of snow in a few days, and may not be hydraulically connected to the basin drainage, leading to a weak, noisy signal. The prominent 24-hour periodicity of the second interval seems to be initiated when almost all of the snow cover has become isothermal and so releasing meltwater in response to radiative energy inputs. This is also the time when a drainage network within and beneath the snowpack has developed, allowing drainage to the stream channel network. The forward phase shift that occurs over this interval is similar to that reported by Jordan (1983) and occurs at the rate of about 1 hr/wk consistently during years of near- or above-average snow accumulation (Caine 1992a). It is best accounted for as a response to the reduction in average snow depth as the season progresses, that is, a reduced thickness of the porous medium through which melt (and rain) water is transmitted. The decline in the amplitude of the daily flow cycle occurs as flow volumes themselves are declining, apparently as a reflection of the shrinking area of snow in the basin, which constrains the volume of meltwater produced each day.

At NAV, a clear pattern of diurnal variations in flows occurs after the seasonal peak, and this pattern continues later into the autumn than at MAR (Fig. 5.5a) because a surface flowpath from the perennial ice and snow of Arikaree Glacier is maintained at NAV until freezing temperatures and a low sun angle reduce discharges to below the level of the NAV gauge.

At a larger spatial scale, the 24-hour periodicity at GL4 and ALB is only intermittently evident in the flow records (Fig. 5.5b, d, e). At ALB, it is most marked during the early part of those seasons when the Lake Albion reservoir is recharged in the early season (e.g., in 1992; Fig. 5.5d). At these times, flows at ALB are generated from the 1.6 km$^2$ area below Lake Albion. This reduction in the effective catchment size combined with a relatively uniform snow cover and a saturated valley-floor allows the development of a strong diurnal signal (Fig. 5.5d). At other times, when the entire basin is contributing to flows at ALB, such a periodicity is not strong. At GL4, the 24-hour periodicity is usually only evident at time of high flow (Fig. 5.5b), with an extensive snow cover in the upper basin. Thus, it seems to be associated with snowmelt on the valley-floor, for the signal is lost when the snow cover there becomes intermittent and sparse.

## Transient Effects on Streamflow

Transient patterns in streamflow that occur relatively frequently within this system are associated with snowmelt and the drainage of meltwater through the snow cover. The release of water and water-saturated snow as slush flows (Clark 1988; Clark and Seppala 1988) occurs almost annually over the valley floor below Green Lake 5. These events are associated with the impoundment of meltwater in Green Lake 5 when snow and ice block its outlet. This temporary storage raises the lake level by more than 1 m and retains 40,000 m$^3$ of water, which is released in less than 8 hours

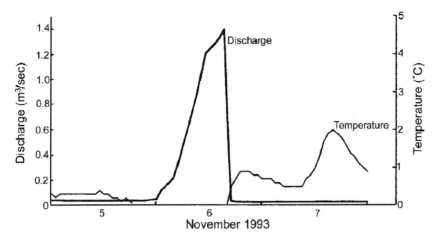

Figure 5.6. Hydrograph at Albion, November 1993. On November 6, a snow dam upstream of the recording site failed. The plot shows the resulting flows (Q) and the associated water temperatures (T) (from Caine 1995).

when drainage through the blockage is established. These releases frequently provide the annual peak flow at GL4.

Similar releases have been recorded in winter from beaver ponds on the flat valley-floor below Lake Albion. The resulting flows produce water levels equivalent to those of the peak flows at this site. Converting these flows to volume discharges remains a problem since they occur when the channel is snow covered, and they produce temporary ice and snow dams within it. The event of November 6, 1993 (Fig. 5.6) appears to have drained 1.5–2.0 m of water from the ponds on the valley-floor 250 m upstream from the ALB gauge.

Precipitation in the summer months (June–September) amounts to more than 200 mm (Table 5.3), i.e., more than 20% of the annual precipitation in Green Lakes Valley. It occurs largely as rainfall from convective cells of a few kilometers in diameter that develop on a quasi-daily frequency. These are occasionally intense but rarely prolonged enough to induce a flow response for more than a few hours (e.g., Fig. 5 in Caine 1992a). Such local storms are, however, important in sediment transport since they provide a mechanism for sediment entrainment from exposed bare soil areas and transport to the channel (e.g., Caine 1992b; chapter 4).

In the 15 years of record in Green Lakes Valley, more widespread rainfall events have provoked marked hydrologic responses on only two occasions, in late July 1982 and late June 1988. The first of these provided 43 mm of rain over a 3-day period and a sixfold flow increase at the ALB site. The effect at GL4 was not so great, though the doubling of the flow did provide the peak flow for the year (Table 5.1; Fig. 5.7). In contrast, flows at MAR responded only to the reduction in solar radiation by declining during the early part of the event before recovering and maintaining the 24-hour periodicity (Fig. 5.7). On June 26, 1988, a rainstorm yielding 52 mm in 8 hours (with a recurrence interval of about 100 years) in the upper Green Lakes Valley provoked similar responses at the same three sites, the only ones for which records are available (Fig. 5.7b). As in 1982, marked increases in flow occurred at

Table 5.3. Green Lakes Valley Hydroclimate

| | Jan | Feb | Mar | Apr | May | Jun | Jul | Aug | Sep | Oct | Nov | Dec |
|---|---|---|---|---|---|---|---|---|---|---|---|---|
| Temperature[a] | −13.2 | −12.3 | −11.2 | −7.0 | −0.9 | 4.6 | 8.2 | 7.1 | 3.4 | −2.1 | −8.9 | −11.8 |
| Precipitation[b] | 102.5 | 80.1 | 127.8 | 100.1 | 87.5 | 59.5 | 50.5 | 62.1 | 49.1 | 40.3 | 80.3 | 90.1 |
| ALB runoff[b] | 17.3 | 13.5 | 43.3 | 23.5 | 35.0 | 211.7 | 126.4 | 86.5 | 33.4 | 14.5 | 30.3 | 34.7 |
| GL4 runoff[b] | | | | 45.9 | 232.2 | 290.6 | 190.1 | 103.1 | 31.6 | | | |
| MAR runoff[b] | | | | 72.6 | 534.0 | 441.1 | 142.5 | 22.6 | 0.2 | | | |

[a] In °C.
[b] In mm.

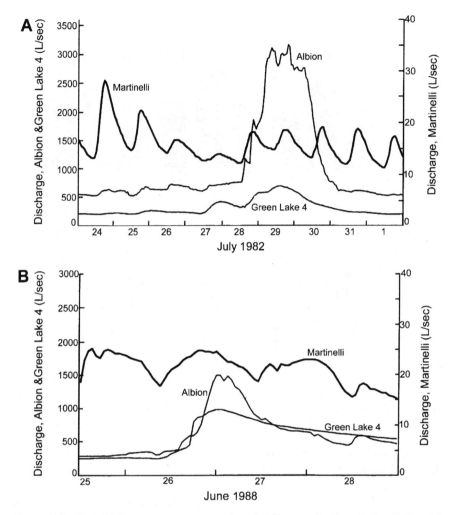

Figure 5.7. Hydrologic responses to two major rainfall events in Green Lakes Valley. (a) Flow responses to 43 mm of rainfall over the period July 27–29, 1982. (b) Flow responses to about 52 mm of rainfall in 8 hours on June 26, 1988 (the storm which produced the debris flow in the upper valley described in Figs. 4.12 and 4.13).

ALB and GL4 while flows at MAR continued at about the same level, retaining the 24-hour periodicity.

These differentiated responses seem to reflect two influences of major rainfall events in the basin. Quick flows from saturated ground in the riparian zone and lake surfaces, with focusing along the main flowpaths, amplify the hydrologic response. In contrast, most summer rainstorms do not have the intensity, duration, or extent to produce equivalent effects on the main drainage. At MAR, a small headwater basin that might be expected to give a quicker response to rainfall than the main one,

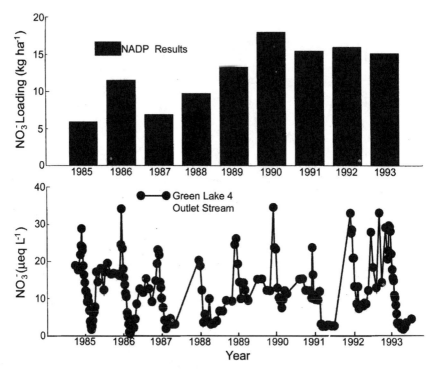

Figure 5.8. Time series of $NO_3^-$ concentrations from the outlet stream of Green Lake 4 and annual $NO_3^-$ loading (kg ha$^{-1}$ yr$^{-1}$) measured at the Niwot Ridge NADP site (redrawn from Williams et al. 1996a).

some meters of snow thickness in the basin center retain large volumes of rainwater and modulate any response, while continuing to respond to radiation inputs despite cloudy conditions.

Later in the season, following melting of the snowcover on the basin floor at MAR, rainstorm responses change. Then, rainstorms of more than 10-mm precipitation provoke a brief quickflow response that ends within a few hours of the termination of the rainstorm. These events often yield a flow volume approximating the product of the depth of precipitation and the area of the saturated zone close to the channel in lower part of the basin.

## Hydrochemical Responses to Atmospheric Deposition

Nitrogen saturation is occurring in the Green Lakes Valley and is associated with episodic acidification of surface waters. We define N-saturation as occurring when $NO_3^-$ concentrations in surface waters during the growing season are about the same as volume-weighted mean $NO_3^-$ concentrations in annual precipitation. This response occurs when the system is no longer capable of removing or sequestering inorganic N as organic forms within the system. Estimates of the forms and amounts of N currently being deposited at our site are discussed in chapter 3. Annual inorganic N loading in wet deposition to the Front Range of about 4 kg ha$^{-1}$ yr$^{-1}$ is about twice that of Pacific states and approaches that of the northeastern United States (Williams et al. 1996b). At GL4 in 1985 and 1986, $NO_3^-$ concentrations during the growing season were near or below detection limits (Fig. 5.8). Biotic and abiotic sinks apparently sequestered all available N from atmospheric deposition during the growing season. Starting in 1987, $NO_3^-$ began to leak out of the basin in surface waters during the growing season, reaching annual minimum concentrations of about 10 $\mu eqL^{-1}$ in 1990. This increase in annual minimum concentrations of $NO_3^-$ in surface waters parallels the large increase in $NO_3^-$ loading from wetfall in the late 1980s (Fig. 5.8).

There has been a concurrent shift in the relationship of seasonal discharge to $NO_3^-$ concentrations at GL4. We examined the relationship between $NO_3^-$ concentrations and discharge during the growing season, investigating the recession limb of the discharge curve from maximum discharge in early June through August. In 1985 and 1986, the regression of $NO_3^-$ concentration as a function of discharge had a y-intercept of zero $NO_3^-$ (Fig. 5.9a). In 1989 and 1990, the zero discharge intercept increased to 9 $\mu eqL^{-1}$, indicating potentially increased $NO_3^-$ in the soil solution in excess of uptake capacity. We interpret this increase in $NO_3^-$ concentrations during the growing season to the leaching of N to deeper hydrologic flowpaths that maintain summer base low. This retention and subsequent release of N may seem inconsistent with the flush of ions is released during the early stages of snowmelt. While this occurs to some extent, a substantial portion of the inorganic N in this system does appear to be retained by biota both within the snow and on the soil surface (chapter 12).

Annual $NO_3^-$ yield from GL4 increased by 50% from the period 1985–1988 (5.0 kg ha$^{-1}$ yr$^{-1}$) to 1989–1992 (7.5 kg ha$^{-1}$ yr$^{-1}$). There is a significant correlation ($p < 0.05$) between N loading in wet deposition and N export, with $r^2 = 0.53$ ($n = 8$) (Fig. 5.9b). Most of this increase in N export from GL4 occurred during the growing season of June, July and August, consistent with the change from an N-limited system to an N-saturated system. Mass balance analysis shows that currently about 50% of the $NO_3^-$ loading from annual wet deposition is exported in stream waters.

Episodic acidification is occurring in the headwaters of the Green Lakes Valley (Fig. 5.10). We define acidification as occurring when acid neutralizing capacity (ANC) values decrease below 0 $\mu eqL^{-1}$. On June 9, 1994, ANC values were $-6.1$ $\mu eqL^{-1}$ at the 9-ha ARK site and $-1.6$ $\mu eqL^{-1}$ at the 42-ha NAV site (Fig. 5.10). ANC values increased with basin area but were still below 50 $\mu eq\ L^{-1}$ at the inlet to Lake Albion on this date, an area of 380 ha. ANC values were below 0 $\mu eqL^{-1}$ at the ARK site for much of June–September in 1994, and at the NAV site there were five sampling dates with negative ANC values. The pH measurement on this date at the ARK site was 4.86 and pH at the NAV site was 5.37. The pH values then increased rapidly to about 6.0 at the 135-ha Green Lake 5. The decrease in pH and the low values for ANC were associated with concentrations of $NO_3^-$ as great as 35 $\mu eqL^{-1}$ (Fig. 5.10). Caine (1995b) has shown that there has been a decrease in ANC of these headwater streams over the last decade, coincident with the increased loading of strong acid anions. If atmospheric deposition of strong acid anions continues

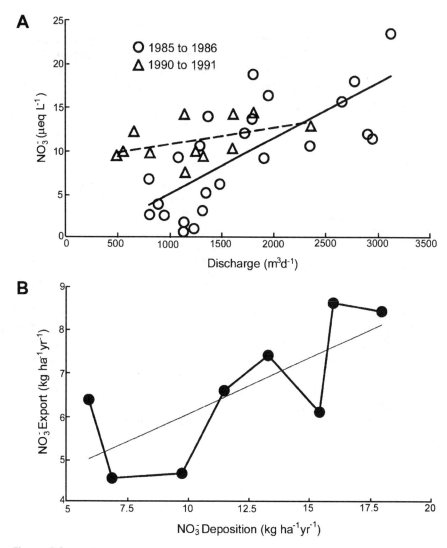

Figure 5.9. (a) Changes in the relationship between streamwater $NO_3^-$ concentrations and discharge at GL4 during the growing season. Growing season is defined as the recession limb of the hydrograph from maximum discharge in early June–August in 1985–1986 (circles) and 1989–1990 (squares). In 1985–1986, the $NO_3^-$ intercept at zero discharge was not significantly different than zero (from Williams et al. 1996b). However, in 1989–1990, the zero-discharge intercept for $NO_3^-$ of 9 $\mu eqL^{-1}$ was significantly greater than zero ($p < 0.01$). (b) Regression analysis between annual $NO_3^-$ input from wet deposition and output in surface waters at GL4. There is a significant increase in $NO_3^-$ export with increasing deposition.

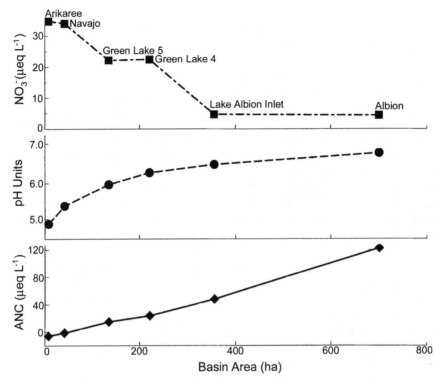

Figure 5.10. Acid neutralizing capacity (ANC), pH, and $NO_3^-$ concentrations as a function of basin area, June 9, 1994 in Green Lakes Valley. Episodic acidification is occurring at present deposition levels in the headwater catchment (redrawn from Williams et al. 1996a).

or increases, episodic acidification will continue downstream to include larger basin areas. Furthermore, surface waters that are now undergoing episodic acidification (ARK, NAV) will become chronically acidified.

Current concepts of critical loads need to be reconsidered since only modest atmospheric loadings of N are sufficient to induce N leaching to surface waters of high-elevation catchments in the western United States. An important result of this analysis is the demonstration that high-elevation ecosystems may be nearly N-saturated at current levels of N deposition. These findings have important implications for policy relating to N emissions. The rapidity of the increase in $NO_3^-$ flux from these test basins was unexpected given the historically low rates of N deposition in the western United States, the modest increase in N loading, and the generally N-limited status of montane ecosystems in the western United States (chapter 9). Determining critical loads for N deposition has been problematic because of the inability to quantify ecosystem storage capacity of N (Aber 1992). Critical load estimates for N deposition have been set at 10 kg ha$^{-1}$ yr$^{-1}$ for northern Europe, based on empirical results that showed no N leaching to surface waters below this value (Dise and Wright 1995). Clearly, leakage of N to surface waters in the Colorado

Front Range occurs at N deposition values well below 10 kg ha$^{-1}$ yr$^{-1}$ of N, perhaps because the percentage of surface area of alpine catchments occupied by biota capable of retaining this material, as opposed to inert rock, is much less than in other ecosystems.

Our ability to effectively manage high-elevation catchments is constrained by a lack of process-level understanding of the nitrogen cycle. At present, we cannot quantify the sources of $NO_3^-$ in surface waters of the Colorado Front Range and other high-elevation catchments in the western United States (chapter 3). Snowmelt runoff is the dominant hydrologic event in these catchments. Nitrate concentrations in surface waters are generally consistent with the release of $NO_3^-$ from storage in the seasonal snowpack in the form of an ionic pulse (Williams et al. 1995), and the $NO_3^-$ in stream waters is often assumed to be from wet and dry deposition (Williams and Melack 1991). However, experiments conducted in 1993 at the Niwot Ridge Saddle research site indicate that microbial activity in snow-covered soils plays a key role in N-cycling in alpine ecosystems prior to snowmelt runoff, with net mineralization rates from March 3 to May 4 ranging from 2.2 to 6.6 g N m$^{-2}$ (Brooks et al. 1996; chapter 12). Furthermore, addition of $NH_4Cl$ to the snowpack as a nonconservative tracer showed that $NH_4^+$ is not nitrified in the snowpack but is retained in underlying soil (Williams et al. 1996c). Preliminary analysis tracing sources of streamwater $NO_3^-$ at the Loch Vale Watershed in nearby Rocky Mountain National Park during snowmelt runoff, using both the oxygen and nitrogen isotopic composition of $NO_3^-$, suggests that atmospheric $NO_3^-$ eluted from the snowpack was a minor source of $NO_3^-$ in streamflow and that microbial activity was an important source of $NO_3^-$ in streamwaters (Kendall et al. 1995). Innovative experiments and tools such as the dual isotope analysis of $NO_3^-$ are needed for a mechanistic understanding of N dynamics in high-elevation ecosystems.

References

Aber, J. D. 1992. Nitrogen cycling and nitrogen saturation in temperate forest ecosystems. *Trends in Ecology and Evolution* 7:220–223.

Bales, R. C. 1992. Snowmelt and the ionic pulse. In *The Encyclopedia of Earth Science*. Vol. 1. Orlando, FL: Academic Press.

Bales, R. C., R. E. Davis, and M. W. Williams. 1993. Tracer release in melting snow: diurnal and seasonal patterns. *Hydrologic Processes* 7:389–401.

Berg, N. H. 1986. Blowing snow at a Colorado alpine site: measurements and implications. *Arctic and Alpine Research* 18:147–161.

Borman, F. H., and G. E. Likens. 1969. The watershed-ecossytem concept and studies of nutrient cycles. In *The ecosystem concept in resource management,* edited by G. M. Van Dyne. New York: Academic Press.

Brooks, P. D., M. W. Williams, and S. K. Schmidt. 1996. Microbial activity under alpine snowpacks, Niwot Ridge, Colorado. *Biogeochemistry* 32:93–113.

Caine, N. 1989. Diurnal variations in the quality of water draining from an alpine snowpack. *Catena* 16:153–162.

Caine, N. 1992a. Modulation of the diurnal streamflow response by the seasonal snowcover of an alpine basin. *Journal of Hydrology* 137:245–260.

Caine, N. 1992b. Sediment transfer on the floor of the Martinelli snowpatch, Colorado Front Range. *Geografiska Annaler* 74A:133–144.

———. 1995a. Snowpack influences on geomorphic processes in Green Lakes Valley, Colorado Front Range. *Geographical Journal* 161:55–68.

———. 1995b. Temporal trends in the quality of streamwater in an alpine environment: Green Lakes Valley, Colorado Front Range, USA. *Geografiska Annaler* 77a:207–220.

Caine, N., and E. M. Thurman. 1990. Temporal and spatial variations in the solute content of an alpine stream, Colorado Front Range. *Geomorphology* 4:55–72.

Chorley, R. J. 1967. The drainage basin as the fundamental geomorphic unit. In *Water, Earth, and Man*. London: Methuen.

Clark, M. J. 1988. Periglacial hydrology. In *Advances in periglacial geomorphology*, edited by M. J. Clark. Chichester, England: Wiley.

Clark, M. J., and M. Seppala. 1988. Slushflows in a subarctic environment, Kilpisjarvi, Finnish Lappland. *Arctic and Alpine Research* 20:97–105.

Cline, D. 1993. Modeling the redistribution of snow in alpine areas using geographical information processing techniques. Forty-ninth Proceedings of the Western Snow Conference, U.S. Forest Service, Fort Collins, CO.

———. 1995. Snow surface energy exchanges and snowmelt at a continental alpine site. In *Biogeochemistry of seasonally snow covered catchments*, edited by K. Tonnessen, M. W. Williams, and M. Tranter. IAHS-AIHS Publication 228. Wallingford, England: International Association of Hydrological Sciences.

———. 1997. Snow surface energy exchanges and snowmelt at a continental, mid-latitude alpine site. *Water Resources Research* 33:689–701.

Dise, N., and R. Wright. 1995. Nitrogen leaching from European forests in relation to nitrogen deposition. *Forest Ecology and Management* 71:153–161.

Furbish, D. J. 1988. The influence of ice layers on the travel time of meltwater flow through a snowpack. *Arctic and Alpine Research* 20:265–272.

Jarett, R. D. 1989. Paleohydrologic techniques used to define the spatial occurrence of floods. *Geomorphology* 3:181–195.

Jordan, P. 1983. Meltwater movement in a deep snowpack. 1. Field observations. *Water Resources Research* 19:971–978.

Kendall, C., D. H. Campbell, D. A. Burns, J. B. Shanley, S. R. Silva, and C. C. Y. Chang. 1995. Tracing sources of nitrate in snowmelt runoff using the oxygen and nitrogen isotopic compositions of nitrate: Pilot studies at three catchments. In *Biogeochemistry of seasonally snow covered basins,* edited by K. A. Tonnessen, M. W. Williams, and M. Tranter. IAHS-AIHS Publication 228. Wallingford, England: International Association of Hydrological Sciences.

Lewis, W. M., Jr., and M. C. Grant. 1980. Relationships between snow cover and winter losses of dissolved substances from a mountain watershed. *Arctic and Alpine Research* 12:11–17.

Marsh, P., and M. -k. Woo. 1984. Wetting front advance and freezing of meltwater within a snow cover. Part 1, Observations in the Canadian Arctic. *Water Resources Research* 20:1853–1864.

Martinelli, M., Jr. 1975. Water-yield improvement from alpine areas: The status of our knowledge. US Department of Agriculture, Forest Service Research Paper RM-138.

Psenner, A. R. 1989. Chemistry of high mountain lakes in siliceous catchments of the Central Eastern Alps. *Journal of Aquatic Sciences* 51:108–128.

Reddy, M. M., and N. Caine. 1988. A small alpine basin budget: Front Range of Colorado. In *Proceedings of the International Mountain Watershed Symposium, Lake Tahoe,* edited by I. G. Poppoff, C. R. Goldman, S. L. Loeb, and L. B. Leopold. Lake Tahoe, NV: Tahoe Resource Conservation District.

Schmidt, R. A. 1982. Vertical profiles of wind speed, snow concentrations and humidity in blowing snow. *Boundary Layer Meteorology* 23:223–246.

Slaymaker, H. O. 1974. Alpine hydrology. In *Arctic and alpine environments*, edited by J. D. Ives and R. G. Barry. London: Methuen.

Walker, D. A., J. C. Halfpenny, M. D. Walker, and C. A. Wessman. 1993. Long-term studies of snow-vegetation interactions. *BioScience* 43:287–301.

Williams, M. W. 1993. Snowpack storage and release of nitrogen in the Emerald Lake Watershed, Sierra Nevada. *Journal of the Proceedings of the Eastern Snow Conference* 50:239–246.

Williams, M. W., R. C. Bales, A. Brown, and J. Melack. 1995. Fluxes and transformations of nitrogen in a high-elevation catchment, Sierra Nevada. *Biogeochemistry* 28:1–31.

Williams, M. W., J. Baron, N. Caine, R. Sommerfeld, and R. Sanford. 1996a. Nitrogen saturation in the Colorado Front Range. *Environmental Science and Technology* 30:640–646.

Williams, M. W., P. D. Brooks, A. Mosier, and K. A. Tonnessen. 1996b. Mineral nitrogen transformations in and under seasonal snow in a high-elevation catchment, Rocky Mountains, USA. *Water Resources Research* 32:3161–3171.

Williams, M. W., M. Losleben, N. Caine, and D. Greenland. 1996c. Changes in climate and hydrochemical responses in a high-elevation catchment, Rocky Mountains. *Limnology and Oceanography* 41:939–946.

Williams, M. W., and J. M. Melack. 1991. Solute chemistry of snowmelt and runoff in an alpine basin, Sierra Nevada. *Water Resources Research* 27:1575–1588.

# Part II

# Ecosystem Structure

# 6

# The Vegetation
## Hierarchical Species-Environment Relationships

Marilyn D. Walker
Donald A. Walker
Theresa A. Theodose
Patrick J. Webber

## Introduction

The vegetation of Niwot Ridge has a rich history of study, beginning with phytosociological studies directly on the Ridge and in the surrounding mountains and incorporating more experimental and dynamic approaches in later years. This chapter provides an overview of the spatial patterns of Niwot Ridge plants and plant communities relative to the primary controlling environmental gradients at scales from the individual to the landscape.

The spatial patterns of vegetation at all scales are dominated by physical forces, particularly the interaction of wind, snow, and topography (Fig. 6.1). The controls of biotic factors on the distribution and abundance of plant species on Niwot Ridge have received considerably less attention than have physical factors, but recent studies have revealed the importance of competition and certain mutualisms in structuring community composition.

Community research on Niwot Ridge has been organized around a hierarchy of spatial scales, from the plot to the region (Fig. 6.2). Plot-based studies have focused on physiological and ecological dynamics of specific species and communities, and more spatially extensive studies have provided a hierarchical framework for the plot studies. In this chapter, we first present an overview of the broader patterns in the vegetation, followed by descriptions of the communities, and then the specifics of physical and biotic controls on species and plant growth that drive the community patterns.

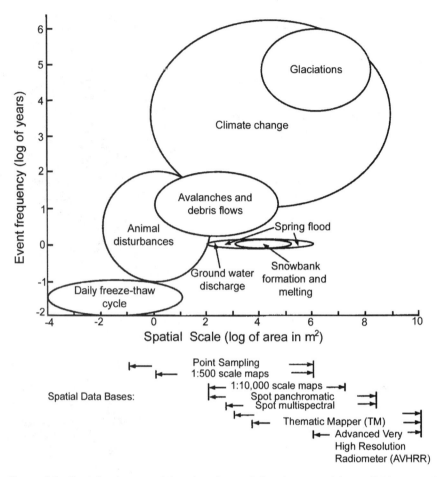

Figure 6.1. Spatial and temporal domains of natural disturbances and the available types of data for examining phenomena at each scale (modified from Walker et al. 1993).

## Landscape and Regional Patterns

The landscape-scale patterns in the Niwot vegetation are driven by a complex elevation gradient, which is a combination of temperature and snow regime, with wind modifying and interacting with temperature and snow at all points along the gradient (chapter 2). Certainly the most critical boundary in the system is the upper tree limit, which defines the alpine system and which lies roughly between 3400 and 3600 m elevation on Niwot Ridge.

Billings (1988) provided a climatic-floristic-physiographic review of major North American alpine systems that helps to place Niwot Ridge into a larger perspective. Climatically, Niwot is intermediate between the dry Sierras, which have

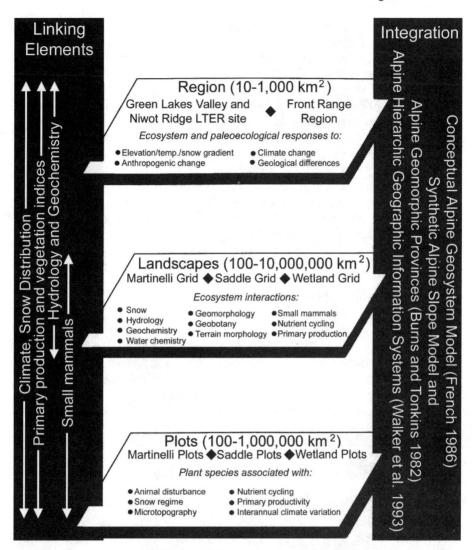

Figure 6.2. Conceptual framework for the Niwot hierarchic GIS, showing scales of research (plot, landscape, region), linking elements between scales, and types of integration (modified from Walker et al. 1993).

greater precipitation but almost none of it falling during the summer, and the wetter northern Appalachians (Mt. Washington), which have fairly even annual precipitation and no drought. Because of this intermediate position with respect to annual water balance, Niwot has the potential for some of the greatest ecological heterogeneity among North American alpine areas.

The origin of the flora is another key element in defining the vegetation. The flora has linkages to and elements of other North American and Asian alpine areas, to the

Arctic, to a lesser degree to the other major life zones that surround it, and also has a significant endemic element (Weber 1976; Komárková 1979). Weber and Willard (1967) listed 313 vascular taxa in the Colorado alpine, of which 67 were circumpolar and 130 were restricted to the alpine. Although the Rocky Mountain Cordillera provides the opportunity for northerly and southerly migration of species, there are few or no northern outliers of the Mexican highland flora present in the Colorado alpine, probably because of predominantly colder conditions than present during most of the last 20,000 years. Billings (1988) pointed out that although past climate and geographic connections are certainly important in explaining the importance of arctic species in the Colorado alpine and Appalachian alpine floras, the presence of adequate summer moisture is likely equally important in keeping them there, because those elements are largely missing in the drier Sierras. The discontinuity in the alpine zone in Wyoming may also play an important role in shaping the flora and therefore the vegetation of Niwot Ridge. The rich ericaceous flora of the Arctic and the Northern Rockies is largely missing in the Colorado alpine. Ericaceous species are particularly important in arctic snowbed vegetation, where they form dwarf-shrub heaths at the shallow end of the snow gradient. Therefore the snow-dominated communities of Niwot Ridge have little floristic affinities with either their Northern Rocky or arctic counterparts or with the snowbed vegetation of the alpine northern Appalachians, which maintains a strong ericaceous element and does have dwarf shrub vegetation (Bliss 1963).

The first landscape-scale view of Niwot Ridge was represented by the vegetation map of Komárková and Webber (1978), who mapped the alpine areas of the ridge at a scale of 1:10,000 using floristically defined units of the Braun-Blanquet system. The hierarchical nature of that system allowed them to map at different levels of floristic resolution depending on the spatial resolution of the units, so the polygons are coded as associations, alliances, orders, or classes depending on the situation. The map indicates a complex spatial mosaic across the entire ridge, but a trend toward drier communities dominated by *Kobresia myosuroides* at the eastern edge.

Walker et al. (1993) examined the landscape-scale patterns in more detail on remotely sensed images of normalized-difference vegetation index (NDVI). They looked at changes in NDVI with elevation on the east- and west-facing slopes on the two sides of the Continental Divide and found that while elevation explained significant variance in NDVI on the West Slope, on the East Slope the pattern only held true on east-facing slopes. The extremely high westerly winds apparently reduce the potential heterogeneity of these sites, acting as a stronger control on production and pattern than the complex elevation, temperature, and moisture gradient (80% of the winds are out of the west or northwest; see chapter 2). Walker et al. (1993) also mapped just the Saddle research area (Fig. 1.5) to the association level at a scale of 1:500, which also showed the east- and west-facing slope differentiation very clearly. The east-facing slope of the Saddle consists of a complex mosaic of communities differentiated by snow regime, moisture, and animal disturbance, whereas the west-facing slope is strongly dominated by a single community type.

## Plant Communities and Their Controls

### Classification of Niwot Vegetation

The first descriptions of the Niwot Ridge vegetation were by John Marr and his students as part of a general description of Eastern Slope ecosystems along a transect from the grasslands to the alpine (Marr, 1961). Marr's "D-1" site, at the western end of Niwot Ridge, served as the alpine representative along this major climate transect. Following Marr, work by Patrick Webber and students broadened the understanding of the vegetation patterns by developing formal classifications of the vegetation.

Two main systems of description have been applied to the Niwot vegetation, one based primarily on habitat, but with a floristic basis, the other on floristics and species (Table 6.1). May and Webber (1982) described six physiognomic-habitat-based units (noda) that were identified from their positions within a polar ordination diagram. Although this work was based on only 30 samples from the Saddle research site, the habitat-based approach was familiar to ecologists and physical scientists working on Niwot and thus became the standard for differentiating plant communities in other studies. Komárková (1979, 1980) used the Braun-Blanquet approach to classify the vegetation in the Indian Peaks region, of which Niwot Ridge is a part. Her detailed study was based on a total of 545 relevés (plots), of which 482 were used to produce the classification. Komárková recognized 52 plant associations (*sensu* Braun-Blanquet 1965, with class > order > alliance > association) in the Indian Peaks—20 of snowbeds, 7 of wind-blown sites, 6 of rock crevices, 12 of wet meadows, 8 of shrublands, and 4 of springs. Willard (1979) developed a classification for the alpine vegetation of Rocky Mountain National Park, approximately 20 miles to the north of Niwot. Table 6.1 shows the correspondence between the Komárková and May-Webber approaches.

The Niwot Ridge vegetation is most easily explained relative to a snow gradient, similar to the mesotopographic approach of Billings (1973, 1988) and recognized by Komáková (1979) and Burns and Tonkin (1982) (Fig. 6.3; Table 6.1). Here we present the major tundra habitats and their communities along a gradient of moisture and snow depth, excluding the vegetation associated with krumholz tree islands.

### Fellfield, Extremely Windblown

Fellfield communities are one of the May-Webber noda, and they are well defined as a distinct physiognomic and habitat type. They occur on windblown slopes and ridge crests and are characterized by an open plant canopy and a high diversity of cushion plants. Crustose lichens reach their greatest diversity in these sites, and bryophytes are uncommon. About 10–50% of the ground is covered by cobbles or exposed gravel. Fellfields remain snow free throughout the year, although they may get a thin crust of snow in slight depressions. Important species, based on abundance, include *Carex rupestris, Paronychia pulvinata* (Fig. 6.4), *Minuartia obtusiloba, Trifolium dasyphyllum, Eritrichium aretioides,* and *Silene acaulis.* Komárková

Table 6.1. Summary of Front Range Vegetation Classification with Equivalent Noda (May and Webber 1982) and Typical Habitats

| Syntaxonomic Unit and Author (Komárková 1979) | Habitat | Nodum (May and Webber 1982) |
|---|---|---|
| **CLASS** *Asplenietea rupestria* Meier et Br.-Bl. 1934 | Rock crevices | |
| **ORDER** *Heuchero-Saxifragetalia* Komárková 1976 | Rock crevices | |
| **ALLIANCE** *Heucherion bracteato-parvifoliae* Komárková 1976 | Low-altitude, dry, south-facing rock crevices | |
| **ASSOCIATION** *Heucheretum bracteato-parvifoliae* Komárková 1976 | Low-altitude, dry, south-facing rock crevices | |
| **ALLIANCE** *Saxifrago-Claytonion megarhizae* Komárková 1976 | High-altitude, mesic, northwest-facing crevices | |
| **ASSOCIATION** *Besseyo alpinae-Caricetum nardinae* assoc. prov | Rocky, exposed ridges above 3800 m | |
| **ASSOCIATION** *Saxifragetum serpyllifoliae* Kiener 1939 *em.* Komárková 1976 | Northwest-oriented high peak faces in the subnival belt | |
| **ASSOCIATION** *Sagino saginoidis-Claytonietum megarhizae* Komárková 1976 | Stable, snow-free, subxeric ridges | |
| **ASSOCIATION** *Besseyo alpinae-Saxifragetum rivularis* Komárková 1976 | Rock faces of sheltered and mesic ravines | |
| **ASSOCIATION** *Polytrichastro alpini-Poetum lettermanii* Komárková 1976 | Wet rocks or scree at high altitude | |
| **CLASS** *Thlaspietea rotundifolii* Br.-Bl. 1948 | Scree and gravel deposits | |
| **ORDER** *Aquilegio-Cirsietalia scopulorum* Komárková 1976 | Scree and gravel deposits | |
| **ALLIANCE** *Aquilegio-Cirsion scopulorum* Komárková 1976 | High-altitude scree and gravel | |
| **ASSOCIATION** *Oxyrio digynae-Ligularietum taraxocoidis* Komárková 1976 | North-facing, fine-moving scree | |
| **ASSOCIATION** *Cirsietum scopulorum* Kiener 1939 | South-facing, stable, medium-to-large scree | |
| **ASSOCIATION** *Minuartio biflorae-Caricetum araphoensis* Komárková 1976 | East-facing, stable, small-to-medium scree | |
| **ALLIANCE** *Cirsio-Phacelion sericeae* Komárková 1976 | Warm, xeric, scree and gravel | |
| **ASSOCIATION** *Aquilegio coeruleae-Rubetum idaei* Komárková 1976 | Warm, xeric, scree near treeline | |
| **ASSOCIATION** *Aquilegio coeruleae-Ribesetum montigeni* Komárková 1976 | Warm, xeric, stable scree near treeline | |
| **ASSOCIATION** *Phacelio sericeae-Polemonietum viscosi* Komárková 1976 | Steep, south-facing, fine-moving scree | |
| **CLASS** *Elyno-Seslerietea* Br.-Bl. 1948 | Well-drained, basophilous to weakly acidophilous alpine areas | |
| **ORDER** *Kobresio-Caricetalia rupestris* Komárková 1976 | Climax habitats on well-drained gently sloping ridge tops of the Front Range | |
| **ALLIANCE** *Caricion foeneo-elynoidis* Komárková 1976 | South-facing, warm, stable scree | |
| **ASSOCIATION** *Cerastio arvensi-Caricetum foeneae* Komárková 1976 | Steep, stable, fine scree | |
| **ASSOCIATION** *Caricetum elynoidis* Willard 1963 | South- and east-facing, moderate, stable slopes | |
| **ALLIANCE** *Kobresio-Caricion rupestris* Komárková 1976 | Windy, stable, cool, well-drained broad interfluves and ridges | |
| **ASSOCIATION** *Trifolietum dasyphylli* Willard 1963 | Subxeric, snow-free >200 d | Fellfield |
| **ASSOCIATION** *Potentillo-Caricetum rupestris* Willard 1963 | Xeric to subxeric, south-facing slopes, snow-free >200 d | Fellfield |
| **ASSOCIATION** *Eritricho aretioidis-Dryadetum octopetalae* Kiener 1939 *corr.* Komárková 1976 | Terrace sides on terraced ground | |
| **ASSOCIATION** *Sileno-Paronychietum* Willard 1963 | Xeric, extremely wind-exposed fellfields, snow-free >200 d | Fellfield |
| **ASSOCIATION** *Selaginello densae-Kobresietum myosuroidis* Cox 1933 *corr.* Komárková | Subxeric to mesic turfs on gentle slopes, snow-free 150–200 d | Dry meadow |

| Taxonomy | Description | Habitat |
|---|---|---|
| **CLASS** *Salicetea herbaceae* Br.-Bl. 1948 | | |
| **ORDER** *Trifolio-Deschampsietalia* Komárková 1976 | Alpine snow patches | |
| **ALLIANCE** *Deschampsio-Trifolion parryi* Komárková 1976 | Earlier-melting snow patches of the Front Range, snow-free 100–150 d | |
| **ASSOCIATION** *Acomastylidetum rossii* Willard 1963 | Shallow mesic depressions and broad leeward hill slopes | |
| **ASSOCIATION** *Deschampsio caespitosae-Trifolietum parryi* Komárková 1976 | Mesic, early-melting snow cover | Moist meadow |
| **ASSOCIATION** *Stellario laetae-Deschampsietum caespitosae* Willard 1963 corr. Komárková 1976 | Subxeric to mesic, early-melting snow patches | Moist meadow |
| **ALLIANCE** *Vaccinio-Danthonion intermediae* Komárková 1976 | Mesic, early-melting snow patches | Moist meadow |
| **ASSOCIATION** *Vaccinietum scoparii-cespitosi* Komárková 1976 | Subxeric to mesic, early-melting snow patches at low elevations | |
| **ASSOCIATION** *Solidagini spathulatae-Danthonietum intermediae* Komárková 1976 | Subxeric to mesic, sloping snow patches at treeline | |
| **ASSOCIATION** *Artemisietum arcticae saxicolae* Willard 1963 | Subxeric to mesic, southeast- and northeast-facing prolonged snow patches | |
| **ORDER** *Sibbaldio-Caricetalia pyrenaicae* Komárková 1976 | Upper alpine belt on southeast-facing slopes | |
| **ALLIANCE** *Poo-Caricion haydenianae* Komárková 1976 | Later-melting snow patches of the Front Range, snow-free <75 d | |
| **ASSOCIATION** *Sileno acaulis-Caricetum perglobosae* Komárková 1976 | Subxeric to mesic, very prolonged snow patches, eroded | |
| **ASSOCIATION** *Oxyrio digynae-Poetum arcticae* Komárková 1976 | Rocky snow patches | |
| **ASSOCIATION** *Poo arcticae-Caricetum haydenianae* Komárková 1976 | Snow patch centers | |
| **ALLIANCE** *Sibbaldio-Caricion pyrenaicae* Komárková 1976 | Southeast-facing, very late-melting snow patches | |
| **ASSOCIATION** *Toninio-Sibbaldietum* Willard 1963 | Late-melting snow patches of the low alpine | |
| **ASSOCIATION** *Caricetum pyrenaicae* Willard 1963 | Subxeric to subhygric, margins of late-melting snow | Snowbed |
| **ASSOCIATION** *Juncetum drummondii* Willard 1963 | Subxeric to mesic, late-melting snow patches | Snowbed |
| **ASSOCIATION** *Epilobio anagallidifolii-Antennarietum alpinae* Komárková 1976 | Mesic, moderately late-melting snow patches | Snowbed |
| **ASSOCIATION** *Epilobio anagallidifolii-Rorippetum alpinae* Komárková 1976 | Subhygric, late-melting snow cover, upper alpine | |
| **ASSOCIATION** *Phleo commutati-Caricetum nigricantis* Komárková 1976 | Lakeshores and streambanks with late-melting snow, accumulation of fine material | |
| **ASSOCIATION** *Sibbaldio procumbentis-Lewisietum pygmaeae* assoc. prov. | Mesic to subhygric depressions below deep snow | Snowbed |
| **ALLIANCE** *Anthelio-Pohlion obtusifoliae* Komárková 1976 | North-facing, stable sites, very prolonged snow cover | |
| **ASSOCIATION** *Solorino croceae-Polytrichetum piliferi* Komárková 1976 | Bryophyte-dominated, very late melting snow patches | |
| **ASSOCIATION** *Polytrichastro alpini-Anthelietum juratzkanae* Komárková 1976 | Late-melting snow patches in subnival belt | |
| **ASSOCIATION** *Polytrichastro alpini-Pohlietum obtusifoliae* Komárková 1976 | Hygric flat depressions on lakeshores | |
| **ASSOCIATION** *Bryo turbinati-Philonotidetum tomentellae* Komárková 1976 | Mesic to subhygric springs and late-melting snow | Snowbed |
| | Very wet, late-lying snow | |
| **CLASS** *Scheuchzerio-Caricetea fuscae* Tx. 1937 | | |
| **ORDER** *Pediculari-Caricetalia scopulorum* Komárková 1976 | Alpine bogs and marshes | |
| **ALLIANCE** *Bistorto-Caricion capillaris* Komárková 1976 | Marsh communities of the Front Range, snow-free period varies | |
| **SUBALLIANCE** *Salicenion arctico-reticulatae* Komárková 1976 | Mesic to hygric, cold, prolonged snow cover | |
| **ASSOCIATION** *Salicetum arcticae* Kiener 1939 em. Willard 1963 | Moist depressions | |
| **ASSOCIATION** *Bistorto viviparae-Salicetum reticulatae* Komárková 1976 | Early-melting snow patches with abundant moisture | |
| **SUBALLIANCE** *Junco-Caricenion capillaris* Komárková 1976 | Mesic, snow-covered, stable sites | |
| **ASSOCIATION** *Carici capillaris-Bistortetum viviparae* Komárková 1976 | Wet rocks with fine soil accumulation | |
| **ASSOCIATION** *Clementsio rhodanthae-Rhodioletum integrifoliae* Komárková 1976 | Wet rocks, marsh margins, frost boils | |
| **ASSOCIATION** *Bistorto viviparae-Caricetum microglochinis* Komárková 1976 | Hydric to hygric sites with a continuous moisture supply | |
| **ASSOCIATION** *Koenigietum islandicae* Willard 1963 | Depressions with abundant moisture | |
| | Lakeshores and wet rocks | |

*(continued)*

Table 6.1. (continued)

| Syntaxonomic Unit and Author (Komárková 1979) | Habitat | Nodum (May and Webber 1982) |
|---|---|---|
| **ALLIANCE** *Pediculari-Caricion scopulorum* Komárková 1976 | Wetter, warmer, shallower snow marsh communities of the Front Range | |
| ASSOCIATION *Drepanoclado exannulati-Caricetum tripartitae* Komárková 1976 | Small depressions, lake shores, wet rocks | |
| ASSOCIATION *Caricetum scopulorum* Kiener 1939 em. Willard 1963 | Subhygric to subhydric marshes on mineral soils | Wet meadow |
| ASSOCIATION *Pediculari groenlandicae-Eleocharitetum* Komárková 1976 | Marshes in lower alpine belt | |
| ASSOCIATION *Clementsio rhodanthae-Caricetum vernaculae* Komárková 1976 | Moist lakeshores, solifluction terraces | |
| ASSOCIATION *Philonotido tomentellae-Caricetum illotae* Komárková 1976 | Lakeshores, springs below snow patches | |
| ASSOCIATION *Pediculari groenlandicae-Caricetum aquatilis* Komárková 1976 | Marshes near springs and lakes | |
| **CLASS** *Betulo-Adenostyletea* Br.-Bl. et. Tx. 1943 | Tall herb, grass, and shrub communities of the lower alpine and subalpine belt | |
| **ORDER** *Salici-Trollietalia* Komárková 1976 | Tall herb and shrub communities of the Colorado subalpine | |
| **ALLIANCE** *Salicion planifolio-villosae* Komárková 1976 | Subxeric to subhygric willow shrublands, snow-free 100–150 d | |
| ASSOCIATION *Bistorto viviparae-Salicetum villosae* Komárková 1976 | Subxeric to mesic shrublands | Shrub tundra |
| ASSOCIATION *Rhodiolo integrifoliae-Salicetum planifoliae* Komárková 1976 | Mesic to subhygric shrublands | Shrub tundra |
| **ALLIANCE** *Agropyro-Calamagrostion canadensis* Komárková 1976 | Mesic screes and streams near treeline | |
| ASSOCIATION *Agropyro trachycauli-Calamagrostietum canadensis* Komárková 1976 | Mesic meadows near treeline | |
| ASSOCIATION *Carici scopulorum-Mertensietum ciliatae* Komárková 1976 | Streamlets and creeks | |
| **ALLIANCE** *Ligustico-Trollion laxi* Komárková 1976 | Springs, streams, and other mesic locations near treeline | |
| ASSOCIATION *Adoxo moschatellinae-Mertensietum ciliatae* Komárková 1976 | At base of wet rocks, mesic scree | |
| ASSOCIATION *Ligustico filicini-Trollietum laxi* Komárková 1976 | Streambanks near treeline | |
| ASSOCIATION *Ligustico filicini-Senecietum triangularis* Komárková 1976 | Streams and wet scree, lower alpine | |
| ASSOCIATION *Ligustico filicini-Athyrietum distentifolii* Komárková 1976 | Southeast-facing, coarse, mesic scree | |
| **CLASS** *Montio-Cardaminetea* Br.-Bl. et. tx.1943 | Mountain to subalpine springs | |
| **ORDER** *Primulo-Cardaminetalia* Komárková 1976 | Spring communities of the Rocky Mountains, snow-free period varies | |
| **ALLIANCE** *Cardamino-Primulion parryi* Komárková 1976 | Spring communities of the Colorado alpine and subalpine | |
| ASSOCIATION *Epilobio anagallidifolii-Cardaminetum cordifoliae* Komárková 1976 | West-facing, stable sites | |
| ASSOCIATION *Philonotido tomentellae-Saxifragetum odontolomae* Komárková 1976 | Springs in subalpine, lower alpine | |
| ASSOCIATION *Epilobio anagallidifolii-Primuletum parryi* Komárková 1976 | Subhydric to hydric springs, streams, and snow patches | Snowbed |
| ASSOCIATION *Clementsio rhodanthae-Calthietum leptosepalae* Komárková 1976 | Springs and wet areas along streams, lower alpine | |

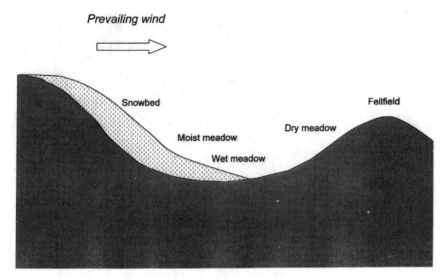

Figure 6.3. Idealized depiction of an alpine mesotopographic gradient, showing the position of major vegetational noda in relation to snow distribution and topography (after Billings 1973).

Figure 6.4. Cushion plant growth form, characteristic of fellfield communities, exemplified by *Paronychia pulvinata* (photo by William D. Bowman).

Figure 6.5. Fellfield dominated by *Dryas octopetala*, a circumboreal community (photo by William D. Bowman).

placed these communities in the Alliance Kobresio-Caricion rupestris, which includes some associations in the slightly more mesic windblown dry meadows. The alliance is the most extensive and well developed in the region. Typical fellfield associations include Eritricho aretioidis-Dryadetum octopetalae and Sileno-Paronychietum. Sileno-Paronychietum is the most common on Niwot, particularly in the Saddle area. *Dryas*-dominated types occur more frequently toward the eastern end of Niwot Ridge, and are some of the most microclimatically extreme types (Fig. 6.5). Komárková (1979) noted a high degree of homogeneity within all of these stands, indicating very similar among-stand composition. Slightly more mesic fellfield sites, in minor depressions or transitional to dry meadows, are in the associations Potentillo-Caricetum rupestris and Trifolietum dasyphylli. These types have a higher canopy cover than the drier types, but still have a primarily open canopy.

The Vegetation 109

Figure 6.6. Dry meadow community dominated by *Kobresia myosuroides* (photo by William D. Bowman).

## *Dry Meadow, Windblown*

Communities within the dry meadow noda described by May and Webber are dominated by the sedge *Kobresia myosuroides,* which forms extensive turfs (Fig. 6.6). Other important species include *Carex rupestris, Selaginella densa,* and *Thlaspi arvense.* These sites maintain a thin winter snow cover, which melts early, resulting in a growing season length of between 150 and 200 days. The dominance of the densely caespitose *K. myosuroides,* which is apparently unable to withstand the harsh winter conditions of the fellfield environment, allows for the development of thick soils and a complete plant cover (Bell and Bliss, 1979). Diversity is lower than in the fellfield stands, most likely due to the strong dominance of *K. myosuroides.* The Saddle dry meadow stands are placed in the association Selaginello densae-Kobresietum myosuroidis. Two other dry meadow types that do not occur in the Niwot Ridge Saddle include Cerastio arvensi-Caricetum foenea and Caricetum elynoidis, both of which occur primarily on south-facing slopes, often in association with pocket gophers (*Thomomys talpoides*). Caricetum elynoidis and Cerastio-Caricetum are in a separate alliance, Caricion foeneo-elynoidis, from the other dry types. All are united into the circumpolar arctic-alpine class Kobresio-Caricetalia rupestris. A key difference between the slightly more mesic Caricion foeneo-elynoidis and the drier Kobresio-Caricion rupestris is a greater predominance and diversity of erect forbs in the former and an abundance and importance of lichens as characteristic taxa in the latter. Caricetum elynoidis is common on Niwot's southern slope and is often intermixed with small patches of Cerastio-Caricetum, which is much more common on the western side of the Continental Divide (Komárková 1979).

An "unsolved puzzle" in the Niwot Ridge vegetation is how *Kobresia myosuroides* can maintain such high dominance; about 30% of the live biomass of the dry meadows is this species (Walker et al. 1994). Viable seeds are rare in the Colorado population, and Humphries et al. (1996) reported that no one has yet found seedlings of this species at this latitude. However, Ebersole (unpublished data) observed seedlings on 20+-year-old disturbance plots on Niwot Ridge. *Kobresia myosuroides* has an arctic-alpine circumpolar distribution and is near the southern limit of its distribution in Colorado (it does extend into northern New Mexico, http://plants.usda.gov/). *Kobresia myosuroides* is globally important in dry arctic and alpine vegetation and is a key character taxa for this class of vegetation. Its persistence is apparently a function of life history traits that are "invisible" within the time frame of research. *Kobresia myosuroides* produces abundant viable seed in northern latitudes and often behaves as a colonizing species in those situations (Walker 1990). These lower latitude populations disperse their seeds before they are fully ripened, apparently using a degree-day sum as a dispersal cue. This results in inviable seeds in most years. *Kobresia myosuroides* is a tillering species, and a single individual may live upwards of 200 years, based on growth rates of individual tussocks (Bell and Bliss 1979). Humphries et al. (1996) demonstrated, using a life-history-based simulation model, that the species can persist in this environment as long as there is an occasional extremely cold year. Thus, *K. myosuroides* should continue to persist and dominate as long as there is adequate summer climatic variability to successfully reproduce on occasion. *Kobresia myosuroides* behaves more like a woody plant than an herbaceous one in this environment, both in its life history and growth characteristics.

### Moist Meadow, Early-Melting Snowbank

The moist meadow noda is at the shallow end of the gradient of communities whose dynamics are controlled primarily by winter snow cover and melt patterns (Fig. 6.7). These sites get a modest snow cover and are usually released from snow relatively early in the growing season, for a total season length of 100–150 days. The combination of abundant moisture and reasonably long growing season makes these sites relatively rich and productive. They are predominated by a lush cover of forbs and grasses. Komárková (1979) placed all of the moist meadow associations in class Salicetea herbaceae, the circumpolar arctic-alpine snowbed class. The moist meadow noda recognized by May and Webber is dominated by the forb *Acomastylis rossii* and the grass *Deschampsia caespitosa*. Komárková's Deschampsio-Trifolion parryi alliance encompasses the noda. Komárková recognized three separate associations within the broader alliance: Acomastylidetum rossii, Deschampsio casepitosae-Trifolietum parryi, and Stellario laetae-Deschampsietum caespitosae. Acomastylidetum matches the concept of moist meadow that has been mainly used on Niwot. These stands are dominated by *A. rossii,* but the composition of individual stands is

Figure 6.7. Moist meadow in the Saddle research site, dominated by the tillering grass *Deschampsia caespitosa* and the clonal forb *Acomastylis rossii* (photo by William D. Bowman).

extremely variable. This association has one of the greatest total numbers of vascular species and one of the lowest average numbers of species (i.e., it has low alpha diversity and high beta diversity). Deschampsio-Trifolietum occurs in sites with slightly longer-lying snow and more mesic conditions, and Stellario-Deschampsietum is more mesic still. The latter type is far less variable than the first two.

A key difference between the *Deschampsia caespitosa-* and *Acomastylis rossii-*dominated associations is their linkages to other regions. *Acomastylis rossii* is a Beringian species limited to the Rocky Mountains, Alaska, and far northeast Siberia. Thus *A. rossii*-dominated alpine communities are unique to the Rocky Mountain Cordillera, occurring as far south as northern Arizona. *Deschampsia caespitosa*, on the other hand, has a circumpolar arctic-alpine distribution, occurring in many different ecological situations, although always associated with at least a moderate snow cover. Thus, distinguishing between these associations in field research would greatly aid in the strength of comparative studies with other alpine regions.

The alliance Vaccinio-Danthonion intermediae contains another group of three moist meadow associations that were not included within the May-Webber moist meadow concept. This alliance occurs primarily in the lower alpine belt, often in the tundra-forest ecotone, and thus could be classified as either tundra or subalpine meadow. However, explicit recognition of these types is essential, because they are quite distinct from the Deschampsio-Trifolion. Vaccinio-Danthonion stands are characterized by *Pedicularis parryi, Carex brevipes, Solidago spathulata, Dantho-*

*nia intermedia, Penstemon whippleanus,* and *Vaccinium scoparium* or *Vaccinium caespitosum.* Snow and moisture conditions are similar between the two alliances, but the former has a slightly longer growing season and presumably a higher degree-day sum due to its lower slope position.

### Late-Melting Snowbank

Communities of late-melting snowbanks were identified by May and Webber (1982) as the snowbed noda, and Komárková (1979) recognized 14 distinct associations within this broad group, many of them infrequent. Because of the extreme conditions associated with late-melting snow, there is rapid turnover in species composition along a rather narrow environmental band. The late-melting snowbeds all fall within Order Sibbaldio-Caricetalia pyrenaicae. Komárková (1979) recognized the allegiance of this vegetation to the circumpolar snowbed class Salicetea herbaceae, and the nomenclatural species for the order, *Sibbaldia procumbens* and *Carex pyrenaica,* also have circumpolar arctic-alpine distributions. Thus, this vegetation is strongly related to other circumpolar arctic-alpine snowbed types.

Three alliances within the class represent a gradient of increasingly late-lying snow, with the most extreme communities having only a partial coverage of nonvascular plants and few or no vascular representatives. Disturbed, dry sites near the center of snowpatches are in Alliance Poo-Caricion haydenianae (Komárková 1979). Of the three associations within the alliance, only Poo arcticae-Caricetum haydenianae, named after the same species as the alliance, is frequent, occupying very late-melting snowpatches with southern exposures. These sites frequently have standing water during the melt period and have a high number of species tolerant to saturated soils. Stand composition is highly variable; important species include *Polytrichum piliferum* and *Toninia* spp. in addition to the nomenclatural species.

Alliance Sibbaldio-Caricion pyrenaicae, named after the same species as the order, is the most abundant snowbed group and encompasses the May-Webber snowbed type. Komárková recognized seven associations, corresponding with a soil moisture gradient. Toninio-Sibbaldietum, Caricetum pyrenaicae, Juncetum drummondii, Epilodia anagallidifolii-Antennarietum alpinae, Epilodia anagallidifolii-rorippetum alpinae, Phleo commutati-Caricetum nigricantis, and Sibbaldio procumbentis-Lewisietum pygmaea define an approximate moisture-snowmelt gradient within this large alliance. The drier end of the spectrum has affinities with the moist meadow types, particularly the Deschampsio-Trifolion, and the wet end intergrades with another snowbed alliance, Anthelio-Pohlion obtusifoliae.

### Wet Meadow

May and Webber's wet meadow noda is represented by subhygric sites, which are downslope from snowbeds and thus receive inputs of surface water throughout most of the growing season (Fig. 6.8). Komárková (1979) placed these into the circumpolar arctic-alpine mire class Scheuchzerio-Caricetea fuscae. She also noted, however, that the Indian Peaks communities were lacking in many of the character taxa for the class and were related to it more by habitat than by floristics. They are also

Figure 6.8. Wet meadow, dominated by *Caltha leptosepala* and *Carex* species, on the north flank of Niwot Ridge, looking toward (left to right) Pawnee Peak, Mount Toll, and Paiute Peak (photo by William D. Bowman).

fairly distinct from arctic North American fens and bogs, which are dominated by *Carex aquatilis* and species of *Eriophorum*. There is a single order, Pediculari-Caricetalia scopulorum, within the class, 2 alliances, and 12 associations, all newly named by either Komárková (1976) or Willard (1963) from the Colorado Front Range.

The two alliances have distinct floristic composition and habitat. Communities of the Pediculari-Caricion scopulorum are found around small pools and have some affinities to spring communities. These habitats best typify the primary coverage of the wet meadow noda within the Niwot Ridge Saddle, but the Saddle communities are unusual because the pools are quite small and dry up during most growing seasons. Communities of the Bistorto-Caricion capillaris have relatively high cover and occur on small, gently-sloping seepage sites (Komárková 1979). This alliance grades into the moist meadow associations, primarily the Trifolio-Deschampsietalia, and thus also has representatives within the Saddle. There are many small pools in close physical approximation to snow flush sites in the Saddle, and thus these two distinct types are often found side-by-side and have been confused as a single distinct, but variable type.

The primary wet meadow associations in the Saddle are those dominated and characterized by *Carex scopulorum*, *Pedicularis groenlandica*, *Caltha leptosepala*, and *Rhodiola integrifolia*. These belong primarily to the associations Caricetum scopulorum and Clementsio rhodanthae-Rhodioletum integrifoliae, each of which belongs to different alliances (Table 6.1).

Because many of these types are found downslope of snowbanks, they share many characteristics controlled by these snowbanks. Cold air drainage, even off relatively small snowbanks, cools the area immediately downslope, resulting in cooler mean temperatures and shorter thaw seasons along snowbank margins (chapter 2). Billings and Bliss (1959) noted an average of 1.5°C cooling from the top to the bottom of a 100-m long snow drift in the Medicine Bow mountains. The combined effect of the cool air drainage and cold water from the drift can create an exceptionally cold environment at the base of very-late-lying snow drifts, and soil temperatures can be considerably colder than the shoot temperature in areas marginal to snowbeds (Holway and Ward 1965). This can cause delayed phenology and reduced water and nutrient uptake at a time when photosynthetic potential is at a maximum. This can be important for phenological development. For example, Holway and Ward (1965) noted that *Acomastylis rossii* consistently flowers only after the soil temperature at 3-cm depth exceeds 10°C, regardless of the aboveground temperatures. Thus its phenology is delayed in these sites relative to others (Walker et al. 1994). Conversely, these meadows senesce last in the autumn, which results in these sites being preferred by grazers such as elk (chapter 14).

The wet meadow associations reach their best expression in the lower alpine zone, and there are some rare associations, dominated by equally rare plants, that occur outside the Saddle. An example is the Koenigietum islandicae, known from only three locations in the Colorado Front Range, at the base of rock glaciers where there is a season-long moisture supply. *Koenigia islandica* is an extremely diminutive annual arctic-alpine species. Similar communities dominated by this plant have been found at the base of permanent snowfields on Alaska's North Slope and in the Chukotka region of eastern Siberia (Walker 1990; Razzhivin 1994).

## *Shrub Tundra*

Shrub tundra dominated by erect *Salix planifolia* and *Salix villosa* is common in lower alpine areas along the margins of ponds and streams. They occur in similar situations in the alpine zone, such as the relatively protected innermost part of the Niwot Ridge Saddle. This area, which was used as the basis for the shrub tundra noda, has a mixed stand of these two species, with areas approximately 5–7 m across of fairly continuous shrub cover intermixed with open regions without shrubs. Maximum heights are about 45 cm, and the stands are associated with a region of inactive low-centered polygons.

The shrub tundra noda corresponds very well with the single alliance Salicio planifolio-villosae, which contains two associations, Bistorto viviparae-Salicetum villosae and Rhodiolio integrifolia-Salicetum planifoliae. *Salix villosa* reaches its optimum in the first, which occurs in slightly drier habitats, and *Salix planifolia* reaches its optimum in the second, which occurs in more mesic habitats. The shrub tundra communities grade into the Clementsio rhodanthae-Rhodioletum integrifoliae on the mesic end (i.e., the Rhodiolio-Salicetum) and into the Kobresio-Caricetalia and Trifolia-Deschampsietea on the dry end.

Komárková (1979) placed these shrub communities provisionally into the European Class Betulo-Adenostyletea but felt that a new North American class should be named. These same willow taxa predominate alpine shrub vegetation throughout the Rocky Mountain chain and are found in similar situations in the Arctic (Viereck et al. 1992). Although the mix of *Salix* species increases in the overall landscape changes along this transect, the importance of *Salix planifolia* and *Salix glauca* (a close relative or synonym of *S. villosa*) remains constant in these situations. Komárková (1979) felt *Salix glauca* to be a distinct species and ecologically distinct from the alpine stands of *Salix villosa*.

## Other Associations and Types Not Defined by the Saddle Noda

### Rock Crevices and Scree Slopes

Habitats dominated by bare rock include rock crevices, boulder fields, and scree slopes. Communities of these sites are dominated by a great abundance of bare rock, with plants limited to narrow zones in crevices or in thin soil between or on rocks. These habitats do not occur within the Saddle and have received little attention from ecologists studying ecosystem processes because they have little soil and are not spatially extensive. These are important communities from a conservation perspective, however, containing some of the most rare species. Komárková (1979) placed all of these communities into the alpine class Asplenietea rupestris, described by Braun-Blanquet (1965), but felt that a North American class was required, because only one characteristic species for the class, *Cystopteris fragilis*, occurs in the Indian Peaks flora, and thus the primary association is by habitat. Komárková (1976) named a new order, the Heuchero-Saxifragetalia, which has the vascular taxa *Claytonia megarhiza, Saxifraga caespitosa, Draba fladnizensis, Draba lonchocarpa,* and *Saxifraga bronchialiis* ssp. *austromontana* as characteristic taxa as well as the mosses *Tortula norvegica, Pohlia cruda,* and *Bryum algovicum* and the lichens *Physcia caesia, Pelitgera malacea,* and *Orthotrichum indet.*

The order and class are further subdivided into two alliances and six associations, illustrating the diversity even in these rather depauperate sites. Many of these types have floristic connections to the nival and subnival zones of the Alps and to High Arctic and Polar Desert stands of the Arctic.

### Springs

Komárková (1979) placed associations of springs and seeps into Class Montio-Cardaminetea, Order Primulo-Cardaminetalia, and Alliance Cardamino-Primulion parryi. There are four associations. Some of the key taxa for class, order, and alliance are *Caltha leptosepala, Primula parryi, Epilobium angallidifolium, Ranunculus escholtzii, Cardamine cordifolia,* and *Saxifraga odontoloma*. These types have some similarities to the wet meadow communities of small ponds (Pediculari-Caricion scopulorum) and to certain snowbed communities. The showy *P. parryi* is a striking visual clue to these communities, which do have some stands on the northern flank of the Saddle and in the lower alpine belt of Niwot Ridge.

## Physical Factors Controlling Plant Communities

Physical controls of alpine vegetation on Niwot Ridge have been examined using ordination methods (Komárková, 1979, 1980; May and Webber 1982). A number of factors related to topography, wind, redistributed snow, and meltwater runoff are the main factors that correlate with species and vegetation patterns. Billings's (1973) classic alpine mesotopographic gradient portrays the typical vegetation associated with windward, ridge crest, leeward, and meltwater accumulation areas. An idealized adaptation to the Niwot Ridge situation shows the sequence of microsites, vegetation, and snow depth along the gradient (Fig. 6.3). The following discussion focuses on soil and hydrology as they relate to the plant environment along this gradient.

### Growing Season Length: Effects on Phenology and Production

The amount of snow on a site is the single most important factor governing the length of the growing season and hence the total amount of warmth available for plant development and growth (Billings and Bliss 1959; Holway and Ward 1963, 1965; Walker et al. 1993; Stanton et al. 1994). At the plant level, the presence of a winter snow cover offers plants protection against frost damage, dehydration, and physical damage from wind and wind-blown particles (Wardle 1968; Tranquillini 1979). Early snow cover limits intensive and deep freezing of the soil, thereby lowering soil instability caused by frost action and weathering. However, the protection that snow provides against winter climate extremes also results in a belated, shortened growing season. Conversely, snow-free places are exposed to severe winds, and plants in these places are subject to high rates of evapotranspiration and high vapor pressure deficits (LeDrew 1975; Bell and Bliss 1979; Isard 1986; chapter 2). Very deep snow patches melt out in late summer or not at all in some years. In these extreme sites, the short growing season combined with other negative influences, such as wet, poorly developed, and unstable soils results in areas completely devoid of vegetation. In contrast, early-melting areas have the advantage of a relatively warm and protected winter environment combined with a long growing season, adequate soil moisture, and a relatively moderate summer microclimate (Billings and Bliss 1959; Stanton et al. 1994).

The Niwot mesotopographic gradient illustrates the effect of a variety of intercorrelated site factors. Species composition, biomass, phenology, and site factors have been monitored in five vegetation types representing the portion of the gradient from exposed ridge tops to shallow snowbeds (Fig. 6.3; May 1976, 1982; Walker et al. 1994, 1995). Snow depths were monitored year round in 1972–1974. The mean maximum depths ranged from less than 10 cm in the fellfield community (Sileno-Paronychietum) to 120 cm in the shallow snowbeds (Toninio-Sibbaldietum). The corresponding growing season length varied from 109 days in the fellfield community to 52 days in the snowbed.

In the already short alpine growing season, delayed snow melt strongly affects patterns of vegetative devlopment, flowering, seed set, and total primary production. It is not uncommon to see the flowering of snow-margin species, such as

*Ranunuculus adoneus,* occurring along the melting edge of snow patches and moving as a wave across the tundra through the growing season. Delayed vegetative development occurs in small depressions where the snow melts late, and conversely, delayed senescence may occur in the fall in snowbeds that are wetter than the surrounding dry tundra. In some alpine and arctic plants, nutrient uptake, shoot growth, and flowering may commence under the snow before melt (chapters 9, 12). Good examples of such early season growth under snow at Niwot Ridge include *Bistorta bistortoides, Caltha leptosepala* (Rochow 1969), *Oxyria digyna* (Mooney and Billings 1961), *R. adoneus* (Caldwell 1968; Salisbury 1985; Galen and Stanton 1991; Galen et al. 1993; Galen and Stanton 1993; Mullen and Schmidt 1993; Scherf et al. 1994; Mullen et al. 1998), and *Acomastylis rossii* (Spomer and Salisbury 1968; Chambers 1991). Reradiation of heat from standing dead shoots from the previous season often melts slender holes in snow cover as deep as 20–30 cm. All of these species and many others can leaf out and flower so quickly because they have leaf primordia and flower buds preformed during prior growing seasons (Billings and Mooney 1968; Billings 1974; Mark 1970; Diggle 1997; Aydelotte and Diggle 1997).

## Wind

Wind is a primary factor determining the dominance of cushion-plant and tussock-graminoid growth forms in fellfields and exposed tundra turfs. The tightly packed stems and leaves of these growth forms minimize winter abrasion from wind-transported particles and reduces drought stress during the summer. In contrast, many plants that are protected by snow during winter have erect growth forms, soft leaves, and are not drought resistant (Billings and Mooney 1968).

One of the most thorough winter studies of plant physiology in relation to severe winter wind conditions was that of Bell, who studied *Kobresia myosuroides* during winter in Rocky Mountain National Park, Colorado (Bell 1974; Bell and Bliss 1979). Although *K. myosuroides* is the dominant plant in large areas of the Colorado Front Range, it occurs only in a narrow range of snow accumulation regimes. It is a tussock-forming sedge that forms dense turfs in areas that are largely snow free during much of the winter, except for microdrifts that form leeward of the *Kobresia* tussocks. *Kobresia* does not occur in extreme wind-blown fellfields nor in areas of even shallow snow accumulation. Bell compared behavior of undisturbed *Kobresia* with that of transplants into habitats with more and less winter snow accumulation. She found that *Kobresia*'s success in snow-free meadows is related to rapid summer growth and to its use of an extended period for development, from about April 1, well before snowmelt over much of the tundra, to October 20, after the beginning of drift development in snowbeds. Wintergreen *Kobresia* leaves can even elongate during warm periods ($> -4°C$) in midwinter, an apparently unique phenomenon in tundra plants. New leaves begin elongation in the autumn and complete growth the following summer. Most carbohydrates are stored aboveground in leaves, primarily as oligosaccharides, sugars that likely contribute to frost hardiness of the evergreen leaves. Storage of carbohydrates in the leaves obviate the need for translocation from the roots in frozen soils in winter. Transplanted *Kobresia* do not survive in fellfields because of mechanical damage by windblown snow and sand and low soil water potentials. Early-spring melting of shallow snow cover (about 15 cm) in *Kobresia*'s preferred habitat permits leaf elongation in saturated soils. In sites of moderate to deep snow accumulation (>75 cm), autumn dieback is incomplete before drifts first form in September. A long snow-free period after early September is apparently necessary for proper onset of normal winter carbohydrate status in the leaf shoots. Winter freezing destroys the apparently unhardened leaf tissues and meristems, resulting in loss of carbohydrate reserves.

## Temperature

The temperature environment within an alpine plant community is strongly influenced by microtopography and slope position (chapter 2). May and Webber (1982) monitored mean monthly temperatures in five noda, at 5 cm above the surface and 10 cm below the surface for 1 year (Fig. 6.9). Only during the late-summer-to-fall period (Aug.–Oct.) was the temperature comparable in all of the communities. The temperature contrast between the fellfields and snowbeds was the greatest during winter. In December, aboveground mean temperature of the fellfield leaf environment was $-15.5°C$, whereas aboveground mean temperature of the snowbed community was a relatively warm $-1°C$. Mean aboveground monthly temperature of the snowbed did not drop below $-3.5°C$ during the winter, but it also did not warm above the freezing point until mid-June, six full weeks later than the fellfield, and remained cooler than the fellfield throughout the summer until September, when senescence had begun in all the communities. Aboveground temperature of the dry and moist meadow communities was intermediate during both the winter and early summer (June–August) periods. Belowground temperatures were a few degrees warmer than the air temperature during the winter and few degrees cooler in the summer for all communities (Fig. 6.9). In winter, the greatest contrast between winter aboveground and belowground temperatures occurs in fellfields and the least contrast occurs in snowbeds. During the summer the situation is reversed, with the greatest contrast occurring in the snowbeds.

The winter climate within the plant canopy is strongly affected by the depth of snow cover. Bell (1974) found that in the Colorado alpine, plant canopy temperatures beneath >50 cm of snow are very stable. In windblown areas, temperatures are less variable during winter than summer due to the higher convective cooling of consistent winter winds. Differences between the soil surface temperature and the air temperature at 120 cm is rarely greater than $1-3°C$. Frequent storms with strong winds slow diurnal heating at the soil surface and accelerate soil cooling in the afternoon.

In spring, wind speeds are much reduced, and the snow cover becomes the primary determinant of temperature profiles. Snowfree areas warm rapidly on sunny days. Bell (1974) found that fellfields and *Kobresia* meadows were $12-15°C$ warmer than air above on clear and overcast days. If the areas are inundated with water after the snow melts, the temperature around the plants remain close to $0°C$ as long as the water remains. Bell (1974) found that this common situation paradoxically leads to cooler plant temperatures in fellfields and *Kobresia* meadows dur-

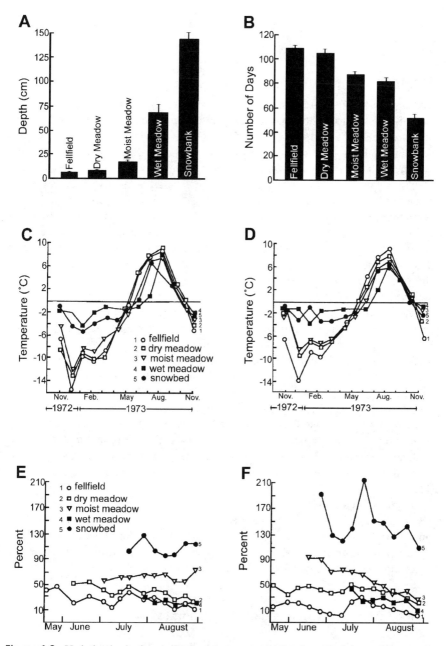

Figure 6.9. Variation in site factors for five plant communities along an alpine toposequence, Niwot Ridge, Colorado. 1972–1973. (a) Average snow depths. (b) Length of the growing season. (c) Foliosphere temperatures at 5 cm above the soil surface for the same five plant communities. (d) Rhizosphere temperatures at 10 cm below the soil surface. (e) Gravimetric soil moisture in 1973 (a dry year). (f) Gravimetric soil moisture in 1974 (a wet year) (adapted from May 1976).

ing May after the melt begins than during the time in April when only a small part of the tundra is snow free.

### Soil Water

In temperate alpine areas, water is released from melting snow over a prolonged period of time due to cold night-time temperatures and deep snow drifts (Caine and Thurman 1990; Caine 1992; chapter 5). Abundant well-defined snow-bed vegetation communities, cold seeps, spring communities, and small wetlands are consequences of late-melting snow in alpine areas (Komárková, 1979). The relationship of snow depth to soil moisture conditions is illustrated by the five plant communities along the Niwot Ridge snow gradient (Fig. 6.3). Soil moisture values are lowest in the fellfield and highest in the wet meadow at the lower margin of the Niwot Ridge snow drift. The central part of the snowdrift has relatively low gravimetric soil moisture, comparable to that of the fellfield, due to rocky course-grained soils, and a tendency to drain rapidly once the snow melts. The summer soil moisture in the fellfield and dry meadow generally decline through the summer but respond somewhat to rainfall events, whereas soil moisture in the moist meadow and wet meadow are related more to the supply of meltwater from snow (Fig. 6.9). The lower margins of snowbanks melt early and are provided with a continuous supply of water as long as the snowbank persists.

Soil moisture, in combination with timing of release from snow cover, governs plant water relations and influences nutrient availability (Fisk et al. 1998; chapter 12). Long-term spatial patterns are an important control on community composition, and interannual variation is an important control on productivity. Oberbauer and Billings (1981) found a high correlation between the seasonal courses of plant water potentials and stomatal conductances along the mesotopographic gradient and the distribution of plant species and vegetation patterns along this same transect. Similar results were found for alpine tundra in the Sierra Nevada (Beyers 1983).

Winter drought stress may be particularly important in areas with little snow cover. Bell (1974) found that wintergreen plants in the Colorado *Kobresia* meadows and fellfields needed at least a small amount of protection provided by microdrifts. Wintergreen *Kobresia* leaf shoots elongated in winter only on warm days when the soil water was available and water potentials rose above $-2.0$ MPa.

## Biotic Factors Controlling Plant Communities

The influence of interspecific interactions in climatically stressful environments has historically invited much debate, focused primarily on negative interactions such as competition (Grime 1977; Tilman 1987; Grace 1991). Ecologists have paid less attention to the importance of positive interactions in structuring plant communities. However, in environments where levels of physical stress are high, such as alpine tundra, facilitation may be a more important structuring force than competition is (Bertness and Callaway 1994).

Only a few studies have specifically examined interspecific competition in arctic or alpine tundra (Griggs 1956; McGraw and Chapin 1989; Wilson 1993; Theodose and Bowman 1997). All indicate that competition is an important structuring force in these communities. Theodose and Bowman (1997) experimentally demonstrated the role of interspecific competition in influencing the distribution of an alpine graminoid. Although the dry meadow dominant *Kobresia myosuroides* is absent from the moist meadow community due to physical factors, as discussed earlier (Bell and Bliss 1979), the moist meadow codominant *Deschampsia caespitosa* is apparently absent from the dry meadow due to competitive displacement by *Kobresia*. Several other studies have alluded to the possibility that interspecific competition may influence species composition on Niwot Ridge. Alpine graminoids from different communities differ in traits thought to influence competitive ability, potentially affecting distribution patterns across communities (Theodose 1995). Within the moist meadow, species differ greatly in several aspects of competitive ability, including growth rates and N uptake capacity, correlating with dominance relationships within that community (Theodose et al. 1996). In addition, mortality of transplanted species into communities differing from community of origin may have been due in part to interspecific competition (May 1976).

Herbivores have been studied extensively on Niwot Ridge, but little research has examined any direct effect of herbivory on alpine plant species composition (chapter 14). Belowground herbivory by gophers (*Thomomys talpoides*) has received much more attention than aboveground herbivory has (Halfpenny and Southwick 1982; Litaor et al. 1996; Cortinas and Seastedt 1996). Gopher activity is greatest in communities with high snowcover, especially snowbed communities (Stoecker 1976; Thorn 1978, 1982; Burns 1979; Willard 1979; Litaor et al. 1996). Direct feeding by gophers may reduce the abundance of preferred species, potentially reducing the degree of competitive exclusion (Halfpenny and Southwick 1982; Tilman 1983; Andersen 1987; Huntley and Inouye 1987; Inouye et al. 1987), although much of the gopher influence on species composition is indirect and dependent on changes in physical conditions of the soil. Gopher activity increases nutrient availability, particularly N, which favors some alpine plant species over others (Bowman et al. 1995; Cortinas and Seastedt 1996; Litaor et al. 1996). Plant community composition surrounding gopher mounds is different from that of undisturbed areas (Davies 1994).

Most Niwot Ridge herbivores seem to favor forb species over grasses (chapter 14). Thus herbivory may contribute to the abundance of graminoids on Niwot Ridge, exacerbating effects of increased atmospheric N deposition by increasing rates of N mineralization (Steltzer and Bowman 1998; Bowman and Steltzer 1998). Exclosure studies need to be conducted to establish these trends.

The influence of positive species interactions on the species composition of plant communities has received less attention than that of negative interactions (Thomas and Bowman 1998). This is particularly true for facilitation, an interaction that is important for plants in the seedling stage. Although sexual reproduction is considered less important than vegetative propagation in alpine tundra, seedling recruitment may be a colonization mechanism in disturbed sites (Chambers 1991). Seedling recruitment is considered an important colonizing mechanism in arctic tundra (Freedman et al. 1983; Gartner et al. 1983). In alpine tundra, mulch prevents small seeds from being blown away by the wind (Chambers 1991), suggesting some plant cover may facilitate establishment. Facilitation may increase seedling survival in intact alpine communities as well. Alpine vegetation cover decreases the formation of needle ice, which contributes to seedling mortality (Bliss 1971; Roach and Marchand 1984). In alpine fellfields, cushion plants act as nurse plants, facilitating seedling establishment, although cushion plants may compete with the seedlings as well (Griggs 1956).

How plants are arranged in space and time within alpine communities is in part dependent on physical factors, but may also be directly or indirectly influenced by presence and type of mutualism. The influence of mutualisms such as symbiotic $N_2$ fixation and mycorrhizae on species composition in alpine tundra is just beginning to be understood. A few species having mutualistic relationships with $N_2$-fixing bacteria occur on Niwot Ridge. These include three species of *Trifolium,* which constitute 12% of the projected cover on Niwot Ridge (May and Webber 1982; Bowman et al. 1996). The presence of $N_2$-fixing plant species influences the species composition of the surrounding vegetation, increasing the abundance of some species but decreasing the abundance of others. Areas adjacent to patches of *Trifolium* have a species composition similar to that of N fertilization plots (Theodose and Bowman 1997; Thomas and Bowman 1998). The abundance of these $N_2$-fixing species within a community may be related to soil P availability. In the dry meadow, the abundance of *Trifolium dasyphyllum* increased significantly following P fertilization but not following N fertilization. This response was in contrast to most other forb species whose abundance was N limited (Theodose and Bowman 1997).

Although mycorrhizal status of many alpine plants has been determined (Haselwandter and Read 1980; Read and Haselwandter 1981; Lesica and Antibus 1985; Gardes and Dahlberg 1996), the role of this mutualism at the plant community level has not been thoroughly examined. The alpine tundra has VA mycorrhizal species, non-mycorrhizal species, and is unique among herbaceous plant communities in its high number of herbaceous ectomycorrhizal species. Variation in the types of mycorrhizal infection may influence alpine plant distribution and abundance. In the dry meadow community, mycorrhizal species abundance increased following N additions, whereas non-mycorrhizal species abundance increased following P additions (Theodose and Bowman 1997). In the snowbed community, which has the shortest growing season, the early season *Ranunculus adoneus* depends on mycorrhizal infection for nutrient uptake. VA mycorrhizal fungi enable *Ranunculus* to obtain and later store P, whereas dark septate fungi may be a means of obtaining organic N without having to rely on soil N mineralization (Mullen and Schmidt 1993; Mullen et al. 1998). Thus, *Ranunculus* may be an important component of the snowbed community in part because of the symbiotic relationship with these fungi.

## Summary and Conclusions

The vegetation of Niwot Ridge has a long history of study, with a particularly strong focus on spatial patterns and phytosociology. Most of the Niwot Ridge vegetation falls into a single class, the Elyno-Seslerietea, which encompasses the fellfield and dry meadow noda, but includes additional associations not included in the original conceptions of those noda. The spatial patterns of Niwot Ridge vegetation is predominated by physical factors, but biotic interactions are particularly important in shaping microscale patterns and short-term dynamics.

The rich history of spatially focused study sets a precedent for work on temporal dynamics at a variety of scales. Because of the short growing seasons and long life spans of most alpine species, temporal studies are a particular challenge. A combination of modeling and carefully selected experimentation is necessary to understand dynamics and trends in the system and to make reasonable predictions about how it might change in response to a changing environment.

References

Andersen, D. C. 1987. Below-ground herbivory in natural communities: A review emphasizing fossorial animals. *Quarterly Review of Biology* 62:261–286.

Aydelotte, A. R., and P. K. Diggle. 1997. Analysis of developmental preformation in the alpine herb, *Caltha leptosepala*. American Journal of Botany 84:1646–1657.

Bell, K. L. 1974. Autecology of *Kobresia* bellardii: Why winter snow accumulation limits local distribution. Ph.D. diss., University of Alberta.

Bell, K. L., and L. C. Bliss. 1979. Autecology of *Kobresia* bellardii: Why winter snow accumulation limits local distribution. *Ecological Monographs* 49:377–402.

Bertness, M. D., and R. Callaway. 1994. Positive interactions in communities. *Trends in Ecology and Evolution* 9:191–193.

Beyers, J. L. 1983. Physiological ecology of three alpine plant species along a snowbank gradient in the northern Sierra Nevada. Ph.D. diss., Duke University.

Billings, W. D. 1973. Arctic and alpine vegetations: Similarities, differences, and susceptibility to disturbances. *BioScience* 23:697–704.

———. 1974. Arctic and alpine vegetation: Plant adaptations to cold summer climates. In *Arctic and alpine environments,* edited by R. G. Barry and J. D. Ives. London: Methuen

———. 1988. Alpine vegetation. In *North American terrestrial vegetation,* edited by M. G. Barbour and W. D. Billings. Cambridge: Cambridge University Press.

Billings, W. D., and L. C. Bliss. 1959. An alpine snowbank environment and its effects on vegetation, plant development, and productivity. *Ecology* 40:388–397.

Billings, W. D., and H. A. Mooney. 1968. The ecology of arctic and alpine plants. *Biological Review* 43:481–529.

Bliss, L. D. 1963. Alpine plant communities of the Presidential Range, New Hampshire. *Ecology* 44:678–697.

———. 1971. Arctic and alpine plant life cycles. *Annual Review of Ecology and Systematics* 2:405–438.

Bowman, W. D., Schardt, J. C., and Schmidt, S. K. 1996. Symbiotic N-2 fixation in alpine tundra: Ecosystem input and variation in fixation rates among communities. *Oecologia* 108:345–350.

Bowman, W. D., and H. Steltzer. 1998. Positive feedbacks to anthropogenic nitrogen deposition in Rocky Mountain alpine tundra. *Ambio* 27:514–517.

Bowman, W. D., T. A. Theodose, and M. C. Fisk. 1995. Physiological and production responses of plant growth forms to increases in limiting resources in alpine tundra: Implications for differential community response to climate change. *Oecologia* 101: 217–227.

Braun-Blanquet, J. 1965. *Plant sociology: The study of plant communities,* edited and translated by C. D. Fuller and H. S. Conard. London: Hafner.

Burns, S. F. 1979. The northern pocket gopher (*Thomomys talpoides*): A major geomorphic agent on the alpine tundra. *Journal of the Colorado-Wyoming Academy of Science* 2:86.

Burns, S. J., and P. J. Tonkin. 1982. Soil-geomorphic models and the spatial distribution and development of alpine soils. In *Space and time in geomorphology,* edited by C. E. Thorn. London: Allen and Unwin.

Caine, N. 1992. Modulation of the dirunal streamflow response by the seasonal snowcover of an alpine snowpatch. *Catena* 16:153–162.

Caine, N., and E. M. Thurman. 1990. Temporal and spatial variations in the solute content of an alpine stream, Colorado Front Range. *Geomorphological Abstracts* 4:55–72.

Caldwell, M. M. 1968. Solar ultraviolet radiation as an ecological factor for alpine plants. *Ecological Monographs* 38:243–268.

Chambers, J. C. 1991. Patterns of growth and reproduction in a perennial tundra forb (Geum rossii): Effects of clone area and neighborhood. *Canadian Journal of Botany* 69:1977–1983.

Cortinas, M. R., and T. R. Seastedt. 1996. Short- and long-term effects of gophers (*Thomomys talpoides*) on soil organic matter dynamics in alpine tundra. *Pedobiologia* 40:162–170.

Davies, E. F. 1994. Disturbance in alpine tundra ecosystems: The effects of digging by Northern pocket gophers (*Thomomys talpoides*). Master's thesis, Utah State University, Logan.

Diggle, P. K. 1997. Extreme preformation in alpine *Polygonum viviparum:* An architectural and developmental analysis. *American Journal of Botany* 84:154–169.

Fisk, M. C., S. K. Schmidt, and T. R. Seastedt. 1998. Topographic patterns of above- and belowground production and nitrogen cycling in alpine tundra. *Ecology* 79:2253–2266.

Freedman, B., J. Svoboda, C. Labine, M. Muc, G. Henry, M. Nams, J. Stewart, and E. Woodley. 1983. Physical and ecological characteristics of Alexandra Fiord, a High Arctic oasis on Ellesmere Island, Canada. Proceedings of the Fourth International Conference on Permafrost (Fairbanks, Alaska). Washington, DC: National Academy Press.

Galen, C., T. E. Dawson, and M. L. Stanton. 1993. Carpels as leaves: Meeting the carbon cost of reproduction in an alpine buttercup. *Oecologia* 95:187–193.

Galen, C., and M. L. Stanton. 1991. Consequences of emergence phenology for reproductive success in *Ranunculus adoneus* (Ranunculaceae). *American Journal of Botany* 78:978–988.

———. 1993. Short-term responses of alpine buttercups to experimental manipulations of growing season length. *Ecology* 75:1546–1557.

Gardes, M., and A. Dahlberg. 1996. Mycorrhizal diversity in arctic and alpine tundra: An open question. *New Phytologist* 133(1):147–157.

Gartner, B. L., F. S. Chapin III, and G. R. Shaver. 1983. Demographic patterns of seedling establishment and growth of native graminoids in an Alaskan tundra disturbance. *Journal of Applied Ecology* 20:965–980.

Grace, J. B. 1991. A clarification of the debate between Grime and Tilman. *Functional Ecology* 5:583–587.

Griggs, R. G. 1956. Competition and succession on a Rocky Mountain fellfield. *Ecology* 37:8–20.

Grime, J. P. 1977. Evidence for the existence of three primary strategies in plants and its relevance to ecological and evolutionary theory. *American Naturalist* 111:1169–1194.

Halfpenny, J. C., and C. H. Southwick. 1982. Small mammal herbivores of the Colorado alpine tundra. In *Ecological studies in the Colorado alpine, a festschrift for John W.*

*Marr,* edited by J. C. Halfpenny. Occasional Paper 37, Institute of Arctic and Alpine Research, University of Colorado, Boulder.
Haselwandter, K., and D. J. Read. 1980. Fungal associations of roots of dominant and subdominant plants in high-alpine vegetation systems with special reference to mycorrhizae. *Oecologia* 45:57–62.
Holway, J. G., and R. T. Ward. 1963. Snow and meltwater effects in an area of Colorado alpine. *America Midland Naturalist* 69:189–197.
———. 1965. Phenology of alpine plants in northern Colorado. *Ecology* 46:73–83.
Humphries, H. C., D. P. Coffin, and W. K. Lauenroth. 1996. An individual-based model of alpine plant distributions. *Ecological Modeling* 84:99–126.
Huntly, N., and R. S. Inouye. 1987. Small mammal populations of an old-field chronosequence: Successional patterns and associations with vegetation. *Journal of Mammalogy* 68:739–745.
Inouye, R. S., N. J. Huntly, D. Tilman, and J. R. Tester. 1987. Pocket gophers (*Geomys bursarius*), vegetation, and soil nitrogen along a successional sere in east central Minnesota. *Oecologia* 72:178–184.
Isard, S. A. 1986. Factors influencing soil moisture and plant community distribution on Niwot Ridge, Front Range, Colorado, U.S.A. *Arctic and Alpine Research* 18:83–96.
Komárková, V. 1976. Alpine vegetation of the Indian Peaks area, Front Range, Colorado Rocky Mountains. Ph.D. diss., University of Colorado, Boulder.
———. 1979. Alpine vegetation of the Indian Peaks area, Front Range, Colorado Rocky Mountains. Vaduz, Lichtenstein: J. Cramer.
———. 1980. Classification and ordination in the Indian Peaks area, Colorado Rocky Mountains. *Vegetatio* 42:149–163.
Komárková, V., and P. J. Webber. 1978. An alpine vegetation map of Niwot Ridge, Colorado. *Arctic and Alpine Research* 10:1–29.
LeDrew, E. F. 1975. The energy balance of mid-latitude alpine site during the growing season. *Arctic and Alpine Research* 7:301–314.
Lesica, P., and R. K. Antibus. 1985. Mycorrhizae of alpine fell-field communities on soils derived from crystalline and calcareous parent materials. *Canadian Journal of Botany* 64:1691–1697.
Litaor, M. I., R. Mancinelli, and J. C. Halfpenny. 1996. The influence of pocket gophers on the status of nutrients in alpine soils. *Geoderma* 70:37–48.
Mark, A. F. 1970. Floral initiation and development in New Zealand alpine plants. *New Zealand Journal of Botany* 8:67–75.
Marr, J. W. 1961. *Ecosystems of the east slope of the Front Range in Colorado.* Vol. 8 of *University of Colorado Series in Biology.*
May, D. E. 1976. The response of alpine tundra vegetation in Colorado to environmental modification. Ph.D. diss., University of Colorado, Boulder.
May, D. E., and P. J. Webber. 1982. Spatial and temporal variation of the vegetation and its productivity on Niwot Ridge, Colorado. In *Ecological studies in the Colorado alpine: A festschrift for John W. Marr,* edited by J. C. Halfpenny. Occasional Paper 37, Institute of Arctic and Alpine Research, University of Colorado, Boulder.
McGraw, J. B., and F. S. Chapin, III. 1989. Competitive ability and adaptation to fertile and infertile soils in two *Eriophorum* species. *Ecology* 70:736–749.
Mooney, H. A., and W. D. Billings. 1961. Comparative physiological ecology of arctic and alpine populations of Oxyria digyna. *Ecological Monographs* 31:1–29.
Mullen, R. B., and S. K. Schmidt. 1993. Mycorrhizal infection, phosphorus uptake, and phenology in *Ranunculus adoneus:* Implications for the functioning of mycorrhizae in alpine systems. *Oecologia* 94:229–234.

Mullen, R. B., S. K. Schmidt, and C. H. Jaeger. 1998. Nitrogen uptake during snowmelt by the snow buttercup, *Ranunculus adoneus*. *Arctic and Alpine Research* 30:121–125.

Oberbauer, S. F., and W. D. Billings. 1981. Drought tolerance and water use by plants along an alpine topographic gradient. *Oecologia* 50:325–331.

Razzhivin, V. Y. 1994. Snowbed vegetation of far northeastern Asia. *Journal of Vegetation Science* 5:829–842.

Read, D. J., and K. Haselwandter 1981. Observations on the mycorrhizal status of some alpine plant communities. *New Phytologist* 88:341–352.

Roach, D. A., and P. J. Marchand. 1984. Recovery of alpine disturbances: Early growth and survival in populations of the native species, *Arenaria groenlandica, Juncus trifidus,* and *Potentialla tridentata. Arctic and Alpine Research* 16:37–43.

Rochow, T. F. 1969. Growth, caloric content, and sugars in Caltha leptosepala in relation to alpine snowmelt. *Bulletin of the Torrey Botanical Club* 96:689–698.

Salisbury, F. S. 1985. Plant growth under snow. *Aquilo, Series Botanica* 23:1–7.

Scherf, E. J., C. Galen, and M. L. Stanton. 1994. Seed dispersal, seedling survival, and habitat affinity in a snowbed plant: Limits to the distribution of the snow buttercup, Ranunculus adoneus. *Oikos* 69:405–413.

Spomer, G. G., and F. B. Salisbury. 1968. Eco-physiology of Geum turbinatum and implications concerning alpine environments. *Botanical Gazette* 129:33–49.

Stanton, M. L., M. Rejmánek, and C. Galen. 1994. Changes in vegetation and soil fertility along a predictable snowmelt gradient in the Mosquito Range, Colorado, U.S.A. *Arctic and Alpine Research* 26:364.

Steltzer, H., and W. D. Bowman. 1998. Differential influence of plant species on soil nitrogen transformations in moist meadow alpine tundra. *Ecosystems* 1:464–474.

Stoecker, R. 1976. Pocket gopher distribution in relation to snow in the alpine tundra. In *Ecological impacts of snowpack augmentation in the San Juan Mountains of Colorado,* edited by H. W. Steinhoff and J. D. Ives. U.S. Department of Interior, for Division of Atmospheric Water Resources Management, Bureau of Reclamation, Denver, Colorado.

Theodose, T. A. 1995. Interspecific plant competition in alpine tundra. Ph.D. diss., University of Colorado, Boulder.

Theodose, T. A., and W. D. Bowman. 1997. Nutrient availability, plant abundance, and species diversity in two alpine tundra communities. *Ecology* 78:1861–1872.

Theodose, T. A., C. H. Jaeger III, W. D. Bowman, and J. C. Schardt. 1996. Uptake and allocation of $^{15}N$ in plants: Implications for the importance of competitive ability in predicting community structure in a stressful environment. *Oikos* 75:59–66.

Thomas, B. D., and W. D. Bowman. 1998. Influence of $N_2$-fixing *Trifolium* on plant species composition and biomass production in alpine tundra. *Oecologia* 115:26–31.

Thorn, C. E. 1978. A preliminary assessment of the geomorphic role of pocket gophers in the alpine zone of the Colorado Front Range. *Geografiska Annaler* 60A:181–187.

———. 1982. Gopher disturbance: Its variability by Braun-Blanquet vegetation units in the Niwot Ridge alpine tundra zone, Colorado Front Range, U.S.A. *Arctic and Alpine Research* 14:45–51.

Tilman, D. 1983. Plant succession and gopher disturbance along an experimental gradient. *Oecologia* 60:285–292.

———. 1987. Secondary succession and the pattern of plant dominance along experimental nitrogen gradients. *Ecological Monographs* 57:189–214.

Tranquillini, W. 1979. *Physiological ecology of the alpine timberline.* Berlin: Springer-Verlag.

Viereck, L. A., C. T. Dyrness, A. R. Batten, and K. J. Wenzlick. 1992. The Alaska vegetation classification. United States Department of Agriculture, General tech. rep. PNW-GTR-286.

Walker, D. A., J. C. Halfpenny, M. D. Walker, and C. A. Wessman. 1993. Long-term studies of snow-vegetation interactions. *BioScience* 43:287–301.

Walker, M. D. 1990. Vegetation and floristics of pingos, Central Arctic Coastal Plain, Alaska. In *Dissertationes Botanicae, 149*. Stuttgart: J. Cramer.

Walker, M. D., R. C. Ingersoll, and P. J. Webber. 1995. Effects of interannual climate variation on phenology and growth of two alpine forbs. *Ecology* 76:1067–1083.

Walker, M. D., P. J. Webber, E. H. Arnold, and D. Ebert-May. 1994. Effects of interannual climate variation on aboveground phytomass in alpine vegetation. *Ecology* 75:393–408.

Wardle, P. 1968. Engelmann spruce (*Picea engelmannii* Engel) at its upper limits on the Front Range, Colorado. *Ecology* 49:483–495.

Weber, W. A. 1976. *Rocky Mountain flora*. Boulder: University of Colorado Press.

Weber, W. A., and B. E. Willard. 1967. Checklist of alpine tundra species of vascular plants found in Colorado. Unpubl. University of Colorado, Boulder. Institute of Arctic and Alpine Research.

Willard, B. E. 1963. Phytosociology of the alpine tundra of Trail Ridge, Rocky Mountain National Park, Colorado. Ph.D. Diss., University of Colorado, Boulder.

Willard, B. E. 1979. Plant sociology of alpine tundra, Trail Ridge, Rocky Mountain National Park, Colorado. *Colorado School of Mines Quarterly* 74:119.

Wilson, S. D. 1993. Competition and resource availability in heath and grassland in Snowy Mountains of Australia. *Journal of Ecology* 81: 445–451.

# 7

# Vertebrates

David M. Armstrong
James C. Halfpenny
Charles H. Southwick

## Introduction

Vertebrates of alpine tundra are near the limits of their genetic tolerance, and thus the alpine provides a natural laboratory for the study of the ecology of these organisms in a climatically stressful environment. The alpine supports a greater species richness of vertebrate herbivores than does arctic tundra (Halfpenny and Southwick 1982). Hoffmann (1974) provided an extensive review of terrestrial vertebrates of arctic and alpine ecosystems, emphasizing circumpolar patterns. For a variety of reasons, however, vertebrates of alpine tundra are considerably less studied than are those of the Arctic, and much remains to be learned about the physiological and behavioral adaptations of vertebrates that allow this group to exist in this extreme and variable ecosystem.

May (1980) offered some generalizations about the state of knowledge of alpine animals. Terrestrial systems are better known than aquatic systems; the magnitude of environmental variability is better known than its predictability and significance to populations of animals; life histories of animals are better known than their roles and functions; dynamics of single species are better known than interactions between and among species; habitat selection by animals is more often defined in terms of the perception of the investigator than in terms of the perception of the organism; the response of animals to patterns of vegetation is better known than the influence animals have in creating and maintaining those patterns; and densities of animals are better known than are patterns of dispersion and their causes. Those generalizations remain broadly accurate.

The purpose of this chapter is to develop a perspective on the structure and function of the vertebrate fauna of alpine environments of the Southern Rocky Mountains, with an emphasis on the fauna found on Niwot Ridge. It considers the origin and ongoing development of the fauna and its biogeographic and ecological relationships. A pattern of distributions is described that is dynamic in space and time. A principal focus is the role of vertebrates to the structure and function of the tundra ecosystem, including both the biotic and physical impacts of vertebrate populations. Some attention is paid to vertebrate trophic guilds, but plant-animal interactions are detailed in this volume by Dearing (chapter 14).

This chapter is not a review of the natural history of Coloradan alpine vertebrates. A number of other literature sources (e.g., Zwinger and Willard 1972; Armstrong 1975, 1987; Smith 1981; Benedict 1991; Andrews and Righter 1992; Cushman et al. 1993; Fitzgerald et al. 1994; Emerick 1995) facilitate access to the literature on individual species. Halfpenny et al. (1986) listed much of the literature on alpine and subalpine environments of the Rockies prior to mid-1985. Comparative data on vertebrates are available from elsewhere in the Southern Rockies, including Mount Evans (Halfpenny 1980), Loveland Pass (Southwick et al. 1986), Mount Lincoln (Blake and Blake 1969), the vicinity of the Rocky Mountain Biological Laboratory at Gothic (Findley and Negus 1953; Armitage 1991), as well as the Medicine Bow Mountains and Snowy Range of southern Wyoming (e.g., Brown 1967).

This review is not completely restricted to the alpine because the tundra has strong faunal relationships and direct ecological linkages with krummholz, the zone of elfin trees that constitutes the alpine/subalpine ecotone, as well as with subalpine and even lower elevation ecosystems. In many respects, "tundra" is a plant-ecological construct. In the simplest terms, treeline represents a limit in the ecological tolerances of certain plant species populations (Wardle, 1974). As vegetation changes, so do resources (food, cover) for consumers. However, one would not expect distributional limits of vertebrates (especially endotherms) to correlate precisely with those of plant species or plant associations or with the same physical factors that shape the vegetation. To the extent possible, our comments pertain to tundra environments and to krummholz, but linkages among ecosystems on a range of scales of space and time cannot be ignored.

Pioneer studies of Coloradan vertebrates did not neglect the alpine fauna. Early biological work often was an adjunct of mineral exploration and exploitation of the mountainous part of the state, so the alpine zone was of particular interest. Brewer (1871) made notes on alpine wildlife observed in Park County on a Harvard Mining School Expedition of 1869, and Allen (1874) reported on collections made above Montgomery, on Mount Lincoln, Park County, in 1871. Martha Maxwell (reported by Coues 1879) included wildlife from the alpine zone—perhaps taken in Boulder County—in the taxidermic exhibit of Coloradan natural history that she prepared for the Philadelphia Centennial Exhibition of 1876. The U.S. Bureau of Biological Survey worked in Colorado in the early years of the twentieth century (Cary 1911), but their ostensible purpose was to determine the suitability of the state for agriculture, so they spent little time in the high mountains. By contrast, E. R. Warren (1910, 1942) first became acquainted with Coloradan vertebrates in the mining camps of the West Elk Range (Ruby, Irwin, Gothic, Crested Butte) and he made important observations (supported by invaluable specimens, field notes, and photographs—see Armstrong 1986) of alpine species.

## Vertebrate Fauna of the Alpine Zone

### Composition of the Fauna

All vertebrates of Niwot Ridge are endotherms (warm-blooded) and are typical of the vertebrate fauna of alpine areas throughout the Southern Rocky Mountains (Table 7.1). There are no records of reptiles or fish on Niwot Ridge, although trout and possibly some amphibians occur in the alpine of the adjacent Green Lakes Valley (Fig. 1.5). Three species of amphibians—tiger salamander (*Ambystoma tigrinum*), striped chorus frog (*Pseudacris triseriata*), and boreal toad (*Bufo boreas*)—are known from above treeline elsewhere in Colorado (Hammerson 1982), but none of them is known currently from Niwot Ridge.

The alpine avifauna includes dozens of occasional visitors during the warmer months, but the number of breeding birds in most patches of tundra is five or fewer (Kingery 1998). The mammalian fauna of the tundra also is a variable assemblage, including 25–30 or more species (Armstrong 1972). Faunal composition changes dramatically through the year and may differ rather strongly from one year to the next. Also, influential members of the fauna (bison, grizzly bear, gray wolf) have been extirpated within historic time.

Because the fauna of a particular locality tends to change on various scales of time (daily, seasonally, and over ecological and evolutionary time), some qualification of "occurrence" needs to be made. A rough hierarchy of occurrence is defined as follows. First, the fauna may be historical (no longer present) or contemporary. If the species is contemporary, then its presence can be defined as a permanent or seasonal resident. Resident animals may either breed in the alpine or have a nonbreeding status. Finally, those animals that have been observed but are not residents of the alpine can be classified as occasional or accidental visitors.

Accidental visitors are often excluded from site lists. Here, we have not attempted hard distinctions, in part because of the opportunism of many alpine vertebrates, which may be present one year and not the next, or which may wander occasionally into the tundra but are not typical residents. For example, on Guanella Pass, Clear Creek County, Colorado, Conry (1978) observed three nesting passerines (Horned Lark, Water Pipit, and White-crowned Sparrow—the first two nesting on open tundra and the third in shrub tundra and krummholz). American Robins and Mountain Bluebirds fed frequently on the tundra in all four field seasons of her study, and in two seasons, Brewer's Sparrows and Vesper Sparrows were important foragers as well. Our general, sometimes subjective, criterion for inclusion in this review was "Does a particular species regularly utilize alpine environments and exploit their resources (or influence energy flow in alpine food webs)?" In other words, "Is the species a typical participant in alpine ecosystems?"

Once a faunal list is assembled, it is still difficult to make reasonable generalizations about vertebrates. The life histories and ecology of birds and mammals tends to differ markedly. Consider the constraints and opportunities of oviparity versus viviparity, for example, or the importance to the ecosystem of the differential dispersal capablilities of birds and terrestrial mammals. Further, generalizations about the avifauna or the mammalian fauna are difficult. For example, just one

Table 7.1. Vertebrate Fauna of Niwot Ridge, Boulder County, Colorado

| Mammals | Breeding Birds |
|---|---|
| Insectivora: Soricidae | Galliformes: Phasianidae |
|   Masked Shrew—*Sorex cinereus* |   White-tailed Ptarmigan—*Lagopus leucurus* |
|   Dwarf Shrew—*Sorex nanus* | Passeriformes: Alaudidae |
|   Montane Shrew—*Sorex monticolus* |   Horned Lark—*Eremophila alpestris* |
| Lagomorpha: Ochotonidae | Passeriformes: Motacillidae |
|   American Pika—*Ochotona princeps* |   American Pipit—*Anthus rubescens* |
| Lagomorpha: Leporidae | Passeriformes: Emberizidae |
|   Snowshoe Hare—*Lepus americanus* |   White-crowned Sparrow—*Zonotrichia leucophrys* |
|   White-tailed Jackrabbit—*Lepus townsendii* | Passeriformes: Fringillidae |
| Rodentia: Sciuridae |   Rosy Finch—*Leucosticte arctoa* |
|   Least Chipmunk—*Tamias minimus* | |
|   Yellow-bellied Marmot—*Marmota flaviventris* | |
|   Golden-mantled Ground Squirrel—*Spermophilus lateralis* | |
| Rodentia: Geomyidae | |
|   Northern Pocket Gopher—*Thomomys talpoides* | |
| Rodentia: Muridae | |
|   Deer Mouse—*Peromyscus maniculatus* | |
|   Bushy-tailed Woodrat—*Neotoma cinerea* | |
|   Southern Red-backed Vole—*Clethrionomys gapperi* | |
|   Heather Vole—*Phenacomys intermedius* | |
|   Montane Vole—*Microtus montanus* | |
|   Long-tailed Vole—*Microtus longicaudus* | |
| Rodentia: Zapodidae | |
|   Western Jumping Mouse—*Zapus princeps* | |
| Rodentia: Erethizontidae | |
|   Porcupine—*Erethizon dorsatum* | |
| Carnivora: Canidae | |
|   Coyote—*Canis latrans* | |
|   Gray Wolf—*Canis lupus*[1] | |
|   Red Fox—*Vulpes vulpes* | |
| Carnivora: Ursidae | |
|   Black Bear—*Ursus americanus* | |
|   Grizzly Bear—*Ursus arctos*[1] | |
| Carnivora: Mustelidae | |
|   American Marten—*Martes americana* | |
|   Short-tailed Weasel—*Mustela erminea* | |
|   Long-tailed Weasel—*Mustela frenata* | |
|   American Badger—*Taxidea taxus* | |
| Carnivora: Felidae | |
|   Mountain Lion—*Felis concolor* | |
|   Bobcat—*Felis rufus* | |
| Artiodactyla: Cervidae | |
|   Wapiti, or American Elk—*Cervus elaphus* | |
|   Mule Deer—*Odocoileus hemionus* | |
| Artiodactyla: Bovidae | |
|   Bison—*Bison bison*[1] | |
|   Bighorn Sheep—*Ovis canadensis* | |

[1] Extirpated within historic time.

Figure 7.1. White-tailed Ptarmigan (*Lagopus leucurus*), the only bird which is a year-round resident of the alpine. Ptarmigan rely heavily on cryptic coloration (camouflage) for predator avoidance (top photo by William Ervin; bottom photo by William D. Bowman).

species of bird, the White-tailed Ptarmigan (Fig. 7.1), can be considered a year-round resident of Niwot Ridge, and even the ptarmigan tends to move to subalpine willow thickets in winter. To complicate the picture further, males often winter in shrub tundra, whereas hens are more likely to move to the subalpine (Hoffman and Braun 1977). Likewise, a number of mammals are resident on Niwot Ridge year-round, but the larger herbivores spend the winter at lower elevations as do their predators and the species that scavenge their carrion. Thus, although summary statements must be qualified, they do highlight the principal features of the alpine vertebrate fauna: opportunism and pronounced seasonal dynamism driven by the rigorous physical environment.

No Coloradan vertebrate is restricted to alpine tundra. The closest approach to a tundra obligate may be the White-tailed Ptarmigan, which breeds on open tundra, especially areas that are snow free early in the season. Females rear broods on wet tundra and fellfield. White-crowned Sparrows breed in alpine and subalpine willow thickets (Hubbard 1978), but their winter range is brushy steppe and riparian habitats at lower elevations (Andrews and Righter 1992). Horned Larks breed on the alpine tundra and also in similarly open habitats in the mountain parks, western valleys, and eastern plains (Andrews and Righter 1992). Among mammals, the pika (*Ochotona princeps*, Fig. 7.2) may be the tundra mammal that comes to mind first, but it occurs in suitable talus/meadow mosaics well below treeline, to as low as 8500 ft (2590 m) in Grand County (Armstrong 1972). The yellow-bellied marmot (Fig. 7.3) ranges into the foothills of the Front Range and locally onto the Colorado Piedmont (Fitzgerald et al. 1994).

Figure 7.2. The pika (*Ochotona princeps*) is one of the most characteristic mammals of the alpine. Pikas must collect all of their winter food supply during the short growing season, as they do not hibernate (see chapter 14) (photo by William Ervin).

Figure 7.3. Yellow-bellied marmot (*Marmota flaviventris*) (photo by William D. Bowman).

Armstrong (1972) reported mammalian similarity between alpine tundra and various other biotic community-types in Colorado (calculated as $2C/[N_1 + N_2]$, where C = number of species in common, and $N_1$ and $N_2$ are numbers of species in the two community-types under comparison) as follows: subalpine forest 0.697, aspen woodland, 0.690, highland streambank, 0.519, subalpine meadow 0.738. Faunal similarity for birds probably is higher still but is impossible to calculate because of the high dispersal and opportunism observed in many species.

The list of birds known or expected on Niwot Ridge is fairly extensive (Table 7.2); however, the number of breeding birds is low, and it surely is the resident breeders, not the occasional visitors or accidentals, that are of greatest importance to the functioning biotic community over ecological time. Nonetheless, visitors can have a significant impact in a particular season.

The avifauna that breeds on or near the tundra and habitually forages on the tundra to feed the young is of interest here. These are the species that are dependent on the primary and secondary production of the tundra for food and often for shelter. This fauna is composed of just five species: White-tailed Ptarmigan, Horned Lark, American Pipit, White-crowned Sparrow, and Rosy Finch. Of those species, only the White-tailed Ptarmigan, American Pipit, and Horned Lark actually nest on the open tundra; White-crowned Sparrows nest in krummholz and Rosy Finches nest in rocky cliffs adjacent to tundra (as do Rock Wrens in many areas, but apparently not on Niwot Ridge). May (1980) concluded that tundra birds probably play a rather minor role in the overall function of alpine ecosystem.

Table 7.2. Birds of Tundra Habitats on Niwot Ridge and Vicinity, Boulder County, Colorado

Falconiformes: Accipitridae—Hawks
  Northern Harrier—*Circus cyaneus* (n)
  Red-tailed Hawk—*Buteo jamaicensis* (n)
  Swainson's Hawk—*Buteo swainsoni* (n)
  Rough-legged Hawk—*Buteo lagopus* (n)
  Ferruginous Hawk—*Buteo regalis* (n)
  Golden Eagle—*Aguila chrysaetos* (n)
Falconiformes: Falconidae—Falcons
  Prairie Falcon—*Falco mexicanus* (n)
  Peregrine Falcon—*Falco peregrinus* (m)
Galliformes: Phasianidae—Grouse
  White-tailed Ptarmigan—*Lagopus leucurus* (r)
Columbiformes: Columbidae—Doves
  Mourning Dove—*Zenaida macroura* (n)
Passeriformes: Alaudidae—Larks
  Horned Lark—*Eremophila alpestris* (b)
Passeriformes: Corvidae—Crows and allies
  Common Raven—*Corvus corax* (n)
  Clark's Nutcracker—*Nucifraga columbiana* (n)
Passeriformes: Troglodytidae—Wrens
  Rock Wren—*Salpinctes obsoletus* (p)
Passeriformes: Turdidae—Thrushes
  American Robin—*Turdus migratorius* (n)
  Mountain Bluebird—*Sialia currucoides* (n)
Passeriformes: Motacillidae—Pipits
  American Pipit—*Anthus rubescens* (b)
Passeriformes: Emberizidae, Emberizinae—Sparrows and Allies
  Vesper Sparrow—*Pooecetes gramineus* (n)
  Savannah Sparrow—*Passerculus sandwichensis* (m)
  White-crowned Sparrow—*Zonotrichia leucophrys* (b)
Passeriformes: Fringillidae—Finches
  Brown-capped Rosy Finch—*Leucosticte arctoa* (b)

*Sources:* Data from Kingery and Graul (1978), Andrews and Righter (1992), and Kingery (1998).

*Notes:* (Status: r = year-around resident, breeder; b = summer resident, breeder; p = possible breeder; n = summer visitor, nonbreeder; m = migrant.)

The list of mammals of Niwot Ridge and vicinity (Table 7.1) is deliberately conservative. Species that are known to have occurred on Niwot Ridge within historic time are included (although they may have been extirpated locally or statewide). The list does not include the wolverine (*Gulo gulo*) or the lynx (*Lynx lynx*), either or both of which may once have occurred occasionally on the Ridge. Both of these species are restricted mostly to subalpine habitats in the Rockies, and neither apparently has ever been particularly abundant in Colorado (Fitzgerald et al. 1994). An ungulate recently introduced to the mountains of Colorado, the mountain goat (*Oreamnos americanus*), has been observed in the vicinity (W. D. Bowman, personal communication) and may eventually reach Niwot Ridge. Bison are known to have occurred on Niwot Ridge (Beidleman 1955), but they have been absent from the area since the mid-nineteenth century (Armstrong 1972; Halfpenny and Southwick 1982; Meaney and Van Vuren 1993). Bighorn sheep (*Ovis canadensis*) once were present on the Ridge but are absent at the present time. Their restoration deserves consideration. Earlier in this century, domestic sheep (*Ovis aries*) were grazed on Niwot Ridge under a Forest Service permit, but that activity ceased after 1949 (Halfpenny and Southwick 1982). Cows were not routinely allowed to graze in alpine areas (Marr 1964), but limited incursions by this species have occurred.

## Development of the Fauna

The vertebrate assemblage of Niwot Ridge consists of species with quite divergent biogeographic affinities, suggesting a dynamic assemblage over the past 10–15,000 years. Direct evidence for faunal development is virtually nonexistent; the fossil and subfossil record consists only of occasional fragments. No actual alpine local fauna has been discovered from the Pleistocene of the Southern Rockies (Walker 1987).

The alpine and subalpine fauna of the Southern Rocky Mountains represents a relict of the last (Pinedale) full glacial interval (Armstrong 1972). At the present time, many species of higher elevations of the Southern Rockies are "marooned" on a relatively mesic island of montane woodland, subalpine forest, and alpine tundra, surrounded by a "sea" of steppe environments. Analysis of the distribution of mammalian species and subspecies (Findley and Anderson 1956; Armstrong 1972) indicates that strongest faunal relationships are to the west and north, with the Middle Rocky Mountains of northern Utah and western Wyoming, rather than directly to the north along the Continental Divide in Wyoming.

Armstrong (1972) argued that areographic analysis—the study of the geographic ranges of species—could provide insight into the history of the Coloradan fauna. Of 33 species of mammals of tundra and krummholz habitats (Table 7.1), 12 (36.4%) are Cordilleran species, 7 (21.2%) are Boreo-Cordilleran species, 13 (39.4%) are widespread species (hence not assignable to any areographic faunal element), and 1 (3.0% of the mammalian fauna, the white-tailed jackrabbit) has a general distribution in steppe habitats of the Great Plains and the Great Basin and so is a Campestrian species. By contrast, percentage contribution of those areographic faunal elements to the Coloradan fauna as a whole is: Cordilleran, 10.9; Boreo-Cordilleran, 10.1; Widespread, 20.3; and Campestrian, 10.9. In other words (and not surprisingly), the fauna of the high mountains is predominated by species of northern and/ or western affinities and by widespread species (eurychores). Species of five areographic elements (Eastern, Great Basin, Yuman, Chihuahuan, and autochothonous Coloradan) are absent from the alpine fauna (Armstrong 1972).

Breeding birds of Niwot Ridge proper include just five species. In terms of faunal elements (in the sense of Armstrong 1972), two (White-tailed Ptarmigan and Rosy Finch) are Cordilleran species, the White-crowned Sparrow exhibits a Boreo-Cordilleran distribution, and the Horned Lark and American Pipit are eurychores.

The nearly equal importance in the Coloradan alpine fauna of western (Cordilleran) and boreal (Boreo-Cordilleran) species provides an important point of contrast with arctic tundra communities. Cordilleran species are absent from arctic tundra. Cordilleran species generally reach northern limits in the mountains of British

Columbia, the Yukon Territory, or southern Alaska. Their southern limits tend to be in California, on isolated ranges in the Great Basin of Nevada and Utah (Brown 1971, 1978), in Colorado (Fitzgerald et al. 1994), or in New Mexico (Armstrong 1995).

At the time of permanent European settlement, the alpine mammalian fauna of the Southern Rocky Mountains included some 39 species (Fitzgerald et al. 1994); 29 species of mammals occur on the North Slope of Alaska (Bee and Hall 1956; Hall 1981). Faunal similarity (estimated using Sørensen's index noted previously) between the Coloradan alpine and the arctic Slope is only 0.324. Seven of the 11 mammalian species in common between the Coloradan alpine and the Alaskan tundra are Holarctic in distribution; the other four (common porcupine, coyote, black bear, and mink) are widespread in North America. None of the species in common between the two areas is a tundra obligate.

## Ecological Distribution of the Vertebrate Fauna

The alpine tundra is by no means a homogeneous landscape, and to some extent various species of vertebrates tend to differentially occupy the patches in the environmental mosaic. Table 7.3 indicates general habitat of vertebrates of Niwot Ridge. The most common small mammals are deer mice, especially prevalent in drier sites and also (late in the season) in snowbed communities (Table 7.4, compiled from 10 years of trapping in the Saddle research area). Species of voles (*Microtus*) predominate on moist and wet tundra. Biomass and density of small mammals are highest in shrub tundra and lowest in wet meadow and snowbed communities. The omnivorous deer mouse is most strongly associated with fellfield and snowbed communities (chapter 6), and absent from moist tundra, wet meadow, and shrub tundra. Species richness of small mammals is low overall, averaging 1.6 in dry meadow, 1.4 in fellfield, and 1.3 in moist tundra. Evenness diversity (Simpson's 1-D) is highest in dry meadow, fellfield, and shrub tundra.

Halfpenny and Southwick (1982) compiled data on ecological distribution of small mammalian herbivores from a variety of studies, mostly from Niwot Ridge. To the extent that description of ecological distribution is comparable with the present study, patterns are similar. Deer mice (*Peromyscus maniculatus*) usually are more abundant on drier sites and in more-disturbed areas (e.g., "gopher gardens") whereas voles (*Microtus*) predominate in moister habitats. The studies summarized by Halfpenny and Southwick (1982) reported *Phenacomys intermedius* (the heather vole) to be relatively more abundant across a wider range of habitats than is indicated in Table 7.4. However, the heather vole is sufficiently difficult to identify that we give no special weight to that difference.

Data in Table 7.4 do not apply to the entire mammalian assemblage, but only to species that are captured consistently in conventional livetraps. Larger animals and species whose habits bias them against standard livetraps are present—and in some habitats are of predominant importance—on Niwot Ridge. Two species of chipmunks (*Tamias minimus, T. umbrinus*) may occur on the Ridge, although only the least chipmunk actually is documented by specimens (Armstrong 1972; the species are not easy to distinguish, except as full adults.) Unidentified chipmunks were taken

Table 7.3. Typical Habitat, Principal Trophic Guild, and Hibernal Activity of Vertebrate Species of Niwot Ridge, Boulder County, Colorado

| Species | General Habitat | Principal Trophic Guild(s) | Hibernal Activity |
|---|---|---|---|
| *Mammals* | | | |
| Masked Shrew | Willow | Insectivore | Subnivean |
| Montane Shrew | Willow | Insectivore | Subnivean |
| Dwarf Shrew | Talus, willow | Insectivore | Subnivean |
| American Pika | Talus/meadow mosaic | Folivore | Subnivean |
| Snowshoe Hare | Willow, krummholz | Browser, folivore | Supranivean |
| White-tailed Jackrabbit | Meadow | Grazer, folivore | Supranivean |
| Least Chipmunk | Talus, meadow, krummholz | Folivore, spermivore, florivore, insectivore, mycovore | Hibernation |
| Yellow-bellied Marmot | Meadow, talus | Folivore | Hibernation |
| Golden-mantled Ground Squirrel | Meadow, talus | Folivore, spermivore, florivore, insectivore, mycovore | Hibernation |
| Northern Pocket Gopher | Meadow | Rhizovore, folivore | Subnivean |
| Deer Mouse | Talus, fellfield | Omnivore | Subnivean |
| Bushy-tailed Woodrat | Talus | Folivore, browser | Subnivean |
| Southern Red-backed Vole | Krummholz, shrub tundra | Spermivore, mycovore | Subnivean |
| Heather Vole | Krummholz, meadow | Spermivore, mycovore, cambivore, folivore | Subnivean |
| Montane Vole | Meadow | Grazer | Subnivean |
| Long-tailed Vole | Meadow, shrub tundra | Spermivore, folivore, grazer | Subnivean |
| Western Jumping Mouse | Willow | Spermivore, insectivore | Hibernation |
| Porcupine | Krummholz | Cambivore, folivore | Elevational migration |

| | | | |
|---|---|---|---|
| Coyote | All | Omnivore | Supranivean |
| Gray Wolf[1] | All | Carnivore | Supranivean |
| Red Fox | Willow, krummholz, meadow | Carnivore | Elevational migration |
| Black Bear | All | Omnivore | Elevational migration, hibernation |
| Grizzly Bear[1] | All | Omnivore | Elevational migration, hibernation |
| American Marten | Krummholz, talus | Carnivore | Supra-, subnivean |
| Short-tailed Weasel | Krummholz, talus, meadow | Carnivore | Supra-, subnivean |
| Long-tailed Weasel | All | Carnivore | Supra-, subnivean |
| Badger | Meadow | Carnivore | Elevational migration |
| Mountain Lion | Krummholz | Carnivore | Elevational migration |
| Bobcat | Krummholz | Carnivore | Supranivean |
| Wapiti, or American Elk | Meadow | Grazer | Elevational migration |
| Mule Deer | Krummholz, willow | Browser | Elevational migration |
| Bison[1] | Meadow | Grazer | Elevational migration |
| Bighorn Sheep | Meadow | Grazer | Elevational migration |
| *Breeding Birds* | | | |
| White-tailed Ptarmigan | Willow | Browser, folivore | Supranivean, elevational migration |
| American Pipit | Meadow | Insectivore | Migration |
| White-crowned Sparrow | Meadow, fellfield | Spermivore, insectivore | Elevational migration |
| Horned Lark | Willow, krummholz | Insectivore, spermivore | Elevational migration |
| Rosy Finch | Meadow, snowbank | Spermivore, insectivore | Elevational migration |

[1]Extirpated within historic time.

Table 7.4. Density and Biomass of Small Mammals (Shrews, Rodents <50 g Other than Pocket Gophers) on Niwot Ridge, 1981–1990, Arrayed by Vegetational Community-Type

| Variable | Fellfield | Dry Meadow | Snowbed | Moist Tundra | Wet Meadow | Shrub Tundra |
|---|---|---|---|---|---|---|
| *Sorex monticolus* | 0.0 | 0.0 | 0.0 | 0.0 | 0.0 | 1.0 |
| *Sorex* sp. | 0.0 | 0.5 | 0.5 | 0.0 | 0.0 | 0.0 |
| *Peromyscus maniculatus* | 13.1 | 4.3 | 5.5 | 0.0 | 0.0 | 0.0 |
| *Clethrionomys gapperi* | 0.0 | 0.0 | 0.0 | 0.0 | 0.0 | 0.0 |
| *Phenacomys intermedius* | 0.2 | 0.3 | 0.0 | 0.0 | 0.0 | 2.0 |
| *Microtus longicaudus* | 0.2 | 0.0 | 0.0 | 0.6 | 0.0 | 1.0 |
| *Microtus montanus* | 0.7 | 1.3 | 0.5 | 2.3 | 2.0 | 0.0 |
| *Microtus* sp. | 0.3 | 1.9 | 0.5 | 5.1 | 2.0 | 9.0 |
| Total small mammal biomass | 296.6 | 156.0 | 121.3 | 223.3 | 103.0 | 385.2 |
| Biomass *P. maniculatus* % | 82.9 | 1.0 | 75.0 | 0.0 | 0.0 | 0 |
| Mean density ha$^{-1}$ | 14.4 | 8.3 | 7.0 | 8.0 | 4.0 | 13.0 |
| Mean species richness | 1.4 | 1.6 | 1.1 | 1.3 | 1.0 | 1.0 |
| Simpson's 1-D (evenness) | 0.1298 | 0.1513 | 0.0714 | 0.0816 | 0.0000 | 0.1212 |
| *n* grids with captures | 53 | 15 | 8 | 7 | 2 | 4 |

*Sources:* Data were summarized from Halfpenny et al. (1984), Auerbach et al. (1987), and unpublished records from the LTER database. Vegetation data from May and Webber (1982).

*Notes:* Mean density by species. Because sampling grids differed in size, all data were scaled per hectare. Density as tabulated is merely numbers of different individuals captured per hectare per trapping period. Biomass is the sum of masses (in g) of individual small mammals upon initial capture. Species richness of small mammals is the average number of different species captured on grids in a particular community-type. Evenness diversity is shown as Simpson's diversity index, expressed as 1-D so that diversity values increase with greater evenness—see Magurran (1988). Data are averaged only from grids where captures were made (hence the low numbers of grids for some community-types).

occasionally on fellfield, dry meadow, and moist meadow grids (LTER unpublished data), but in numbers too erratic to tabulate with confidence, even at the generic level. The golden-mantled ground squirrel was captured rarely (LTER unpublished data).

Yellow-bellied marmots were captured and marked on Niwot Ridge each year from 1981 to 1990, but these animals were not found in the trapping areas summarized in Table 7.4. Marmots favor forb-dominated, productive tundra meadows, adjacent to protective rock outcrops and boulders. Density and biomass vary from year to year, as indicated in Table 7.5.

Pikas are not generally distributed on Niwot Ridge, favoring instead forb-dominated meadow interspersed with talus. Grids were established in 1981 in the Saddle study area explicitly for various studies of pikas (summarized by Halfpenny and Southwick 1982; Brown et al. 1989; chapter 14). Table 7.5 indicates density and biomass of pikas over the period 1981–1987.

Northern pocket gophers (Fig. 7.4) seldom are captured in conventional live-traps. Pocket gophers were captured in 1982–1984 (Halfpenny et al. 1984; LTER unpublished data), but again not on the trapping areas summarized in Table 7.4. The data available do not lend themselves to calculating biomass or density.

Table 7.5. Density and Biomass Estimates of Pikas and Marmots, Niwot Ridge, Boulder County, Colorado, 1981–1987

|  | 1981 | 1982 | 1983 | 1984 | 1985 | 1986 | 1987 |
|---|---|---|---|---|---|---|---|
| Pika |  |  |  |  |  |  |  |
| density, individuals (ha$^{-1}$) | 9.1 | 6.1 | 7.2 | 9.6 | 7.8 | 5.4 | 9.6 |
| biomass, g (ha$^{-1}$) | n/a | 758 | 949 | 1260 | 1112 | 634 | 1325 |
| Yellow-bellied marmot |  |  |  |  |  |  |  |
| density, individuals (ha$^{-1}$) | 3.4 | 2.6 | 3.4 | 4.6 | 2.6 | 1.8 | 2.2 |
| biomass, g (ha$^{-1}$) | 5100 | 7060 | 6300 | 9080 | 7080 | 2630 | 6380 |

*Sources:* Data from Halfpenny et al., 1984 and Auerbach et al., 1987.

Pikas and marmots (and also northern pocket gophers) are particularly important to the structure and function of the communities in which they are involved because (correlated with relatively large body size) they are longer lived than other resident (nonmigratory) mammals of Niwot Ridge. Survivorship (percentage of marked individuals from one year recaptured the next year) of deer mice, voles, marmots, and pikas, respectively, averaged 1.5, 0, 27.7, and 32.3 for the period between 1981 and 1987 (Halfpenny et al. 1984; Auerbach et al. 1987).

Figure 7.4. A young northern pocket gopher (*Thomomys talpoides*) being held by Niwot Ridge researcher Eve Davies. Pocket gophers are important agents of disturbance in the alpine (see chapters 6, 8, 11, 14) (photo by William D. Bowman).

Year-round studies of alpine small mammals to assess intrannual population fluctuations have not been done but have been undertaken in subalpine habitats below Niwot Ridge. In aspen woodland, populations of small mammals were highest in autumn (September–November) and lowest in April and May (Stinson 1977a). In subalpine coniferous forest, populations of Southern Red-backed Voles (*Clethrionomys gapperi*) ranged from 42 ha$^{-1}$ in November to 10/ha in May and those of deer mice (*Peromyscus maniculatus*) ranged from 27/ha in September to 2.5/ha in June (Merritt 1976).

Halfpenny (1980—summarized by Halfpenny and Southwick 1982) compared the life history characters of deer mice near the alpine summit of Mount Evans, Clear Creek County (4300 m) and at the University of Colorado Mountain Research Station (at an elevation of 2730 m in subalpine lodgepole forest below Niwot Ridge). Generally the study substantiated and extended earlier work by Spencer and Steinhoff (1968), which suggested that, given the shorter breeding season at higher elevations, female deer mice tended to have fewer litters compensated by more young per litter. It also indicated that alpine populations tend to have higher life expectancies and lower ratios of juveniles:adults.

## Seasonal Dynamics of the Fauna

One of the most striking aspects of alpine environments is their physical rigor and their intense seasonality. Marchand (1987) outlined three "options" for overwintering success in animals of seasonally cold climates: migration, hibernation, and resistance. Of these strategies, migration and resistance are much more common in the Colorado alpine than is hibernation. Hibernation may be a "mid-latitude strategy" in general (Marchand 1987), but it is not a strategy suitable to most species at high elevations in mid-latitudes.

Four of the five breeding birds of Niwot Ridge migrate in winter, two (White-crowned Sparrow, Horned Lark) to steppe habitats adjacent to the Southern Rocky Mountains, one (American Pipit) to the southern United States and adjacent Latin America, and one (Rosy Finch) irregularly and locally to subalpine or montane habitats. Male White-tailed Ptarmigan often winter in the alpine zone (in shrub tundra), but hens frequently move along stream courses into taller, subalpine willow stands (Hoffman and Braun 1975, 1977).

Table 7.3 indicates typical hibernal activity of vertebrates of Niwot Ridge. (For present purposes, "hibernation" is intended to encompass all forms of extended winter dormancy.) Of 33 mammalian species of Niwot Ridge, only four (12.1%: least chipmunk, golden-mantled ground squirrel, yellow-bellied marmot, western jumping mouse) hibernate on or near the tundra. Larger herbivores (elk and mule deer today, and formerly bison and bighorn sheep as well) migrate to lower elevations. The black bear (only occasional on the tundra even in summer) hibernates in upper montane and subalpine forests, especially in areas of deep snow accumulation (Fitzgerald et al. 1994).

Winter activity on open tundra is minimal, mostly confined to medium- to large-sized mammals (hares, carnivores). Over one half of local mammals (including seven rodents, three shrews, and the pika, and the small carnivores that prey on them) are active through the winter, mostly in the subnivean environment. Pikas subsist on vegetation stored in "haypiles," whereas other species generally forage through the winter. The ecology of the subnivean mammalian assemblage is fairly well studied in the Southern Rockies in the subalpine zone adjacent to Niwot Ridge (Stinson, 1977a; Merritt and Merritt 1978a, 1978b; Merritt 1983, 1984), and the relative stability of the subnivean environment has been described in detail (Pruitt 1984; Marchand 1987; Halfpenny and Ozanne 1989).

## Influences of Vertebrates in the Alpine Ecosystem

### Biotic Impacts: Trophic Guilds

Table 7.3 indicates broad food habits of vertebrates of Niwot Ridge, allowing a first approximation of trophic guilds. Halfpenny and Southwick (1982) reviewed literature on small mammalian herbivores in the Coloradan alpine as background for long-term ecological research on Niwot Ridge. Mammals ranging in size from mice and voles ($10^1$ g) to porcupines ($10^4$ g) were included. (Of course, this is a wider range of sizes than usually is implied by the construct, "small mammals.") The data behind Table 7.3 are mostly derived from studies in the subalpine zone or at even lower elevations. For most species, we have only a general and qualitative idea of diet and no sense of quantitative variation within and between seasons and years or with microhabitat, geography, ontogenic stage, or individual preference.

Available data suggest that narrow food specialization is absent in the alpine zone. The closest to a dietary specialist may be the badger, which feeds mostly on northern pocket gophers, although they are not averse to chipmunks, ground squirrels, an occasional mouse, or carrion. Shrews usually are considered to be insectivorous, but in fact will eat virtually any invertebrate of appropriate size and most will take vertebrate carrion as well. The White-tailed Ptarmigan is a browser in winter, eating mostly buds and bark of willows, and a more generalized herbivore in summer (May and Braun 1972).

There are no specialized fruit- or seed-eaters in the vertebrate fauna. Obligatory spermivory (seed-eating) would seem a poor strategy in an ecosystem where most producers allocate a disproportionate amount of biomass to vegetative growth and storage organs. Chipmunks and ground squirrels often are considered spermivores but actually are nearly omnivorous; in the alpine, they probably eat more reproductive structures in the form of flowers than they do mature fruits or seeds. Carleton (1966) found that sympatric least chipmunks and golden-mantled ground squirrels both were highly dependent on dandelions (*Taraxacum officinale*) on subalpine meadows near Gothic, Gunnison County, Colorado (contributing 80% of the summer diet), but the chipmunk emphasized flowers and fruit (seed heads) whereas the ground squirrel emphasized leaves.

Bears and coyotes are well known as omnivores, as is the deer mouse. The golden-mantled ground squirrel and the least chipmunk might also be called omnivores if the annual diet were simply summed and seasonal specialization ignored.

**144** Ecosystem Structure

Figure 7.5. Short-tailed weasel (*Mustela erminea*) hunting pikas on Niwot Ridge (photo by Denise Dearing).

White-crowned Sparrows, Horned Larks, and Rosy Finches are similarly opportunistic. Depending on habitat, young may be reared on a diet of insects, with a shift later to a diet predominated by seeds.

Broadly speaking, of 38 species tabulated in Table 7.3, 45% are primary consumers, 32% are secondary consumers (Fig. 7.5), and some 24% are omnivores, operating at both trophic levels. Data are not available to describe energy flow between these levels, and annual fluctuations in population numbers might make such description impossible or—considering the mobility of most of the larger-bodied consumers—at least spurious. Perhaps the most interesting arena for further study in terms of diet would be quantitative description of trophic guilds. Based on the rough compilation in Table 7.3 about 18% of the vertebrates of Niwot Ridge are folivores; insectivory includes about 16% of the fauna. Nearly 15% of the fauna are spermivorous, some 13% are carnivorous, and nearly 10% of vertebrates are grazers.

Table 7.6 provides data on some variables pertinent to the impact of vertebrates in the Niwot Ridge ecosystem. The compilation suggests patterns and potential avenues for further investigation. Data are from alpine and subalpine habitats in the Southern Rockies and should be viewed and used conservatively. Some of the parameters tabulated vary widely with geography and habitat and within and between years.

Body mass of extant native vertebrates of Niwot Ridge differs by five orders of magnitude, from dwarf shrews (2–4 g) to half-ton bull elk ($4.5 \times 10^5$ g). Two of the largest vertebrates of the Coloradan alpine zone have been extirpated within historic time, the grizzly bear and the bison. Estimates of total biomass of non-ungulate mammalian herbivores are available from subalpine localities. In the Wasatch Mountains of northern Utah, estimates ranged from 1287 to 3770 g/ha along a subalpine sere (Anderson et al. 1980); in a meadow near Rabbit Ears Pass, Grand County, Colorado, estimates were 3277–3476 g/ha (Vaughan 1974). In the alpine zone, pikas may achieve densities of 15–20 individuals per ha, and biomass of >3 kg/ha. Northern pocket gophers may achieve biomass of >5 kg/ha.

To gauge potential impact on the ecosystem, one needs to know more than mere size. Home range and crude density (let alone ecological density—numbers per useable habitat) have seldom been studied in alpine communities. Data from a variety of sources are compiled in Table 7.6. Broadly speaking, density varies inversely with body size and, as one would expect, primary consumers of a given body size tend to have higher densities (and smaller home ranges) than do secondary consumers.

## Physical Impacts: Vertebrates as Geomorphic Influences

Billings (1979) noted that many high mountain ecosystems developed in the absence of hoofed mammals, but that is not the case for Niwot Ridge or for the alpine of the Southern Rockies in general. The physical effects of ungulates (presently elk and mule deer, but formerly bighorn sheep and bison as well) on Niwot Ridge have not been studied, but surely they are (and have been) important. Trails along and across contours contribute to microrelief. Trampling of vegetation must influence productivity and succession of communities and the persistence of individual plants.

Perhaps the nonhuman mammals most important as geomorphic agents on Niwot Ridge in particular (and the alpine of western North America in general) are pocket gophers (Geomyidae), especially the northern pocket gopher. Thorn (1978) reviewed literature on pocket gopher activity in alpine tundra and emphasized the importance of gophers in the evolution of "terracette" microtopography (see chapter 11).

Pocket gophers are of moderate size (ca. 100 g) and tend to maintain dense populations (up to 50/ha). Effects of northern pocket gophers on numerous aspects of community and ecosystem ecology have been investigated in detail and widely over their geographic range in western North America (e.g., see Ellison 1946; Hansen et al. 1960). Both biotic and physical effects of pocket gophers on alpine tundra have been studied on Niwot Ridge for many years (Litaor et al. 1996; Cortinas and Seastedt 1996; chapters 6 and 8).

Osburn (1958) investigated plant succession on the Ridge and was particularly cognizant of the influence of pocket gophers: "Any process which significantly reduces the amount of humus will also alter the plant community and change the type of stand. The combined effects of gopher digging, which kills plants, and the strong wind which blows away humus and fine soil is one of these processes." Pocket gophers increase the extent of fellfield communities, and recovery of fellfield to continuous tundra vegetation is rare and/or quite slow, occurring on a timescale of "many decades" (Webber and May 1977).

Table 7.6. Body Mass, Home Range, and Density of Vertebrate Species of Niwot Ridge, Boulder County, Colorado

| Species | Body Mass (g) | Home Range (ha) | Density (individuals/ha) | References |
|---|---|---|---|---|
| *Mammals* | | | | |
| Masked Shrew | 3–6 | | | Vaughan (1974) |
| Montane Shrew | 4–7 | 0.1–0.4 | 1–4 | |
| Dwarf Shrew | 2–4 | | | |
| American Pika | 120–250 | 0.07–0.13 | 1.3–15 | Brown et al. (1989), Golian (1985), Meaney (1983), Smith and Weston (1990), Southwick et al. (1986) |
| Snowshoe Hare | $1 \times 10^3$–$1.5 \times 10^3$ | 8.1 | | Dolbeer and Clark (1975) |
| White-tailed Jackrabbit | $2 \times 10^3$–$3 \times 10^3$ | 300–700 | 2.2 | Lim (1987) |
| Least Chipmunk | 25–60 | | 11–22 | Vaughan (1974) |
| Yellow-bellied Marmot | $1.6 \times 10^3$–$5.2 \times 10^3$ | 0.5–0.7[a] | 3.5–8.5 | Johns and Armitage (1979), Frase and Hoffmann (1980) |
| Golden-mantled Ground Squirrel | 170–290 | | 1–12 | Fitzgerald et al. (1994) |
| Northern Pocket Gopher | 100–145 | 0.0054 | 15–25, 40–46 | Hansen et al. (1960), Vaughan (1974), Thorn (1978) |
| Deer Mouse | 16–22 | 0.01–0.03 | 1–20 | Vaughan (1974), Merritt and Merritt (1978b), Halfpenny and Southwick (1982) |
| Bushy-tailed Woodrat | 270–300 | | | |
| Southern Red-backed Vole | 20–40 | 0.01–0.02 | 2–48 | Merritt and Merritt (1978b) |
| Heather Vole | 30–50 | | | Hoffmann (1974), Halfpenny and Southwick (1982) |
| Long-tailed Vole | 35–60 | | 5–50 | Smolen and Keller (1987) |
| Montane Vole | 35–60 | | 15–32 | Vaughan (1974) |
| Western Jumping Mouse | 20–40 | 0.24–0.35 | 28–35 | Stinson (1977b) |
| Porcupine | $4 \times 10^3$–$18 \times 10^3$ | 5–15 | 0.01–0.1 | Dodge (1982) |
| Coyote | $9 \times 10^3$–$16 \times 10^3$ | $1$–$4 \times 10^3$ | 0.002–0.004 | Bekoff (1977, 1982) |

| Species | Body mass | Home range | Density | Source |
|---|---|---|---|---|
| Gray Wolf[b] | $18 \times 10^3 – 80 \times 10^3$ | | $0.00002–0.0005$ | Paradiso and Nowak (1982) |
| Red Fox | $3 \times 10^3 – 7 \times 10^3$ | 58–160 | | Samuel and Nelson (1982), Larivière and Pasitschniak-Arts (1996) |
| Black Bear | $90 \times 10^3 – 225 \times 10^3$ | $200–20 \times 10^3$ | | Pelton (1982) |
| Grizzly Bear[b] | $135 \times 10^3 – 275 \times 10^3$ | $38.4 \times 10^3 – 82.8 \times 10^3$ | | Pasitschniak-Arts (1993) |
| American Marten | 500–1,200 | 100–300 | 0.01–0.02 | Buskirk and Ruggiero (1994) |
| Short-tailed Weasel | 30–200 | 4–200 | 0.02–0.06 | King (1983) |
| Long-tailed Weasel | 100–300 | 12–16 | 0.06–0.07 | Svendsen (1982) |
| Badger | $6 \times 10^3 – 14 \times 10^3$ | 160–1,700 | 0.004 | Lindzey (1982) |
| Mountain Lion | $35 \times 10^3 – 100 \times 10^3$ | $10 \times 10^3 – 30 \times 10^3$ | $2–4 \times 10^{-4}$ | Currier (1983) |
| Bobcat | $5 \times 10^3 – 14 \times 10^3$ | $60–20.1 \times 10^3$ | 0.0004–0.03 | McCord and Cordoza (1982) |
| Wapiti (American Elk) | $220 \times 10^3 – 450 \times 10^3$ | Migratory | | |
| Mule Deer | $30 \times 10^3 – 70 \times 10^3$ | 30–800 | 0.005–0.05 | Mackie et al. (1982) |
| Bison[b] | $400 \times 10^3$ | $~900 \times 10^3$ | Migratory | |
| Bighorn Sheep | $50 \times 10^3 – 125 \times 10^3$ | $1 \times 10^3 – 4 \times 10^3$ | | Shackleton (1985) |
| *Breeding Birds* | | | | |
| White-tailed Ptarmigan | 300–500 | | | Terres (1995) |
| American Pipit | 21 | 1.5–2.2[a] | | Conry (1978) |
| White-crowned Sparrow | 28–30 | 0.4–1.0[a] | | Hubbard (1978) |
| Horned Lark | 30 | 1.0–1.6[a] | 1.1–1.7 | Conry (1978) |
| Rosy Finch | 28–32 | | | Terres (1995) |

*Sources:* Data on body mass of mammals from Armstrong (1972), Fitzgerald et al. (1994); sources of data on other variables as noted.

[a] territory.
[b] extirpated within historic time.

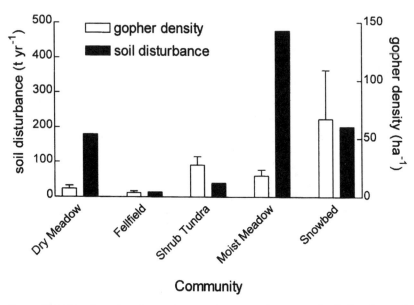

Figure 7.6. Number of pocket gophers per hectare and amount of soil disturbance for five vegetative noda (see chapter 6 for full description) on Niwot Ridge, Colorado (from Thorn 1982). Values for pocket gopher densities are means ±1 s.e. ($n = 3–26$).

Thorn (1982) was particularly interested in the geomorphic effects of pocket gophers on Niwot Ridge and differential use by gophers of patches of the mosaic of the tundra landscape. He calculated that about 55% of gopher disturbance occurred on just 21% of the area of tundra. Mean rate of soil movement was $40 \pm 7.1$ t ha$^{-1}$ yr$^{-1}$. Disturbance was greatest on southern and eastern aspects, minimal on northern aspects. Pocket gopher activity differed widely among plant associations, being least in dry fellfield (which the gophers may have helped to create—see Osburn 1958) and greatest in moist meadows (Fig. 7.6). Density of pocket gophers is greatest in areas of long-lying snowpack, although gophers do disturb less soil in those areas. This seeming incongruity is traceable to the animals' habit of following retreating snowbanks well into the summer. Young are born and reared in the subnivian environment.

Burns (1979, 1980) reported pocket gopher densities of 10.6 ha$^{-1}$ on his Niwot Ridge study area, with 91% of burrowing activity on downwind slopes (areas of snow accumulation). There were seasonal differences in soil movement, averaging 72,900 cm$^{-3}$ individual$^{-1}$ during the warmer months (June–October) and 28,040 cm$^{-3}$ individual$^{-1}$ during the period November–May, for an annual total of 100,940 cm$^{-3}$ individual$^{-1}$ pocket gopher, resulting in an average lowering of the surface of 0.0037 cm yr$^{-1}$. Nivation processes are more effective agents than pocket gophers, lowering the surface 0.009 cm/yr, whereas wind and water processes are much less effective, lowering the surface just 0.0001 cm yr$^{-1}$ (see chapter 4). Of pocket gophers, Burns (1979) concluded conservatively, "their long-term effect on the development of alpine landscape in North America may be great." It would be of real interest to evaluate that statement quantitatively through comparative study of carefully matched alpine areas with and without pocket gophers.

## Humans in the Vertebrate Fauna of Niwot Ridge

The biota of any place and time is essentially a "progress report" on processes that continue on evolutionary and ecological timescales. It is essential to recall that humans are a species in the biota of nearly all ecosystems on Earth, a species that is increasingly prevalent and influential. Billings (1979) observed that "all natural high mountain ecosystems evolved in the absence of man," but that is decidedly not the case with the alpine tundra of Colorado (Fitzgerald et al. 1994). Humans are in important respects a remnant of the Pleistocene megafauna of Niwot Ridge. Evidence of Paleoindian occupation—from stone flakes to walls constructed to hunt large game—is scattered widely in the alpine, dating back 11,000 years (Ives 1980; Benedict 1992a, 1992b). Occupation of the tundra of the Front Range was necessarily opportunistic and seasonal, influenced by a unique combination of environmental factors (Benedict 1992a, 1992b).

The Niwot Ridge alpine site is accessible from all directions, being located at the headwaters of four major river systems of the semiarid continental interior. The abundant, late-lying and perennial snowbanks made the region virtually immune to drought, and convenient access to mild wintering environments and other low-elevation resources was available. The extent of alpine tundra in the Front Range attracted bighorn sheep, elk, and other large ungulates in summer.

Factors that worked against exploitation of the tundra included lack of appropriate stone materials for tools at high elevations, a general scarcity of edible plants that could be harvested in large quantities, and a severe periglacial climate that restricted human habitation to warm seasons and to intervals without excessive snow. More than 50 game-wall systems have been located above timberline (Benedict 1992a, 1992b), ranging in age from Paleoindian (ca. 7650 years BP) to Late Prehistoric (ca. 500 years BP). The major game-drive systems are concentrated along a 70-km section of the Continental Divide east of Middle Park (i.e., from about Rollins Pass north into Rocky Mountain National Park). Cultural features include butchering sites, hunting camps, and vision-quest localities, but no long-term camps or specialized sites for processing plant foods have been located above 3000 m (Benedict 1992a, 1992b).

From Paleolithic origins, human exploitation of—and impact on—the tundra has continued to the present day. Niwot Ridge was too high for trappers and other market-hunters of the nineteenth century (Fitzgerald et al. 1994). However, the Ridge was grazed by domestic sheep for the first half of the twentieth century (Halfpenny and Southwick 1982). The Industrial Revolution at first ignored the Ridge. Lying just to the north of the Colorado mineral belt (Benedict 1991), it was spared the ravages of mining. Today, Niwot Ridge is part of a Man in the Biosphere Reserve of the United Nations, its vertebrates and their environments largely protected from direct impacts of humans other than scientific researchers, but even these seemingly pristine environments are not immune from continuing and expanding anthropogenic changes in the atmosphere (see chapter 16).

## Future Research

Despite well over a century of formal scientific study, much remains to be learned about alpine vertebrates of the Southern Rocky Mountains at the levels of ecosystems, communities, populations, and individuals. Billings (1979) outlined general kinds of needed research on high mountain environments: mapping and censusing permanent sample areas to measure changes in the biota through time in both peopled and unpeopled mountains; studies of primary and secondary productivity and biomass and food webs; and controlled whole-ecosystem measurements before and after perturbation. That list remains a challenge largely unmet, although such research efforts as LTER are a major step in a positive direction.

The composition of the vertebrate assemblage varies through the year and from one year to the next. How tightly structured are communities of such fluid composition? What are quantitative symbiotic relationships (especially trophic relationships and competition) among vertebrates and between vertebrates and other components of the biota?

We recommend more, and more thorough, studies in the alpine zone of fundamental natural history parameters of vertebrate species—diet, reproduction, development, population growth and regulation, behavior—to allow comparison with populations of the same species at lower elevations, where a number of species are rather well known. To be truly revealing, studies in the alpine need to be long term and year around. Techniques will have to improve before that is possible, but there is promise in telemetry and automated data-gathering.

Future studies of alpine vertebrates in general would profit from advice offered by Chew (1978): emphasize responses of vegetation (production, composition, population dynamics) to the measured actions of vertebrates; investigate ways in which vertebrates may regulate processes that are important to plants (nutrient cycling, seed dispersal and predation, feeding, cutting, trampling); conduct more experimental field studies, especially removal studies of vertebrate populations, to reveal unique mutualistic relationships of vegetation and vertebrates; and integrate efforts of various kinds of biologists over long time spans. Halfpenny and Southwick (1982) presented a series of hypotheses regarding small mammalian herbivores that can be extended to the vertebrate fauna in general and subjected to rigorous testing. In all of these efforts, we would be wise to bear in mind sage advice from Johnson and Jehl (1994): "It is time for biologists to face squarely the complexity of the natural world we attempt to interpret."

## Summary and Conclusions

The vertebrate fauna of the alpine tundra of the Colorado Front Range is fairly large, composed of some 50 species of endotherms (but few if any amphibians or fishes and no reptiles) that meet some or all of their resource needs on the tundra at one time or another during the year. Endemism is absent; biogeographic relationships are mostly with regions to the west and/or north.

Alpine tundra is distributed as an archipelago of semi-isolated patches along the highest ranges of the Southern Rockies, but, given the movements of vertebrates on various timescales, adjacent islands of tundra are by no means completely isolated from one another. Significant ecological impacts of vertebrates such as the northern pocket gopher and pika on tundra ecosystems are both biotic (as consumers) and physical (as geomorphic influences).

Alpine environments, and the vertebrate populations that exploit them, are dynamic on scales ranging from the millennia since the recession of Pinedale glaciers, to the decades or centuries of ecological succession following major disturbance (such as elimination of significant areas of vegetation and mass-wasting of the physical substrate, some of it caused by the vertebrates themselves), to the annual and seasonal scales of population fluctuations, and finally to minutes, as a basking marmot disappears from its boulder when a cloud momentarily obscures the summer sun.

Whatever the scale of space (region, landscape, community, population) or time (evolutionary, ecological, or physiological), humans have been and continue to be influential as agents of change on the tundra and important as observers and managers of that change. Our continuing goal on the alpine tundra should be long-term research, in pursuit of deeper understanding essential to more-effective stewardship of this intriguing and beautiful environment.

References

Allen, J. A. 1874. Notes on the mammals of portions of Kansas, Colorado, Wyoming, and Utah. *Bulletin of the Essex Institute* 6:43–66.

Anderson, D. C., J. A. MacMahon, and M. L. Wolfe. 1980. Herbivorous mammals along a montane sere: Community structure and energetics. *Journal of Mammalogy* 61:500–519.

Andrews, R., and R. Righter. 1992. Colorado birds, a reference to their distribution and habitat. Denver, CO: Denver Museum of Natural History.

Armitage, K. B. 1991. Social and population dynamics of yellow-bellied marmots: Results from long-term research. *Annual Review of Ecology and Systematics* 22:379–407.

Armstrong, D. M. 1972. Distribution of mammals in Colorado. Monograph, University of Kansas Museum of Natural History 3:1–415.

———. 1975. Rocky Mountain mammals, a handbook of mammals of Rocky Mountain National Park and Shadow Mountain National Recreation Area, Colorado. Estes Park, CO: Rocky Mountain Nature Association.

———. 1986. Edward Royal Warren (1860–1942) and the development of Coloradan mammalogy. *American Zoologist* 26:363–370.

———. 1987. Rocky Mountain mammals. Rev. ed. Niwot, CO: Colorado Associated University Press.

———. 1995. Northern limits of mammals of northern interior Mexico. In *Contributions in mammalogy: A memorial volume honoring Dr. J. Knox Jones, Jr.,* edited by H. H. Genoways and R. J. Baker. Lubbock, TX: Museum of Texas Tech University.

Auerbach, N. A., J. C. Halfpenny, and M. D. Taylor. 1987. Ecological data on small mammal herbivores from the Colorado alpine tundra, 1984–1987. Institute of Arctic and Alpine Research, University of Colorado, Boulder, *Long-Term Ecological Research Data Report* 87/9:1–67.

Bee, J. W., and E. R. Hall. 1956. Mammals of northern Alaska. Miscellaneous Publications, University of Kansas Museum of Natural History 8:1–309.
Beidleman, R. G. 1955. An altitudinal record for the bison in northern Colorado. *Journal of Mammalogy* 36:470–471.
Bekoff, M. 1977. *Canis latrans*. *Mammalian Species* 79:1–9.
———. 1982. Coyote (*Canis latrans*). In *Wild mammals of North America, biology, management, and economics*, edited by J. A. Chapman and G. A. Feldhamer. Baltimore, MD: Johns Hopkins University Press.
Benedict, A. D. 1991. *A Sierra Club naturalist's guide to the Southern Rockies*. San Francisco: Sierra Club Books.
———. 1992a. Along the Great Divide: Paleoindian archaeology of the high Colorado Front Range. In *Ice Age hunters of the Rockies*, edited by D. J. Stanford and Jane S. Day. Colorado: Denver Museum of Natural History, Denver, and University Press of Colorado, Niwot.
———. 1992b. Footprints in the snow: High-altitude cultural ecology of the Colorado Front Range, U.S.A. *Arctic and Alpine Research* 24:1–16.
Billings, W. D. 1979. High mountain ecosystems, evolution, structure, operation, and maintenance. In *High altitude geoecology*, edited by P. J. Webber. Boulder, CO: Westview Press.
Blake, I. H., and A. K. Blake. 1969. An ecological study of timberline and alpine areas, Mount Lincoln, Park County, Colorado. *University of Nebraska Studies, New Series* 40:1–59.
Brewer, W. H. 1871. Animal life in the Rocky Mountains of Colorado. *American Naturalist* 5:220–223.
Brown, J. H. 1971. Mammals on mountaintops: Nonequilibrium biogeography. *American Naturalist* 105:467–478.
———. 1978. The theory of insular biogeography and the distribution of Boreal birds and mammals. *Great Basin Naturalist Memoir* 2:209–227.
Brown, L. N. 1967. Ecological distribution of mice in the Medicine Bow Mountains of Wyoming. *Ecology* 48:677–680.
Brown, R. N., C. H. Southwick, and S. C. Golian. 1989. Male-female spacing, territorial replacement, and the mating system of pikas (*Ochotona princeps*). *Journal of Mammalogy* 70:622–629.
Burns, S. F. 1979. The northern pocket gopher (*Thomomys talpoides*): A major geomorphic agent. *Journal of the Colorado-Wyoming Academy of Science* 11:86.
———. 1980. Alpine soil distribution and development, Indian Peaks, Colorado Front Range. Ph.D. diss., University of Colorado, Boulder.
Buskirk, S. W., and L. F. Ruggiero. 1994. American marten. In *The scientific basis for conserving forest carnivores, American marten, fisher, lynx, and wolverine in the western United States*. edited by L. F. Ruggiero, K. B. Aubry, S. W. Buskirk, L. J. Lyon, and W. J. Zielinski. General Tech. Rep., Fort Collins, Colorado, USDA Forest Service, RM-254.
Carleton, W. M. 1966. Food habits of two sympatric Colorado sciurids. *Journal of Mammalogy*, 47:91–103.
Cary, M. 1911. A biological survey of Colorado. *North American Fauna* 33:1–256.
Chew, R. M. 1978. The impact of small mammals on ecosystem structure and function. In *Populations of small mammals under natural conditions*, edited by D. P. Snyder. Pymatuning Laboratory of Ecology, University of Pittsburgh, Special Publication Series. 5.
Conry, J. A. 1978. Resource utilization, breeding biology, and nestling development in an alpine tundra passerine community. Ph.D. diss., University of Colorado, Boulder.
Cortinas, M. R., and T. R. Seastedt. 1996. Short- and long-term effects of gophers (*Thomomys talpoides*) on soil organic matter dynamics in alpine tundra. *Pedobiologia* 40:162–170.

Coues, E. 1879. Notice of Mrs. Maxwell's exhibit of Colorado mammals. In *On the plains and among the peaks; or, how Mrs. Maxwell made her natural history collection,* by M. A. Dartt-Thompson. Philadelphia: Claxton, Remsen, and Haffelfinger.
Currier, M. J. P. 1983. Felis concolor. *Mammalian Species* 200:1–7.
Cushman, R. C., S. R. Jones, and J. Knopf. 1993. Boulder County nature almanac. Boulder, CO: Pruett Publishing.
Dodge, W. E. 1982. Porcupine (*Erethizon dorsatum*). In *Wild mammals of North America, biology, management, and economics,* edited by J. A. Chapman and G. A. Feldhamer. Baltimore, MD: Johns Hopkins University Press.
Dolbeer, R. A., and W. R. Clark. 1975. Population ecology of snowshoe hares in the central Rocky Mountains. *Journal of Wildlife Management* 39:535–549.
Ellison, L. 1946. The pocket gopher in relation to soil erosion on mountain range. *Ecology* 27:101–114.
Emerick, J. C. 1995. *Rocky Mountain National Park natural history handbook.* Estes Park, CO: Rocky Mountain Nature Association.
Findley, J. S., and S. Anderson. 1956. Zoogeography of montane mammals of Colorado. *Journal of Mammalogy* 37:80–82.
Findley, J. S., and N. C. Negus. 1953. Notes on the mammals of the Gothic region, Gunnison County, Colorado. *Journal of Mammalogy* 34:235–239.
Fitzgerald, J. P., C. A. Meaney, and D. M. Armstrong. 1994. *Mammals of Colorado.* Colorado: Denver Museum of Natural History, Denver, and University Press of Colorado, Niwot.
Frase, B. A., and R. S. Hoffmann. 1980. Marmota flaviventris. *Mammalian Species* 135:1–8.
Golian, S. H. 1985. Population dynamics of pikas: Effects of snowpack and vegetation. Ph.D. diss., University of Colorado, Boulder.
Halfpenny, J. C. 1980. Reproductive strategies: Intra- and interspecific comparison within the genus *Peromyscus.* Ph.D. diss., University of Colorado, Boulder.
Halfpenny, J. C., K. P. Ingraham, J. Mattysse, and P. C. Lehr. 1986. *Bibliography of alpine and subalpine areas of the Front Range, Colorado.* Occasional Paper 43, Institute of Arctic and Alpine Research, University of Colorado, Boulder.
Halfpenny, J. C., K. Ingraham, C. H. Southwick, S. H. Golian, H. L. Krause, and C. Meaney. 1984. Ecological data on small mammal herbivores from the Colorado alpine tundra, 1981–1983. Long-Term Ecological Research Data Report, Institute of Arctic and Alpine Research, University of Colorado, Boulder, 84/7:1–63.
Halfpenny, J. C., and R. D. Ozanne. 1989. *Winter, an ecological handbook.* Boulder, CO: Johnson Books.
Halfpenny, J. C., and C. H. Southwick. 1982. Small mammal herbivores of the Colorado alpine tundra. In *Ecological studies in the Colorado alpine: A festschrift for John W. Marr,* edited by J. C. Halfpenny. Occasional Paper 37, Institute of Arctic and Alpine Research, University of Colorado, Boulder.
Hall, E. R. 1981. *Mammals of North America.* 2 vols. New York: Wiley.
Hammerson, G. A. 1982. *Amphibians and reptiles in Colorado.* Denver: Colorado Division of Wildlife.
Hansen, R. M., T. A. Vaugn, D. F. Harvey, V. T. Harris, P. L. Hegdal, A. M. Johnson, A. L. Ward, E. H. Reid, and J. O. Kieth. 1960. *Pocket gophers in Colorado.* Bulletin, Colorado State University Experiment Station 508-S:1–26.
Hoffman, R. W., and C. E. Braun. 1975. Migration of a wintering population of white-tailed ptarmigan in Colorado. *Journal of Wildlife Management* 39:485–490.
———. 1977. Characteristics of a wintering population of white-tailed ptarmigan in Colorado. *Wilson Bulletin* 89:107–115.

Hoffmann, R. S. 1974. Terrestrial vertebrates. In *Arctic and alpine environments*, edited by J. R. Ives and R. G. Barry. London: Methuen.

Hubbard, J. D. 1978. Breeding biology and reproductive energetics of Mt. white-crowned sparrows in Colorado. Ph.D. diss., University of Colorado, Boulder.

Ives, J. R. 1980. Introduction: A description of the Front Range. In *Geoecology of the Colorado Front Range: A study of alpine and subalpine environments*, edited by J. R. Ives. Boulder, CO: Westview Press.

Johns, D. W., and K. B. Armitage. 1979. Behavioral ecology of alpine yellow-bellied marmots. *Behavior Ecology and Sociobiology* 5:133–157.

Johnson, N. K., and J. R. Jehl, Jr. 1994. A century of avifaunal change in western North America: Overview. *Studies in Avian Biology* 15:1–3.

King, C. M. 1983. Mustela erminea. *Mammalian Species* 195:1–8.

Kingery, H. E., ed. 1998. *Colorado breeding bird atlas.* Denver: Colorado Bird Atlas Partnership.

Kingery, H. E., and W. D. Graul. 1978. *Colorado bird distribution latilong study.* Denver: Colorado Division of Wildlife.

Larivière, S., and M. Pasitschniak-Arts. 1996. Vulpes vulpes. *Mammalian Species* 537:1–11.

Lim, B. K. 1987. Lepus townsendii. *Mammalian Species* 288:1–6.

Lindzey, F. G. 1982. Badger (*Taxidea taxus*). In *Wild mammals of North America, biology, management, and economics,* edited by J. A. Chapman and G. A. Feldhamer. Baltimore, MD: Johns Hopkins University Press.

Litaor, M. I., R. Mancinelli, and J. C. Halfpenny. 1996. The influence of pocket gophers on the status of nutrients in alpine soils. *Geoderma* 70:37–48.

Mackie, R. J., K. L. Hamlin, and D. F. Pac. 1982. Mule deer (*Odocoileus hemionus*). In *Wild mammals of North America, biology, management, and economics,* edited by J. A. Chapman and G. A. Feldhamer. Baltimore, MD: Johns Hopkins University Press.

Magurran, A. E. 1988. *Ecological diversity and its measurement.* Princeton, NJ: Princeton University Press.

Marchand, P. J. 1987. *Life in the cold, an introduction to winter ecology.* Hanover, NH: University Press of New England.

Marr, J. W. 1964. Utilization of the Front Range tundra, Colorado. In *Grazing in terrestrial and marine environments.* D. J. Crisp, editor. Oxford: Blackwell Scientific.

May, D. E., and P. W. Webber. 1982. Spatial and temporal variation of the vegetation and its productivity on Niwot Ridge, Colorado. In *Ecological studies in the Colorado alpine: A festschrift for John W. Marr,* edited by J. C. Halfpenny. Occasional Paper 37, Institute of Arctic and Alpine Research, University of Colorado, Boulder.

May, T. A. 1980. Animal ecology, overview. In *Geoecology of the Colorado Front Range: A study of alpine and subalpine environments,* edited by J. R. Ives. Boulder, CO: Westview Press.

May, T. A., and C. E. Braun. 1972. Seasonal foods of adult white-tailed ptarmigan in Colorado. *Journal of Wildlife Management* 36:1180–1186.

McCord, C. M., and J. E. Cordoza. 1982. Bobcat and lynx (*Felis rufus* and *F. lynx*). In *Wild mammals of North America, biology, management, and economics,* edited by J. A. Chapman and G. A. Feldhamer. Baltimore, MD: Johns Hopkins University Press.

Meaney, C. A. 1983. Olfactory communication in pikas (*Ochotona princeps*). Ph.D. diss., University of Colorado, Boulder.

Meaney, C. A., and D. Van Vuren. 1993. Recent distribution of bison in Colorado west of the Great Plains. Proceedings of the Denver Museum of Natural History, Series 3, 4:1–10.

Merritt, J. F. 1976. Population ecology and energy relationships of small mammals of a Colorado subalpine forest. Ph.D. diss., University of Colorado, Boulder.

———. 1983. Influence of snowcover on survival of *Clethrionomys gapperi* inhabiting the Appalachian and Rocky Mountains of North America. *Annales Zoologici Fennici* 173:73–74.

———. 1984. Growth patterns and seasonal thermogenesis of *Clethrionomys gapperi* inhabiting the Appalachian and Rocky Mountains of North America. In *Winter ecology of small mammals,* edited by J. F. Merritt. Special Publication 10, Carnegie Museum of Natural History.

Merritt, J. F., and J. M. Merritt. 1978a. Population ecology and energy relationships of *Clethrionomys gapperi* in a Colorado subalpine forest. *Journal of Mammalogy* 59:576–598.

———. 1978b. Seasonal home ranges and activity of small mammals of a Colorado subalpine forest. *Acta Theriologica* 23:195–202.

Osburn, W. S., Jr. 1958. Ecology of winter snow-free areas of the alpine tundra of Niwot Ridge, Boulder County, Colorado. Ph.D. diss., University of Colorado, Boulder.

Paradiso. J. L., and R. M. Nowak. 1982. Wolves (*Canis lupus* and allies). In *Wild mammals of North America, biology, management, and economics,* edited by J. A. Chapman and G. A. Feldhamer. Baltimore, MD: Johns Hopkins University Press.

Pasitschniak-Arts, M. 1993. Ursus arctos. *Mammalian Species* 439:1–10.

Pelton, M. R. 1982. Black bear (*Ursus americanus*). In *Wild mammals of North America, biology, management, and economics,* edited by J. A. Chapman and G. A. Feldhamer. Baltimore, MD: Johns Hopkins University Press.

Pruitt, W. O., Jr. 1984. Snow and small mammals. In *Winter ecology of small mammals,* edited by J. F. Merritt. Special Publication 10, Carnegie Museum of Natural History.

Samuel, D. E., and B. B. Nelson. 1982. Foxes (*Vulpes vulpes* and allies). In *Wild mammals of North America, biology, management, and economics,* edited by J. A. Chapman and G. A. Feldhamer. Baltimore, MD: Johns Hopkins University Press.

Shackleton, D. M. 1985. Ovis canadensis. *Mammalian Species* 230:1–9.

Smith, A. T., and M. L. Weston. 1990. Ochotona princeps. *Mammalian Species* 352:1–8.

Smith, D. 1981. *Above timberline, a wildlife biologist's Rocky Mountain journal.* New York: Alfred A. Knopf.

Smolen, M. J., and B. L. Keller. 1987. Microtus longicaudus. *Mammalian Species* 271:1–7.

Southwick, C. H., S. C. Golian, M. R. Whitworth, J. C. Halfpenny, and R. N. Brown. 1986. Population density and fluctuations of pikas (*Ochotona princeps*) in Colorado. *Journal of Mammalogy* 67:149–153.

Spencer, A. A., and H. W. Steinhoff. 1968. An explanation of geographic variation in litter size. *Journal of Mammalogy* 49:281–286.

Stinson, N., Jr. 1977a. Home range of the western jumping mouse, *Zapus princeps,* in the Colorado Rocky Mountains. *Great Basin Naturalist* 37:87–90.

———. 1977b. Species diversity, resource partitioning, and demography of small mammals in a subalpine deciduous forest. Ph.D. diss., University of Colorado, Boulder.

Svendsen, G. E. 1982. Weasels (*Mustela* species). In *Wild mammals of North America, biology, management, and economics,* edited by J. A. Chapman and G. A. Feldhamer. Baltimore, MD: Johns Hopkins University Press.

Terres, J. K. ed. 1995. *The Audubon Society encyclopedia of North American birds.* Avenel, NJ: Wings Books.

Thorn, C. E.. 1978. A preliminary assessment of the geomorphic role of pocket gophers in the alpine zone of the Colorado Front Range. *Geografiska Annaler* 60A:181–187.

———. 1982. Gopher disturbance: its variability by Braun-Blanquet vegetation units in the Niwot Ridge alpine tundra zone, Colorado Front Range, U.S.A. *Arctic and Alpine Research* 14:45–51.

Vaughan, T. A. 1974. Resource allocation in some sympatric, subalpine rodents. *Journal of Mammalogy* 55:764–795.

Walker, D. N. 1987. Late Pleistocene/Holocene environmental changes in Wyoming: The mammalian record. In *Late Quaternary mammalian biogeography and environments of the Great Plains and prairies,* edited by R. W. Graham, H. A. Semken Jr., and M. A. Graham. Scientific Papers 22, Illinois State Museum.

Wardle, P. 1974. Alpine timberlines. In *Arctic and alpine environments,* edited by J. D. Ives and R. G. Barry. London: Methuen.

Warren, E. R. 1910. *Mammals of Colorado.* New York: G. P. Putnam's Sons.

———. 1942. *The mammals of Colorado.* 2d rev. ed. Norman, OK: University of Oklahoma Press.

Webber, P. J., and D. E. May. 1977. The magnitude and distribution of belowground plant structures in the alpine tundra of Niwot Ridge, Colorado. *Arctic and Alpine Research* 9:157–174.

Zwinger, A. N., and B. E. Willard. 1972. *Land above the trees: A guide to American alpine tundra.* New York: Harper and Row.

# 8

# Soils

Timothy R. Seastedt

## Introduction

This chapter examines alpine soils from a traditional soil science and ecological perspective, with a bias toward the latter. Soil physical and chemical properties are presented, but the soils as a resource for the biota as well as the feedbacks between abiotic and biotic processes are emphasized. Over half a century ago, Hans Jenny (1941) developed a conceptual model of the factors responsible for soil development. Jenny recognized that parent material, climate, topography, and geological and ecological disturbance factors could be viewed as independent phenomena that interact to produce soils. Jenny (1980) subsequently expanded this model to one that was also useful to describe entire ecosystems. To date, I've found no better framework with which to explain soils as true ecosystem characteristics—an entity generated by the interaction of biota with the abiotic environment. Accordingly, the roles that parent materials, topography, climate, biota, and disturbance frequencies have in controlling the structural and functional aspects of alpine soils are discussed. Because each of these five factors of soil formation has the potential to interact with various combinations of the other four factors, the number of possible combinations—and soil types—is surprisingly large, especially when one or more of the five factors exhibits tremendous within-site variability. Certainly the alpine must rank "most heterogeneous" among terrestrial ecosystem types in terms of topography, making this variable particularly important in any discussion of soil characteristics. As will be demonstrated, however, the other four factors also exhibit significant variation that contributes to the complexity of the alpine soil landscape.

Soil characteristics emphasized here include those variables that affect and are affected by biotic processes over time scales ranging from a single growing season to decades to centuries. Hence, cation exchange capacity (CEC), soil acidity (pH), soil water content, nutrient content and flux, and carbon storage and flux, are of primary concern.

Detailed information about the soils of this region comes primarily from two sources, Scott Burns's 1980 dissertation on soil distribution and development in the Niwot Ridge-Green Lakes region, and an extensive series of publications by M. I. Litaor. Burns provided classical soil descriptions based on the analysis of 97 extensive soil pit excavations. Litaor's contributions, which spanned over a decade, were also based on soil pit analyses, but were complemented by additional measurements of soil inputs such as eolian dust and movements of materials within and from the soils by using various types of soil water collectors. Collectively, the information obtained by these two individuals provides an extensive database on the physical and chemical characteristics of alpine soils. These works have been supplemented to some extent by more recent studies by ecologists interested in understanding landscape variation in soil characteristics and how these characteristics influence ecosystem processes of the tundra (e.g., Stanton et al. 1994; Fisk et al. 1998).

There exists a hierarchy of spatial scales relevant to the analysis of soil characteristics. In addition to classifying the tundra soils on the basis of snowpack and moisture conditions, Burns (1980) divided the alpine zone into three geomorphic provinces. The ridge-top surfaces make up 36% of the area and are composed of relatively old and deep soils. Valley-side areas comprise 60% of the region and contain soils that are "young and thin with 70% of the mapping unit being bare rock" (Burns 1980). The valley-floor, which comprises only 4% of the region, is composed of young, thin soils on glacial till and bedrock. These larger geomorphic units constrain characteristics associated with vegetation types and also interact differentially with various disturbance agents. For example, gopher effects on valley-side areas appear more extreme than those observed on ridge-top areas, and tree island and gopher effects along hillslope areas appear to collectively create conditions for the loss of surface horizon soils created under an earlier climate, when these areas may have been well above tree line.

## Soil Classification of Niwot Ridge

Burns (1980) reported that the alpine tundra soils of Niwot Ridge are Inceptisols (young soils with few diagnostic features), mostly Cryochrepts (Inceptisols formed in cold regions with relatively low organic matter in surface horizons) and Cryumbrepts (similar to Cryochrepts, but with darker surface horizons indicating more organic materials). Some Entisols (soils with almost no development) termed Cryorthents (medium-textured Entisols created in cold regions) can be found in areas that have not been colonized by biota. While soils in much of the alpine are shallow relative to most grasslands, these Inceptisols have relatively high-potential fertility. Webber and Ebert-May (1977) described soils on Niwot Ridge as well drained, coarse textured, with thin organic-rich surface horizons. Most of these soils are moderately acidic, ranging in pH from about 4.5 to 5.5 (Fisk 1995). Hence, while the soils may have relatively high cation exchange capacities (e.g., see Burns 1980;

Table 8.1. Niwot Ridge Soil Characteristics

| | Sample Location | | |
|---|---|---|---|
| Variable | Dry Meadow | Moist Meadow | Wet Meadow |
| Depth of A Horizon (cm) | 18 | 34 | 39 |
| Texture (%) | | | |
| Sand | 39 | 46 | 33 |
| Silt | 38 | 33 | 46 |
| Clay | 23 | 21 | 21 |
| pH (1:1 water) | | | |
| Burns (1980) | 5.5 | 5.0 | 4.6 |
| Fisk (1995)[a] | 5.5 | 4.7 | 5.2 |
| Organic Matter (%) | | | |
| Burns (1980) | 22 | 16 | 51 |
| Fisk (1995) | 28 | 31 | 37 |
| CEC (meq/100 g) | 47 | 28 | 58 |
| Average (%) summer soil $H_2O$ (g/g)[b] | 60 | 85 | 144 |

*Note:* Dry meadows dominated by Kobresia, moist meadows dominated by *Acomastylis* and *Deschampsia,* and wet meadows dominated by *Caltha leptosepala* and *Carex* spp. Unless otherwise stated, data are from Burns (1980).

[a] Data from Fisk (1995) include only the top 10 cm of soil.
[b] From Taylor and Seastedt (1994).

Litaor 1987; Table 8.1), hydrogen ions occupy a moderate to large percentage of these exchange sites.

Relative to adjacent forest soils and in contrast to Webber and Ebert-May's (1977) generalizations, surface soils of alpine tundra can have moderate to deep (> 30 cm) A horizons, with surface litter rarely forming litter layers (O horizons) above the A horizon. Burns (1980) did report examples of some thick O horizons in deposition zones of windblown areas, and Litaor (1987) also reported O horizons at some of his wet meadow sites. Litter can accumulate in areas protected from scouring, where decomposition of surface litter is not rapid. Such areas might include dry, protected areas as well as the very wet meadows, where anaerobic conditions could extend to the soil surface, and beneath tree islands. The accumulation of organic matter in the O horizons of tree islands is unusually large, even for forest standards (cf. Holtmeier and Broll 1992), and this accumulation of detritus is responsible for the ability of these trees to sprout adventitious roots from buried branches (Marr 1977).

## Parent Materials of Niwot Ridge and Green Lakes Valley

Osburn and Cline (1967) and Gable and Madole (1976) indicated that this region of the Front Range is composed of Precambrian crystalline rocks, Albion monzonite, Boulder Creek granite, cordierite, and biotite gneisses. The higher elevations of Niwot Ridge were not directly affected by Pleistocene glaciation, whereas the valley bottom was heavily scoured by ice. Hence, valley soils are new, 10,000 years or less in age, while ridge soils may have had a much longer time period to develop (Burns 1980). Moraines and solifluction debris may dominate portions of the lower and higher elevations, respectively. Even relatively flat areas of Niwot Ridge are often dominated by rock debris or patterned ground.

Clays in surface horizons of the soils include mica, smectite, chlorite, and kaolinite. Mica is generally highest; kaolinite abundance is often somewhat less than others (Burns 1980). Litaor (1987) emphasized the significance of wind-derived (eolian) deposits on the characteristics of the soils of this region and the effects that loess can have on soil and streamwater pH. Burns (1980) reported that this loess, originating from unknown regions west of the Continental Divide, was found at depths ranging from 8 cm at sites he termed "extremely windblown" (e.g., fellfields), to 13 cm in depth in areas of "minimal snow cover" (dry *Kobresia* meadows) to 8 cm in "early melting snowbank" (moist *Deschamsia* and *Acomastylis* meadows). Loess was not found elsewhere. Hence, this material is an important component of the surface soils of some but not all of Niwot Ridge and therefore contributes to the spatial heterogeneity of soil characteristics. Analysis from eolian dust collectors indicated that this material was about 20% organic matter and contained significant amounts of calcium carbonate, a major buffer for acidic deposition (Litaor 1987; Caine 1995). Holtmeier and Broll (1992) suggested that calcium sulfate was also found in loess, a compound not reported by Litaor (1987).

## The Role of Climate in Soil Formation

Fahey (1971) suggested that the primary geomorphic processes operating today in tundra soils are freeze-thaw phenomena that occur on a small spatial scale and produce needle ice, frost boils, and frost creep. These relatively small-scaled activities are nested within the solifluction terraces and patterned ground created most recently during the Pleistocene (chapter 4). At a minimum, these Pleistocene features continue to affect drainage and patterns of snowpack and snow duration. Hence, the paleoclimate created legacies in the surface of Niwot Ridge that continue to influence the present-day landscape.

The harsh climate of the alpine (see chapter 2) limits the amount of chemical weathering of parent materials. At the same time, however, freeze-thaw activities may physically break apart rock particles of all sizes and actively create new surface area for enhanced weathering under the existing climate. Diurnal freeze-thaw is a common phenomenon during much of the alpine spring and autumn. In a litterbag decomposition study using crushed rock (rhyodacite), Caine (1979) demonstrated significant losses in mass (ca. 0.08% per year) and was able to demonstrate that the timing and quantity of water flow affected this rate of loss. Hence, weathering of parent materials, accompanied in some areas by loess deposition, has created the potential for a moderately fertile surface soil (Table 8.1).

The wind is a major component of climate at Niwot Ridge. Any surface exposed to the wind is also exposed to scouring by particulates of ice, rock, and organic debris. Windblown sites (often created by disturbance but sometimes generated by scouring alone) may lack soil development or have had soils reduced to gravel and only portions of the initial surface horizons remain (Burns 1980). The wind interacts with topography to produce huge variations in microclimate (chapter 2). Snow depth, snow duration, diurnal temperature fluctuations, soil moisture, soil texture, and others all have large effects on ecosystem processes such as nutrient cycling and energy flow. Hence, while topography is given the credit for the landscape heterogeneity of Niwot Ridge, wind greatly amplifies hillslope effects.

The 100 cm or so of precipitation per year is in excess of soil storage capacity, evaporation, and vegetative demands (Greenland et al. 1984; Greenland 1989). Hence water will move either over the frozen surface or through the alpine soils, and the consequences of this movement is significant to biogeochemical cycles (chapter 5). Litaor (1992) reported that water movement through the alpine soils occurs during both snowmelt and intense summer precipitation events. Macropores can be formed by gopher tunneling activities and by the residual channels created by plants after root decay. The former activity produces noticeable erosion and "terracettes" on steeper slopes (Thorn 1978). Water can often be seen flowing from what are assumed to be gopher tunnels during snowmelt, and the debris associated with this flow often creates very noticeable deposits in the melting snowbanks. At present, we suspect that water movement through micropores (spaces found between soil particles) of alpine soil results in substantial filtering of this water, with a variety of abiotic and biotic processes acting to remove materials such as soluble nutrients. Also, organic material is likely deposited or aggregated with clays under such a filtration system. In contrast, macropore movements directly transport materials in water without benefit of this soil filter. Hence, factors that influence the patterns of micro- and macropore development in soils are important to the terrestrial alpine ecosystem and the subalpine wetlands and aquatic ecosystems. Macropore flow also explains the apparent paradox between streamwater increases in nitrogen content and the tundra's ability to absorb and retain sources of atmospheric nitrogen enrichment (chapter 12). Nitrogen will be removed from percolating waters if filtered through micropores, but this nitrogen can be directly transferred to streams via macropore flow.

## Topographic Effects

Much terrestrial ecosystem research originated as "watershed studies," where ecosystem characteristics were assessed in such terms of ecosystem outputs as stream volume and chemistry. Ecosystems lacking stream outputs required a different conceptual model, and the "catena concept" (Schimel et al. 1985) was useful in describing ecosystem phenomena of semihumid to arid landscapes. Central to this model was the idea that water and solids move within (but not necessarily out of) the landscape and generate ecosystem characteristics. Both the catena and watershed models have substantial relevance and application to ecological studies of alpine tundra, and both concepts are potentially incorporated by employing a landscape or landform analysis. This approach considers the form of the land surface and associated terrestrial and aquatic ecosystems in evaluating factors such as materials transport and energy flow (Swanson et al. 1988). A landform analysis explicitly recognizes both shapes and position of an area. For example, a north-facing hillslope protected from the prevailing winds would generate a unique microclimate and biota that would, in turn, affect soil development and characteristics. A north-facing slope that is wind-scoured would produce a different microclimate and biota and create a different set of soil characteristics. Placement within the landscape is also very significant when evaluating stream or lake characteristics of high-elevation systems (chapter 5). The uppermost lakes receive the highest percentage of direct inputs and likely have the least amount of terrestrial biotic filtering of atmospheric materials. Similar-sized lakes lower in the drainage receive a higher percentage of filtered inputs.

There is little doubt that snowpack amounts, snow duration, and the presence or absence of soil water subsidies are the dominant features determining the landscape patterns of both the biota and soils of Niwot Ridge, a finding emphasized for other alpine areas by Billings and his colleagues (e.g., Billings and Bliss 1959; Billings 1988). This fact was also identified by Burns (1980) and has been a theme of the Niwot Ridge Long-Term Ecological Research (LTER) program since its inception (e.g., Walker et al. 1993). To correctly depict soils (assuming a constant disturbance regime, to be discussed), both the duration of snowpack and soil moisture regime of an area are necessary to determine soil characteristics. An important observation is that two types of wet meadows may be found in the alpine tundra, as shown in Fig. 8.1. Sites just melting out from areas of extensive snow are often described as wet meadows, but the seasonality of soil temperatures and moisture as well as the vegetation of these areas can be drastically different from wet areas with little snowpack, which are naturally irrigated throughout most of the growing season by snowmelt from upslope areas. Indeed, "snowbed communities" versus "wet meadows" represent two extremes in soil characteristics of alpine tundra.

Dry, moist, and wet meadows are more biologically active (e.g., have higher rates of plant productivity) than either fellfields or snowbed communities (chapter 12). These sites have contributed substantial amounts of organic matter to the soils (Table 8.1). Repeated sampling demonstrates that wet sites have the highest amounts of organic matter and highest indices of soil fertility (CEC). However, comparisons of Fisk's (1995) results are not entirely consistent with the earlier sampling by Burns (1980) (Table 8.1). Fisk's data includes south-, west-, and north-facing sites in each of her meadow locations, whereas Burns' results are based on a random sampling approach. Without additional information, it is difficult to attribute patterns to biota, aspect, and slope, or some combination of these factors.

Stanton et al. (1994) measured N, P, K, Ca, and moisture content, organic matter, and pH of soils across a snowpack gradient in the Mosquito Range of Colorado. They found that most soil characteristics exhibited strong patterns dependent on snowpack and snow duration. In particular, nitrogen concentrations varied from 1.0 to 4.8% of soil mass across the landscape, with a very large decline in the sites with the longest snow duration. Soil pH showed an increase with respect to snow duration, suggesting, as did Caine's (1979) study, that higher rates of mineral weathering may occur at such sites.

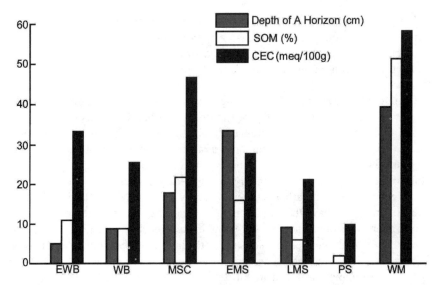

Figure 8.1. Soil organic matter and CEC characteristics as a function of snow duration and soil moisture. Data used in this graph were obtained from tables in Burns (1980). The units shown are depth (cm) of surface horizon, the percentage of dry mass for SOM, and the milliequivalents per 100 g soil for CEC. The landscape symbols include extremely windblown (EWB) and wind blown (WB), which refer to fellfield areas; minimum snow cover (MSC), which refers to *Kobresia* meadows; early melting snow (EMS), referring to moister *Acomastylis* and *Deschampsia* meadows; LMS for *Sibbaldia* meadows; permanent snow (PS) for snowfields; and wet meadows (WM) for *Caltha* and *Carex* meadows.

Burns (1980) reported on the variability in soil temperatures across portions of the alpine snowpack gradient. Areas with substantial snowcover are much warmer in the winter and become warmer still, at least briefly, in the summer relative to areas not receiving much snow. The insulating properties of snow and the reduced plant canopy on snowbed communities explains this seasonal temperature pattern. Snowbed soils are much drier than wet meadow communities once these melt free of snow (LTER, unpublished data). These soils are also very low in organic matter and clay. Despite the fact that they receive substantial snowmelt water relative to other sites, the soils lack the ability to retain this moisture.

## The Role of Biota in Soil Development

The soil is home to most of the biological diversity of the alpine. The vast majority of this diversity is not readily observable and is composed of microarthropods, nematodes, other microinvertebrates, fungi, and bacteria. A few species such as the gopher, *Thomomys talpoides,* are not only visible but are dominant features on large portions of the alpine landscape. These species form an important component of both the biota and disturbance components of Jenny's model of soil development. Here, the discussion of contributions of biota to soil characteristics is limited to some general effects contributed by plant growth forms and microbial and associated detritivore community characteristics. Gophers deserve special treatment and are discussed below.

Graminoid vegetation, if allowed to develop in a favorable climate over extended periods of time, contributes to the development of a Mollisol. This soil is characterized by high fertility and accumulation of vast amounts of carbon and nitrogen (Seastedt 1995). The alpine vegetation is a rather unique variety of grassland, and, to speak anthropomorphically, the same type of vegetation is attempting to create a Mollisol in the alpine. The restricted growing season hinders this, but the "graminoid signature" on soil carbon storage is clearly evident. Niwot Ridge soils from *Kobresia* meadows have the largest amounts of soil carbon per unit area of surface (15 cm deep) soils of any aerobic soil found in the temperate zone (Zak et al. 1994).

The fact that alpine meadow soils are potentially more fertile than are adjacent forest soils is consistent with grassland-forest comparisons made by Jenny (1930) as well as examples provided in soil texts (e.g., Brady 1992). However, actual demonstrations of the mechanisms explaining this phenomenon are few. Why should the upper soil horizons of graminoid vegetation, plants that are not as productive as trees, have more soil carbon, and why are these soils potentially higher in plant nutrients? The presence of moving tree islands in the lower regions of alpine tundra provides an opportunity for direct comparisons and studies directed at understanding how plant life form affects soil characteristics (Fig. 8.2).

Benedict (1984) suggested that soils beneath the tree islands were developing soils similar to Spodosols, which are found under moist, coniferous forests. These soils differ from Mollisols by being more acidic, lower in cation exchange capacity, and with reduced organic matter content. Limited data from Burns (1980), Holtmeier and Broll (1992), and Pauker and Seastedt (1996) support Benedict's contentions. For example, Burns (1980) reported cation exchange capacity beneath krummholz to range from 16.1 to 31.2 meq, while meadow sites had values that ranged from 37.8 to 46.7. Both Burns (1980) and Pauker and Seastedt (1996) demonstrated reduced soil organic matter beneath krummholz. Holtmeier and Broll (1992) also provided data that suggest that clay content of these soils is diminished beneath the trees. If correct, the reduction in soil carbon (some of which is assumed to be humus) and clays means that cation exchange capacity will be reduced. A more recent study (Seastedt and Adams 2001) indicated that soil texture is not altered by tree island passage.

Ongoing studies of tree island and tundra soils are testing the hypotheses that the trees, while having higher rates of NPP, are thought to have lower base cation requirements than do adjacent tundra. Litter is therefore lower in cation concentrations, and decomposition of this material produces more acidic organic matter than does the decomposition of herbaceous material. In the absence of high plant demand for cations (e.g., reduced root uptake), metallic cations are leached from the soil surface into the subsoil. Concurrently, the more acidic leachates from the large mass of coniferous litter results in a destabilization of organic-clay aggregates in the surface layers, and these materials are also leached from the surface horizons.

Soils   165

Figure 8.2. Krummholz tree moving imperceptibly across the alpine, altering soil processes on the way (photo by William D. Bowman).

Because these tree islands move from windward to leeward areas at rates of ca. 2–4 cm per year (Marr 1977), the trees colonize the alpine soils, modify the soil characteristics, then move on, leaving a reduced-fertility soil to be recolonized by alpine species. The trees therefore generate spatial heterogeneity in the soils above that generated by climate and topography. The trees also might qualify as a disturbance, particularly if species in the areas being colonized by the moving trees are the objects of study. These areas first experience increased snowpack due to the snowdrift created by the trees. Such areas appear to receive increased vole and gopher activity and then are subsequently occupied by subalpine fir or Englemann spruce. The original tundra vegetation is completely eradicated in the process.

Biota supply the organic matter of the soil, which represents ca. 10–20% of the mass of surface soils. The biota also have large influences on bulk density, pore sizes, and, as suggested above, on the ability of materials such as carbon and other nutrients to pass through the soil or be stored in either abiotic or biotic forms. Cortinas and Seastedt (1996) were unable to demonstrate that differences in the lifeforms represented by the fibrous-rooted sedge, *Kobresia,* and the tap-rooted forb, *Acomastylis,* produced differences in surface carbon amounts or bulk densities of soils found directly beneath these plants. Phosphorus concentrations (as measured with a persulfate digest) were higher beneath *Acomastylis* (Seastedt, unpublished results). This difference may be related to the plants or, more likely, to higher frequencies of gopher disturbance in the *Acomastylis* areas. Regardless of the mechanism, the results indicate that subtle differences in soil chemistry should be anticipated due to individual species' effects on these and other soil properties. And, while the soil chemical composition may be similar among the diverse species growing in the alpine, plants can strongly influence the amounts and forms in which important plant nutrients such as nitrogen are moving through this matrix (Steltzer and Bowman 1998).

## The Role of Disturbance in the Formation of Alpine Soils

Just as topographic constraints operate on different spatial scales, disturbances operate on different temporal scales, all of which have significance on current soil status. From the Miocene, we have the uplifting processes that created the Front Range. From the Pleistocene, we have large-scale glacier and periglacial activity, producing scours, moraines, solifluction terraces, and patterned ground. Freeze-thaw cycles still affect soil characteristics, but on a more-limited spatial scale (Benedict 1970). From the last millennia, we have subtle but significant shifts in climate that may have allowed periodic advancement and declines of both tree islands and timberline (c.f. Benedict 1984), and these trees clearly modify the tundra soil. For this same period of time, if not substantially longer, we have the ebb and flow of gopher activities with their very measurable effects on hillslope erosion (Thorn 1978, 1982; Sherrod 1999) and more subtle changes in soil chemistry (Litaor et al. 1996; Cortinas and Seastedt 1996; Sherrod 1999). Finally, for the last several decades, we have the anthropogenic modifications, which range from trampling (the "bigfoot effect"), modifications to high-elevation catchments for water management, and changes in the quantity and quality of atmospheric chemical inputs of hydrogen ions, nitrate, ammonia, and other materials (chapters 3, 5).

The following discussion is restricted to current disturbances to the soil. The uplift of the Rockies and past cryoturbation processes are now fixed landscape features. There is no record of fire in the alpine tundra of the Front Range. (Limited amounts of charcoal have been found in soil and lake sediments, but this material is assumed to have originated elsewhere.) This leaves current freeze-thaw phenomena, gopher burrowing activities, and human activities as the major physical disturbance factors of the tundra.

Freeze-thaw processes are of interest at a variety of temporal scales. At the smallest temporal and spatial scale, freezing is assumed to damage biological tissue and perhaps most importantly result in a pulse of nutrients from lysed cells of microbes (Brooks 1995). Freeze-thaw processes, particularly during snowmelt, should make the tundra much more "leaky" in terms of labile compounds such as certain organic materials and inorganics such as nitrate, ammonium, and soluble phosphates.

At an intermediate temporal scale, freeze-thaw creates frost boils and other relatively small-scale disturbances on a subset of the soil surface (Benedict 1970; Fahey 1971). These actively disturbed areas lack plant cover and therefore have reduced organic matter inputs that tend to hold other nutrients in the soil. Such areas, therefore, should have lower soil fertility than unaffected sites.

At the longest temporal scales, freeze-thaw generates substantial surface area that increases weathering rates, thereby replacing lost nutrients and maintaining the moderate buffering capacities of these soils. As previously described, in the absence of loess deposition, the disturbance is critical to the initial steps in converting a rocky surface to a fine-textured soil surface.

As described in the previous chapter, gophers are "keystone species" or "ecological engineers" (Jones et al. 1994) in that their effects on the biogeochemical environment are very large relative to their contributions to energy flux. A substantial literature on gopher effects is available (e.g., Huntley and Inouye 1988), and considerable research has been conducted in the alpine of the Colorado Front Range (Thorn 1978, 1982; Cortinas and Seastedt 1996; Litaor et al. 1996). Work by Sherrod (1999) supplements and synthesizes previous work, particularly in terms of gopher effects on plant species composition and species richness.

As with other disturbances, gophers clearly have short- and long-term effects. Burns (1980) noted that the dry meadows, areas where gophers seldom forage, had the most mature or potentially climax (equilibrium) soil characteristics. Areas with heavy gopher disturbance are maintained in a successional state. While the latter communities likely average less organic matter and lower cation exchange capacity than do undisturbed soils, rigorous direct tests of these predictions are few. Certainly mounding and burrowing often causes large surface losses of both organic matter and clays to wind and water erosion (Fig. 8.3). However, the gophers also increase

Figure 8.3. Approximately a 5-year-old gopher mound, with coarse particles making up a relatively large component of the soil (photo by William D. Bowman).

the depth of the surface horizons, perhaps even doubling this depth in many habitats (Burns 1980). Hence, surface materials are moved deeper into the soil. These may actually increase overall soil storage of such materials as carbon, but the disturbance likely reduces plant-available nutrients in the uppermost regions of the soil or area of maximum rooting density. Burrowing may favor deep rooting species, particularly if these species are either nonpreferred food items or are capable of withstanding moderate gopher herbivory.

Once modest additions of N have been added to the soil, P quickly becomes co-limiting to plant growth. The element also appears to affect plant species composition of the alpine (Theodose and Bowman 1997). Phosphorus is biologically and ecologically interesting because some plants use mycorrhizal symbionts to provide adequate P whereas others do not and because several species of clover (*Trifolium* spp.) are abundant in the alpine and presumably have large P demands associated with their ability to fix atmospheric nitrogen. Phosphorus data were not included in the studies of Burns or Litaor. Detailed work using P fractionation procedures (Hedley et al. 1982) is currently underway, and preliminary data on total persulfate-extractable P indicate some interesting patterns. In paired plots, P concentrations are significantly higher from soils beneath *Acomastylis* than beneath *Kobresia* plants, even though carbon amounts in the surface horizons are not (Fig. 8.4). One inter-

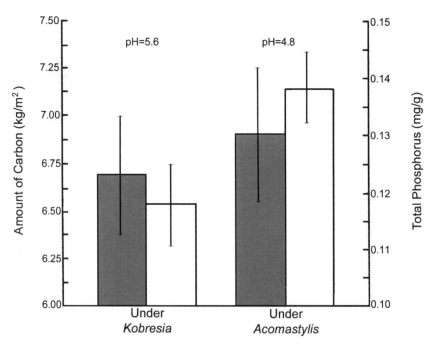

Figure 8.4. Amounts of carbon (shaded bars), persulfate-extractable phosphorus concentrations (open bars), and pH values of soils sampled directly beneath *Acomastylis rossii* and *Kobresia myosuroides*. Fifteen pairs of samples were taken (one beneath each vegetation type) in relatively dry meadow sites.

pretation is that gopher mounding activities tend to be correlated with the former plant, but not the latter, and in any case, the latter species is more common in areas impacted by gophers. The mixing of soil may provide material higher in labile P to the surface horizons, or the difference could be generated by the plants themselves. Additional impacts of gophers on alpine soils are reported by Dearing (chapter 14).

## Human Modifications to Soils

Increased hydrogen ion deposition from atmospheric sulfuric and nitric acids increases the weathering rate of parent materials (e.g., Sollins et al. 1980; Binkley and Richter 1987). The buffering capacity of these soils (the ability to retain current pH levels in the face of increasing acid inputs) is relatively modest but likely adequate for the coming decades. More significant to soil pH and fertility in general are the potential effects of enhanced nitrification from increased N deposition (chapters 12 and 16).

While increased acidity of rainfall along the Front Range of Colorado appears to have been the trend for the last several decades (Lewis et al. 1984; Baron et al. 1994; Williams et al. 1996), the potential for enhanced buffering capacity provided by loess deposition is also a potential future scenario. While speculative, increased loess deposition is possible, particularly if desertification continues in areas that contribute to this loess source (See Sievering et al. 1996; and chapter 3 for origins of the Niwot Ridge airstream).

## Conclusions and Future Studies

Jenny's model of soil development remains a useful conceptual approach to understanding the present status and spatial patterns of soils of alpine tundra. Regretfully, the model provides only guidance, not quantitative rules, regarding interpretations of individual and interactive effects of factors generating the structure and properties of soils. As previously discussed, each of the factors may produce short- and long-term effects, due to the fact that many soil properties (e.g., weathering, mineralization, and decomposition of different substrates) operate at different time scales. The energy and moisture regime created by snowpack, snow duration, and water runoff patterns is undoubtedly the strongest combination of variables influencing the landscape patterns of alpine soils, but these variables themselves both create and are created by the interaction of a specific climate, topography, parent material, biotic component, and disturbance regime.

A comparison of soil characteristics across 10 LTER sites in North America shows that Niwot Ridge soils from *Kobresia* meadows are highest in organic C and N of all sites sampled and among the highest in extractable ammonium and nitrate (only Hubbard Brook, a site also experiencing large anthropogenic inputs of N, was higher in the sum of inorganic N) (Zak et al. 1994). From a plant's perspective, then, these soils would appear to be relatively fertile and concerns about excess atmospheric N inputs in the Front Range (e.g., Williams et al. 1996) are consistent with these high N levels. Yet Bowman et al. (1993) demonstrated that N availability still limits plant productivity. A provocative question is whether the N is limiting the plants or whether the plants are, through some mechanisms of organic N sequestration, assisted perhaps with some microbial and physical properties of soils, thus limiting the N availability. This mechanism has been suggested for prairie soils (Wedin and Tilman 1990; Seastedt et al. 1991), and it is possible that a combination of soil physical and chemical properties, microbial responses, and plant growth dynamics all function to maintain at least a seasonal N limitation in the alpine. Given well-founded concerns with atmospheric N and C enrichment and the fates of these elements (e.g., Vitousek 1994; Asner et al. 1997), the question "What are the mechanisms responsible for the sequestration and storage of N and C in soils?" remains much more than an academic one.

A number of questions relevant to a basic understanding of soils in ecosystem and biogeochemical processes of the alpine remain unresolved. What portion of the carbon in the surface soils of Niwot Ridge is eolian in origin? Can $^{12}C/^{13}C$ ratios of surface carbon in sites with eolian deposits be distinguished from ratios from sites lacking this input? Are surface soils actually at some quasi-equilibrium in terms of carbon and nutrient storage, or are the soils functioning as sources or sinks of these materials to aquatic and lower elevation systems? Finally, the landscape variation observed in the Niwot Ridge soils creates tremendous variation in the biota and site-specific biogeochemical processes controlling carbon, water, nutrients, and trace gas fluxes. Appreciation of this heterogeneity—as well as the integration of this variability into a general model depicting the collective ecosystem processes of the alpine tundra—requires a knowledge base we have only begun to accumulate.

*Acknowledgments* The editorial comments and suggestions provided by Nel Caine, Susan Sherrod, and Diane McKnight were very helpful in preparing this chapter. Roberto Cortinas obtained the soil samples and the pH data reported in Fig. 8.4.

References

Asner, G. P., T. R. Seastedt, and A. R. Townsend. 1997. The decoupling of terrestrial carbon and nitrogen cycles. *BioScience* 47:226–234.

Baron, J. S., D. S. Ojima, E. A. Holland, and W. J. Parton. 1994. Analysis of nitorgen saturation potential in Rocky Mountain tundra and forest. Implications for aquatic systems. *Biogeochemistry* 27:61–82.

Benedict, J. B. 1970. Downslope soil movement in a Colorado alpine region—Rates, processes and climatic significance. *Arctic and Alpine Research* 2:165–227.

———. 1984. Rates of tree-island migration, Colorado Rocky Mountains. *Ecology* 65:820–823.

Billings, W. D. 1988. Alpine vegetation. In *North American terrestrial vegetation.* edited by M. G. Barbour and W. D. Billings. New York: Cambridge University Press.

Billings, W. D., and L. C. Bliss. 1959. An alpine snowbank enviornment and its effects on vegetation, plant development and productivity. *Ecology* 40:388–397.

Binkley, D., and D. Richter. 1987. Nutrient cycles and H+ budgets of forest ecosystems. *Advances in Ecological Research* 16:1–51.

Bowman, W. D., T. A. Theodose, J. C. Schardt, and R. T. Conant. 1993. Constraints of nutrient availability on primary production in two alpine tundra communities. *Ecology* 74:2085–2097.

Brady, N. C. 1992. *The nature and properties of soils.* 10th ed. New York: MacMillan.

Brooks, P. B. 1995. Microbial activity and nitrogen cycling under seasonal snowpacks, Niwot Ridge, Colorado. Ph.D. diss., University of Colorado, Boulder.

Burns, S. F. 1980. Alpine soil distribution and development, Indian Peaks, Colorado Front Range. Ph.D. diss., University of Colorado, Boulder.

Caine, N. 1979. Rock weathering rates at the soil surface in an alpine environment. *Catena* 6:131–144.

———. 1995. Snowpack influences on geomorphic processes in Green Lakes Valley, Colorado Front Range. *Geographical Journal* 161:55–68.

Cortinas, M. R., and T. R. Seastedt. 1996. Short- and long-term effects of gophers (*Thomomys talpoides*) on soil organic matter dynamics in alpine tundra. *Pedobiologia* 40:162–170.

Fahey, B. D. 1971. A quantitative analysis of freeze-thaw cycles, frost heave cycles, and frost penetration in the Front Range of the Rocky Mountains. Ph.D. diss., University of Colorado, Boulder.

Fisk, M. C. 1995. Nitrogen dynamics in an alpine landscape. Ph.D. diss., University of Colorado, Boulder.

Fisk, M. C., S. K. Schmidt, and T. R. Seastedt. 1998. Topographic patterns of above- and belowground production and nitrogen cycling in alpine tundra. *Ecology* 79:2253–2266.

Gable, D. J., and Madole, R. F. 1976. Geologic map of the Ward quadrangle, Boulder County, CO. U.S. Geological Survey Geologic Quadrangle Map GQ-1277.

Greenland, D. 1989. The climate of Niwot Ridge, Front Range, Colorado, USA. *Arctic and Alpine Research* 21:380–391.

Greenland, D., N. Caine, and O. Pollak. 1984. The summer water budget and its importance in the alpine tundra of Colorado. *Physical Geography* 5:221–239.

Hedley, M. J., J. W. B. Stewart, and B. S. Chauhan. 1982. Changes in inorganic and organic soil phosphorus fractions induced by cultivation practices and by laboratory incubations. *Soil Science Society of America Journal* 46:970–976.

Holtmeier, F. K., and G. Broll 1992. The influence of tree islands and microtopography on pedoecological conditions in the forest-alpine tundra ecotone on Niwot Ridge, Colorado Front Range, U.S.A. *Arctic and Alpine Research* 24:216–228.

Huntly, N., and R. Inouye. 1988. Pocket gophers in ecosystems: patterns and mechanisms. *BioScience* 38:786–793.

Jenny, H. 1930. A study of the influence of climate upon the nitrogen and organic matter content of the soil. Univ. of Missouri Agricultural Experiment Station Research Bulletion 152. Columbia.

———. 1941. *Factors of soil formation.* New York: McGraw-Hill.

———. 1980. *The soil resource.* New York: Springer-Verlag.

Jones, C. G., J. H. Lawton, and M. Shachak. 1994. Organisms as ecosystem engineers. *Oikos* 69:373–386.

Lewis, W. W., M. C. Grant, and J. F. Saunders III. 1984. Chemical patterns of bulk atmospheric deposition in the State of Colorado. *Water Resources Research* 20:1691–1704.

Litaor, M. I. 1987. The influence of eolian dust on the genesis of alpine soils in the Front Range, Colorado. *Soil Science Society of America Journal* 51:142–147.

———. 1992. Aluminum mobility along a geochemical catena in an alpine watershed, Front Range, Colorado. *Catena* 19:1–16.

Litaor, M. I., R. Mancinelli, and J. C. Halfpenny. 1996. The influence of pocket gophers on the status of nutrients in alpine soils. *Geoderma* 70:37–48.

Marr, J. W. 1977. The development and movement of tree islands near the upper limit of tree growth in the sourthern Rocky Mountains. *Ecology* 58:1159–1164.
Osburn, W., and A. J. Cline. 1967. Soil survey laboratory data and descriptions of some soils of Colorado. USDA Soil Conservation Service, Soil Survey Investigations Report 10.
Pauker, S., and T. R. Seastedt. 1996. Effects of mobile tree islands on soil carbon storage in tundra ecosystems. *Ecology* 77:2563–2567.
Schimel, D. S., D. C. Coleman, and K. A. Horton. 1985. Soil organic matter dynamics in paired rangeland and cropland toposequences in North Dakota. *Geoderma* 36:201–214.
Seastedt, T. R. 1995. Soil systems and nutrient cycles in the North American prairie. In *The changing prairie,* edited by A. Joern and K. H. Keeler. New York: Oxford University Press.
Seastedt, T. R., and G. A. Adams. 2001. Effects of mobile tree islands on alpine tundra soils. *Ecology* 82:8–17.
Seastedt, T. R., J. B. Briggs, and D. J. Gibson. 1991. Controls of nitrogen limitation in tallgrass prairie. *Oecologia* 87:72–79.
Sherrod, S. K. 1999. A multiscale analysis of the northern pocket gopher (*Thymomys talpoides*) at the alpine site of Niwot Ridge, Colorado. Ph.D. diss., University of Colorado, Boulder.
Sievering, H., D. Rusch, and L. Marquez. 1996. Nitric acid, particulate nitrate and ammonium in the continental free troposphere: Nitrogen deposition to the alpine tundra ecosystem. *Atmosphereic Environment* 30:2527–2537.
Sollins, P., C. C. Grier, F. M. McCorison, K. Cromack Jr., R. Fogel, and R. L. Fredriksen. 1980. The internal element cycles of an old-growth Douglas-fir ecosystem in western Oregon. *Ecological Monographs* 50:261–285.
Stanton, M. L., M. Rejmanek, and C. Galen. 1994. Changes in vegetation and soil fertility along a predictable snowmelt gradient in the Mosquito Range, Colorado, U.S.A. *Arctic and Alpine Research* 26:364–374.
Steltzer, H., and W. D. Bowman. 1998. Differential influence of plant species on soil N transformations within moist meadow alpine tundra. *Ecosystems* 1:464–474.
Swanson, F. J., T. K. Kratz, N. Caine, and R. G. Woodmansee. 1988. Landform effects on ecosystem patterns and processes. *BioScience* 38:92–98.
Taylor, R. V., and T. R. Seastedt. 1994. Short- and long-term patterns of soil moisture in alpine tudnra. *Arctic and Alpine Research* 26:14–20.
Theodose, T. A., and W. D. Bowman. 1997. Nutrient availability, plant abundance, and species diversity in two alpine tundra communities. *Ecology* 78:1861–1872.
Thorn, C. E. 1978. A preliminary assessment of the geomorphic role of pocket gophers in the alpine zone of the Colorado Front Range. *Geografiska Annaler* 60A:181–187.
———. 1982. Gopher disturbance: Its variability by Braun-Blanquet vegetation units in the Niwot Ridge alpine tundra zone, Colorado Front Range, USA. *Arctic and Alpine Research* 14:45–51.
Vitousek, P. M. 1994. Beyond global warming: Ecology and global change. *Ecology* 75:1861–1876.
Walker, D. A., J. C. Halfpenny, M. D. Walker, and C. A. Wessman. 1993. Long-term studies of snow-vegetation interactions. *BioScience* 43:287–301.
Webber, P. J., and D. Ebert-May. 1977. The magnitude and distribution of belowground plant structures in the alpine tundra of Niwot Ridge, Colorado. *Arctic and Alpine Research* 9:157–174.
Wedin, D., and D. Tilman. 1990. Species effects on nitrogen cycling: A test with perennial grasses. *Oecologia* 84:433–441.

Williams, M. W., J. Baron, N. Caine, R. Sommerfeld, and R. Sanford. 1996. Nitrogen saturation in the Rocky Mountains. *Environmental Science and Technology* 30:640–646.

Zak, D. A., D. Tilman, R. R. Parmenter, C. W. Rice, F. M. Fisher, J. Vose, D. Milchunas, and C. W. Martin. 1994. Plant production and soil microorganisms in late-successional ecosystems: A continental-scale study. *Ecology* 75:2333–2347.

# Part III

# Ecosystem Function

# 9

# Primary Production

William D. Bowman
Melany C. Fisk

## Introduction

The production of biomass by plants is of central importance to energy, carbon, and nutrient fluxes in ecosystems. Knowledge of the spatial and temporal variation of production and the underlying biotic and physical controls on this variation are central themes in ecosystem science. The goals of this chapter are to present the estimates of spatial patterns in above- and belowground production associated with the major community types found on Niwot Ridge and other alpine areas of the southern Rocky Mountains and to examine the likely environmental causes and underlying mechanisms responsible for spatial and temporal variation in production as elucidated by experimental and observational studies.

Rates of primary production and standing crops of plant biomass are low in alpine tundra relative to other ecosystem types (Lieth and Whittaker 1975; Zak et al. 1994). However, within communities (i.e., at the plot level), there is large variation in rates of production, the degree of biotic control over response to environmental change, and the principal environmental constraints of primary production. As a result, the alpine is one of the most dynamic ecosystems for research. For example, there is a tenfold difference in annual aboveground production between the most and least productive sites with continuous plant cover on Niwot Ridge. In addition, the high plant diversity is a source of potential variation in physiological and developmental control of plant response to the environment. Dominant species include sedges, grasses, shrubs, and forbs, among which are $N_2$-fixing *Trifolium* species. Nearly all of the dominant species may be mycorrhizal. Soil moisture, a driving force for many biotic processes, may vary by an order of magnitude between wet and dry sites following prolonged periods of drought. Thus the alpine tundra of Niwot Ridge, which might appear superficially homogeneous, in fact has complex physical and biotic gradients. This spatial variation prevents simple generalizations about single limiting resources or climatic driving forces determining spatial and temporal variation in productivity.

## Spatial Patterns of Production

### Community Types, Phenology, and Methodology

Billings (1973) defined the mesotopographic gradient as a working unit for describing the alpine landscape, as it encompasses the full range of snow accumulation and associated microclimates and thus biological diversity. In this chapter, we use five main community types associated with this gradient: fellfield, dry meadow, moist meadow, wet meadow, and snowbed. (For a more complete description of plant community types, see chapter 6 and Komárková 1979.) Fellfield communities typically occupy sites with high winds, low or no snow cover, and poorly developed soils, and are dominated by prostrate shrubs and forbs. Dry meadow communities are dominated by the sedge *Kobresia myosuroides,* and have low (<0.5 m) winter snow cover (see Walker et al. 1993 for description of snow-vegetation interactions). Moist meadow communities are typically found at the base of snowfields, receive moisture input from snowmelt early in the growing season, and are codominated by forbs and grasses, principally *Acomastylis rossii* and *Deschampsia caespitosa.* Wet meadow communities are found in sites of variable snow cover but with low or no topographic relief, therefore accumulating substantial water from melting snow and summer precipitation, and maintain saturated soils through most of the growing season. Wet meadows are dominated by sedges and the forb *Caltha leptosepala* but also have patches of low-growing willows where snow cover is great enough to protect the plants from winter winds. Snowbed communities occur in sites with the highest snowcover (>2 m) and are dominated by forbs, primarily *Sibbaldia procumbens.*

Maximum biomass is usually achieved 2–3 weeks before the end of the growing season (Billings and Bliss 1959; May and Webber 1982). The phenology of biomass production differs among the communities as determined by release from snow (Billings and Bliss 1959; Holway and Ward 1965; May and Webber 1982; Walker et al. 1995). The fellfield and dry meadow communities, which receive low or no winter snowcover, commence and finish vegetative growth earlier than do the other communities, usually reaching maximum biomass by mid- to late July. The moist and wet meadow communities have intermediate vegetative phenologies, reaching peak biomass between late July and early August. The snowbed community, which may not be snow free until late July or early August, reaches peak biomass in mid- to late August.

Virtually all of the alpine plant species on Niwot Ridge and the southern Rockies are perennials with aboveground tissues that senesce to the surface of the ground at the end of the growing season. Exceptions include some grasses (e.g., *Deschampsia*) and forbs (e.g., *Solidago multiradiata*), which will overwinter with some green leaves below the snow, and woody species (e.g., *Salix spp.* and *Pentaphylloides floribunda*), which have woody tissues aboveground. However, the amount of overwinter growth is low compared with summer production, and woody species constitute a small fraction of the total plant cover, except in a few areas where they are locally abundant. Thus, aboveground production consists of the annual production of aboveground shoots and is usually estimated by clip harvest of current year's biomass production at peak growth. This technique may underestimate production consumed by herbivores or that which senesces and decomposes during the growing season (Chapman 1986; Wielgolaski et al. 1981), but repeated harvests of alpine tundra during the growing season indicate that harvest at peak season is a reasonably robust method for estimating primary production (May and Webber 1982).

Efforts to measure belowground production on Niwot Ridge have included the use of both sequential coring and root ingrowth techniques. Sequential coring did not usually reveal a peak in root biomass and hence was not found to be a suitable method for estimating root production in the alpine (Bowman et al. 1993; Fisk et al. 1998). The root ingrowth technique also is subject to recognized problems (Vogt and Persson 1991). Production may be stimulated in ingrowth cores because the soil disturbance creates a favorable soil environment for root proliferation. On the other hand, root production may be underestimated using the ingrowth method because of the disappearance of roots over the measurement period. In northern hardwood forests, rapid disappearance of roots has been observed using minirhizotrons (Hendrick and Pregitzer 1992) and root screens (Fahey and Hughes 1994), and production measured using ingrowth cores was suggested to be underestimated by as much as 50% (Fahey and Hughes 1994). Because decomposition rates are slow in the cool, often dry alpine tundra soils (see chapter 11), we believe root disappearance to be less of a factor than in more-temperate ecosystems. The root ingrowth technique thus should provide a reasonable index of belowground production among alpine tundra communities.

## Variation in Aboveground Production Among and Within Communities

Mean rates of aboveground production for alpine communities on Niwot Ridge range between 100 and 300 g m$^{-2}$ y$^{-1}$ (Table 9.1). The majority of estimates of tundra primary production on Niwot Ridge have been made in or near the Saddle research site (May and Webber 1982; Keigley 1987; Bowman et al. 1993; Walker et al. 1994; Fisk et al. 1998). This site encompasses the full range of snow accumulation and has large areas with relatively uniform species assemblages that are representative of the major plant communities (cf. chapter 6). The rates of production from Niwot Ridge are well within the range of estimates for other alpine sites in the southern Rocky Mountains (Billings and Bliss 1959; Scott and Billings 1964), alpine sites worldwide (Klikoff 1965; Bliss 1966, 1985; Wielgolaski 1975; Rehder 1976; Rehder and Schäfer 1978; Ram et al. 1989), and arctic tundra (Wielgolaski et al. 1981; Webber 1978; Shaver and Chapin 1991). These rates of production also fall within the range of herbaceous-dominated ecosystems of temperate zones, particularly arid grasslands (Lauenroth 1979; Sala et al. 1988).

Table 9.2. Aboveground and Belowground Primary Production Estimates

| | | | Community | | | |
|---|---|---|---|---|---|---|
| Source | Fellfield | Dry Meadow | Moist Meadow | Shrub Tundra | Wet Meadow | Snowbed |
| Aboveground production, Niwot Ridge | | | | | | |
| May and Webber (1982) | 171 ± 16 | 164 ± 13 | 196 ± 13 | 298 ± 87 | 139 ± 28 | 124 ± 10 |
| Bowman et al. (1993) | | 136 ± 8 | | | 299 ± 12 | |
| Walker et al. (1994) | 237 ± 102 | 26 ± 82 | 23 ± 7 | | 162 ± 10 | 97 ± 4 |
| Bowman et al. (1995) | | 134 ± 131 | 167 ± 29 | | | |
| Fisk et al. (1998) | | 155 | 262 | | 291 | |
| Aboveground production, Medicine Bow | | | | | | |
| Scott and Billings (1964) | 129 | 143 | 232 | | 248 | 74 |
| Belowground production (fine roots) | | | | | | |
| Fisk et al. (1998) | | 198 ± 31 | 230 ± 31 | | 364 ± 10 | |

*Note:* Estimates are for Alpine Tundra Communities on Niwot Ridge and a nearby site in the Medicine Bow Mountains of Wyoming. Values are means ± s.e. in g m$^{-2}$ yr$^{-1}$.

What makes the alpine unique with regard to aboveground primary production is the high degree of spatial heterogeneity relative to most other herbaceous dominated ecosystems. At the plot level ($\leq 4$ m$^2$), wet meadow production may be as high as 500 g m$^{-2}$ yr$^{-1}$, while snowbed production may be as low as 50 g m$^{-2}$ yr$^{-1}$ (Bowman et al. 1993; Walker et al. 1994). This range is exclusive of large influences from disturbance and is indicative of the high variation in environmental determinants of production, both biological and climatic, discussed later in this chapter. This heterogeneity is the result of topographic controls on microclimatic variation (chapters 2 and 6), which in turn creates biotic gradients of differing plant dominants, microbial functional types, and rates of animal disturbance (pocket gophers, microtine rodents). Fisk et al. (1998) ascribed much of the variation to landscape gradients in soil water availability, which determined patterns of N cycling. Variation in aboveground production among communities is greater than the temporal variation within a community (Walker et al. 1994).

The lowest rates of aboveground production are found in snowbed communities, and the highest are in moist and wet meadows, particularly those with willow codominants (Table 9.1). The values of fellfield production are probably overestimated, as they include some perennial biomass in the clip harvests (Walker et al. 1994). Estimates of production in the other communities are probably slightly underestimated, as discussed earlier. While the mean production estimates are relatively similar among communities (Table 9.1), there is substantial variation among individual plots within a community and between communities in different sites, as indicated by relatively high standard errors and differences between studies. For example, Walker et al. (1994) selected sites intended to represent mean values for a large landscape, while Bowman et al. (1993) selected dry and wet meadow sites intended to represent extremes in production and site moisture.

Aboveground production in the wet meadow is particularly variable, as it includes sites that are both snow covered and snow free during the winter, and which have different plant growth forms dominating them (sedges, the forb *Caltha leptosepala,* and willows). Wet meadows that are snow free in the winter are particularly susceptible to soil erosion processes (see chapter 8) and may have substantially lower production rates than sites that are snow covered in the winter. Alpine wet meadow communities are significantly different from their counterparts in the Arctic, which have substantially lower rates of production, ca. 50 g m$^{-2}$ y$^{-1}$ (Shaver and Chapin 1991). This difference may be related to the degree of soil anaerobiosis, which may be higher in arctic wet meadow soils, and the extent of water movement through the soil, which due to topographic influences would be higher in alpine soils, facilitating the supply of nutrients (Giblin et al. 1991).

## *Belowground Production and Belowground:Aboveground Ratios*

Belowground production has been measured in only a subset of the plant communities found on Niwot Ridge, but it appears to share the spatial heterogeneity noted for aboveground production in the alpine. Fisk et al. (1998) found an increase in fine root production across topographic sequences of dry, moist, and wet meadows

(Table 9.1). Based on this single estimate of root production and the various estimates of aboveground production listed in Table 9.1, total production (above- plus fine root) varies between 332 and 424 g m$^{-2}$yr$^{-1}$ in dry meadows, between 397 and 492 g m$^{-2}$yr$^{-1}$ in moist meadows, and between 526 and 663 g m$^{-2}$yr$^{-1}$ wet meadows. The domination of annual production by the belowground component leads to a consistent increase in total production from dry to wet meadows, despite lower differences in aboveground production among these communities (Table 9.1).

Belowground:aboveground ratios for production varied from 1 in moist meadows to 1.6 in wet meadows and 2.3 in dry meadows (Fisk et al. 1998). Greater allocation to root production might be expected in the dry meadow compared with the moist or wet meadows as a result of lower availabilities of soil resources (Bowman et al. 1993; Taylor and Seastedt 1994; Fisk et al. 1998; chapter 10). Differences in aboveground: belowground production ratios may also be related to differences in the proportion plant growth forms making up the different communities (Fisk et al. 1998), with greater belowground production in the graminoid-dominated dry and wet meadows relative to the forb-dominated moist meadow. A greater relative proportion of coarse root production probably occurs in moist communities relative to the graminoid-dominated dry and wet meadows; however, coarse root production was not quantified by Fisk et al. (1998).

Temporal variation in root production is unknown but is likely to be high given the high variation among years in aboveground production (Walker et al. 1994) and in root biomass (Fisk et al. 1998; Table 9.1). A more-complete understanding of landscape patterns of production in alpine tundra thus awaits not only quantification of root production in some of the more extreme environments (fellfield, snowbed) but also evaluation of longer term trends of root production, such as has been done for aboveground production (Walker et al. 1994).

In contrast to belowground production, information regarding belowground biomass in alpine tundra is more complete. Belowground biomass has been quantified in several studies on Niwot Ridge (Table 9.2) and in alpine ecosystems in Europe (Rehder 1976; Rehder and Schäfer 1978). Similar to belowground production, belowground biomass increases from dry to wet meadow communities (Table 9.2). Belowground biomass is greatest in shrub tundra and lowest in snowbeds (Table 9.2). The range of living plus dead root biomass on Niwot Ridge is quite large, from 1500 to 5600 g m$^{-2}$ (Table 9.2). Total belowground biomass varied comparatively less, from 1500 to 2000 g m$^{-2}$, among high-elevation sedge and willow mat communities in the Alps (Rehder 1976; Rehder and Schäfer 1978).

Belowground:aboveground ratios are relatively high, offering further evidence that belowground production is an important component of total net primary production (NPP) in the alpine. In addition, high belowground biomass in alpine tundra suggests the critical importance of belowground resources (water, nutrients) as limiting factors for production. Ratios for total (living plus dead) biomass varied within a similar range (2.5–8.0) in two separate studies on Niwot Ridge (Table 9.2). In high-elevation tundra communities of the Alps, total biomass belowground: aboveground varied from approximately 3.5 to 8.8 (Rehder and Schäfer 1978), very similar to the range found on Niwot Ridge (cf. chapter 10). Higher ratios, from 9 to 21, were found for living belowground:aboveground biomass by May and Webber

(1982) and Bowman (1994), whereas Fisk (1995) found ratios varying only from 3.7 to 4.2 (Table 9.2). Differences in live versus dead determinations are likely responsible for these variable results.

## Environmental Control over Primary Production

Several environmental factors contribute to the high degree of spatial and temporal variation in primary production of alpine tundra on Niwot Ridge, both directly and indirectly. These include physical factors such as light (cloudiness), temperature, and water and nutrient availability, and snow-cover-mediated effects such as growing season length and soil temperatures in both winter and summer. Biotic influences include physiological capacities and developmental constraints influencing the ability of plants to respond to variation in resource availability.

### Light

Photosynthesis rates in alpine plants increase with increasing sunlight up to nearly full strength (2200 $\mu$mol m$^{-2}$ s$^{-1}$ on Niwot Ridge). As a result, variation in cloud cover can impact net carbon gain and biomass production (Scott and Billings 1964; Körner 1982; Diemer and Körner 1996). A complete evaluation of the potential decrease in production associated with cloudiness has not been done for alpine plants of the Rocky Mountains, but Körner (1982) and Diemer and Körner (1996) estimated carbon gain during the growing season was approximately 20% lower than the theoretical maximum (estimated from light and temperature response curves) for two alpine species due to cloudiness in the central Alps. This probably represents an upper limit for the limitation on production associated with cloudiness in the southern Rocky Mountains, as cloudiness is lower than the Alps, and a lack of cloud cover would be associated with lower precipitation and greater evapotranspiration, increasing the probability of water availability limiting production.

The decrease in atmospheric density at high elevation is associated with a potential increase in ultraviolet B radiation (UVB 280–320 nm) hitting the surfaces of plants (Caldwell et al. 1989). However, the actual temporally integrated UVB irradiance may not be higher at higher elevations due to the offsetting increase in cloudiness (Caldwell 1968). Alpine plants tend to exhibit lower susceptibilities to elevated UVB damage than do lowland plants (Robberecht et al. 1980; Caldwell et al. 1982; Sullivan et al. 1992; Rau and Hoffman 1996), but there is little evidence to support the contention that UVB limits production in the alpine. Caldwell (1968) experimentally lowered UVB irradiance in plots on Niwot Ridge and found that a few species had slight increases in growth where UVB was screened, but the majority of plants studied were unaffected.

### Temperature

The alpine is characterized by low air temperature (chapter 2), and since photosynthesis is a temperature-sensitive process, it is logical that variation in production

Table 9.2. Belowground Biomass in g m$^{-2}$yr$^{-1}$ and Aboveground:Belowground Biomass Ratios for Alpine Tundra Communities on Niwot Ridge

| | | | Community | | | |
|---|---|---|---|---|---|---|
| Source | Fellfield | Dry Meadow | Moist Meadow | Shrub Tundra | Wet Meadow | Snowbed |
| Root biomass, total (living plus dead) | | | | | | |
| May and Webber (1982) | 1858 | 2929 | 2563 | 5008 | 3583 | 1502 |
| Fisk et al. (1998) | | 2772 | 3039 | | 5674 | |
| Live | | | | | | |
| May and Webber (1982) | 1534 | 1739 | 2090 | 4398 | 2927 | 1208 |
| Bowman et al. (1993) | | 2300 | | | 3300 | |
| Fisk et al. (1998) | | 575 | 1055 | | 1233 | |
| Below:aboveground total | | | | | | |
| May and Webber (1982) | 2.5 | 3.0 | 2.6 | 3.1 | 5.6 | 5.6 |
| Fisk et al. (1998) | | 4.8 | 3.4 | | 8.0 | |
| Live | | | | | | |
| May and Webber (1982) | 9 | 11 | 11 | 6 | 21 | 10 |
| Bowman et al. (1993) | | 12 | | 17 | | |
| Fisk et al. (1998) | | 3.7 | 4.0 | | 4.2 | |

may be related to temperature variation. Although community-level production response has not been measured directly in warming experiments on Niwot Ridge (chapter 16), the responses of individual species' growth is moderate or nil. Temporal variation in aboveground production on Niwot Ridge over a 9-year period was not correlated with temperature (expressed as thawing degree days) (Walker et al. 1994). This insensitivity of production to temperature variation probably reflects the ability of alpine plants to adjust their photosynthetic temperature optima to the ambient temperature, their relatively broad photosynthetic temperature response curves that maintain rates within 5% of the maximum over an 8°C range, and little, if any, apparent photosynthetic inhibition resulting from high light and low temperature (Körner 1999). Variation in environmental temperature is likely to have a greater indirect influence on production via its influence on soil resource availability (chapter 16).

## *Water Availability*

The wide range of soil moistures in time and space (Taylor and Seastedt 1994; chapter 8) and high vapor pressure deficits from leaf to air (Smith and Geller 1980) are indicative of the potential role of water supply in controlling alpine plant growth. Several studies on Niwot Ridge have implicated soil moisture as an important determinant of primary production. Greenland et al. (1984) estimated potential evapotranspiration for alpine tundra on Niwot Ridge and concluded that the replenishment of soil moisture by summer precipitation was inadequate for maintaining plant demands in several communities. Spatially, patterns of NPP appear to coincide with topographic gradients of soil moisture. Walker et al. (1993) proposed that landscape variation in water availability constrains plant biomass production in the alpine tundra. Fisk et al. (1998) found a highly significant correlation between spatial patterns of soil water availability and their estimates of above- plus belowground production over topographic sequences of dry, moist, and wet meadows.

Within any plant community, temporal variation in soil water availability also has the potential to regulate photosynthesis and thus production. An extensive study undertaken by Walker et al. (1994) found that interannual variation in plant biomass was linked with soil moisture availability in four out of five communities studied, but not always within the same growing season. Aboveground production in moist and wet meadows correlated with the current growing season soil moisture, while fellfield and dry meadow production correlated with the previous growing season soil moisture.

Although these studies clearly show a link between soil moisture and aboveground production, Bowman et al. (1995) failed to increase rates of production by adding supplemental water to dry and moist meadow plots for two growing seasons, when summer precipitation was 10–20% below average. Rates of net N mineralization were greater following the watering treatment, and aboveground biomass was greater with N in combination with water. They concluded that during years of near-normal precipitation without prolonged periods of drought, soil moisture was a more-important determinant of plant growth via microbial controls on nutrient supply rather than a direct determinant of plant growth. Fisk et al. (1998) found soil water supply to be a fundamental control on rates of N cycling across alpine communities, correlating with spatial patterns of production.

The temporal distribution of precipitation during the growing season is also potentially an important determinant of production. Periodic summer droughts of multiple days to weeks can occur even in years of near-normal precipitation, producing a condition of pulsed moisture and nutrient supply interspersed with periods when water can limit production. Differences in the timing of water availability are likely to be partly responsible for topographic patterns of net primary production. For example, Fisk et al. (1998) found that fine root production in wet meadows exceeded that of dry and moist meadows as a result of late season root growth, after significant soil drying had occurred in these latter communities.

Low water availability can decrease photosynthesis rates by lowering stomatal conductance to $CO_2$ and impairing biochemical capacity. Water availability also influences growth via internal plant water relations, as positive turgor is required for cell expansion. Ehleringer and Miller (1975) found evidence for reduced stomatal conductance as a result of decreased water potentials in several species along a soil moisture gradient in the Saddle, during a summer with low precipitation early in the growing season. A number of other studies have also found site- and year-specific indications of water limitations of stomatal conductance and/or photosynthesis in alpine tundra plants, including those of Mooney et al. (1965), Klikoff (1965), Kuramoto and Bliss (1970), and Enquist and Ebersole (1994). Oberbauer and Billings (1981), working in the nearby Medicine Bow Mountains of southern Wyoming, found a close correspondence between stomatal conductance and leaf water potential in alpine species, which was influenced by site position (soil moisture) and rooting depth. They reported differential stomatal sensitivities to changes in leaf water potential among species and populations of the same species. Species from sites with higher soil moisture had higher stomatal conductance, which decreased more per-unit change in leaf water potential than did plants from sites with drier soils (Fig. 9.1). The same held true among populations of *Bistorta bistortoides,* which occur along the moisture gradient. These results indicate that production is more sensitive to changes in soil moisture in moist and wet meadow communities than in dry meadow and fellfield communities, consistent with the results of Walker et al. (1994).

Comparisons of plant water relations between arctic and alpine species generally indicate that Rocky Mountain alpine plants are more drought tolerant than arctic plants are (Johnson and Caldwell 1975). This pattern corresponds with the higher incidence of soil water deficits in Rocky Mountain alpine tundra than in arctic tundras.

Vegetative phenology, and thus the length of the growing season, is influenced by soil moisture availability. Failure of summer rains to replenish soil moisture lost to evapotranspiration can result in early senescence of leaves (Jackson and Bliss 1984), decreasing seasonal photosynthetic carbon gain. This influence of soil moisture on growing season length is most pronounced in dry meadows and fellfields, which are more reliant on summer precipitation for soil moisture supply than are moist meadows, wet meadows, and snowbeds, which receive snowmelt water during part or most of the growing season (Taylor and Seastedt 1994).

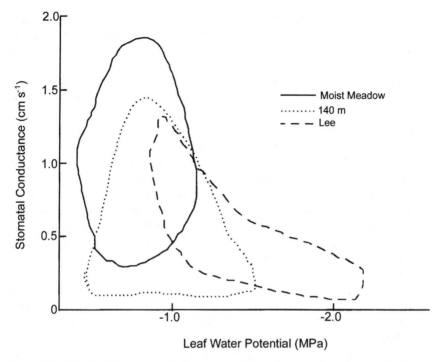

Figure 9.1. Relationship between leaf water potentials and stomatal conductance in *Bistorta bistortoides* from environments differing in soil water availability (moist meadow > 140 m (similar to dry meadow) > lee (similar to fellfield)) (from Oberbauer and Billings 1981).

## Nutrient Availability

The supply of nitrogen to plants is perhaps one of the most ubiquitous constraints on terrestrial primary production, with phosphorus playing a secondary role as a nutrient limiting to production (Vitousek and Howarth 1992). The availabilities of N and P are most often linked with rates of microbially mediated mineralization (Aber and Melillo 1992), which are severely constrained in alpine soils by low temperatures and variable water availability (Rehder and Schäfer 1978; Fisk and Schmidt 1995; chapter 12) and high microbial immobilization of N relative to supply (Fisk et al. 1998). Some arctic and alpine plants may circumvent the mineralization step by using organic solutes for their nutrient requirements (Raab et al. 1996, 1999; chapter 10), but the breadth of the taxa that can do this is unknown.

On Niwot Ridge, Fisk et al. (1998) found a close correlation between topographic position, rates of N cycling, and primary production. They concluded that landscape position was an important control on total production, as soil moisture acts as a strong control on rates of microbial activity, which in turn controls nutrient supply to plants. Although landscape position has also been linked with rates of production in the Arctic, Giblin et al. (1991) and Shaver and Chapin (1991) both provided evidence that production and nutrient cycling do not follow a systematic trend over topographic sequences. Rather, there are relatively large differences among topographic positions that are probably related more to plant chemistry and substrate quality (Nadelhoffer et al. 1991). The soil moisture gradient is more pronounced in alpine tundra (Taylor and Seastedt 1994), resulting in increases in microbial activity, nutrient turnover, and production from dry to wet meadows (Fisk et al. 1998). Similar gradients in NPP generated by topography are also found for prairie sites (Schimel et al. 1991; Benning and Seastedt 1995).

Several experiments indicate that increases in nutrient availability will increase tundra production on Niwot Ridge (Keigley 1987; Bowman et al. 1993, 1995). Fertilization of dry, moist, and wet meadow communities indicate that N and P availability limit plant growth at the community level. However, it is clear that the particular nutrient that limits production varies according to species, and species vary in their capacity to increase biomass when nutrients are added. N most commonly limits production at the community level, while P appears to be secondarily important, except in the wet meadow community, where its importance is equal to that of N.

The production response of alpine tundra to nutrient additions is mediated by a change in community composition rather than by proportional increases in biomass by all species present. This appears to be the result of differential responses among growth forms as well as changes in competitive relations among species. Species-level responses to nutrient additions followed phylogenetic and functional affiliations (Theodose and Bowman 1997). Non-mycorrhizal forbs, $N_2$-fixing forbs, and grasses responded to the addition of P alone, while sedges, mycorrhizal forbs, and grasses in some sites responded to N alone. The response of the dry meadow to N additions was therefore primarily an increase in grass and forb biomass, while the dominant sedge *Kobresia* contributed a substantially lower proportion of total community biomass (Fig. 9.2). The increase in production in this community following 2 years of N fertilization was 10–20% greater than would have been if there were simply a proportional increase in biomass by the existing dominant species. Thus, although nutrient amendments lead to increases in production in alpine tundra, it can be argued that this effect is primarily the result of changes in species composition following fertilization (Körner 1989; Theodose and Bowman 1997), and that in essence the biotic control over variation in production is greater than the influence of nutrient or water supply is. This has important consequences for ecosystem responses to environmental change (Bowman 2000; chapter 16).

Competitive exclusion is one mechanism that may alter community composition and biomass production following fertilization (Theodose and Bowman 1997). For example, forb biomass in fertilized wet meadow plots becomes increasingly negatively related to graminoid biomass through time, as sedges and grasses overtop and shade out forbs or compete more effectively for other belowground resources (Fig. 9.3). Competition may have also lead to the more than 50% drop in *Kobresia* cover in dry meadow communities after 6 years of fertilization.

## *Growing Season Length*

The alpine growing season is extremely short and has long been assumed to be an important determinant of primary production. Phenology of alpine plants is linked

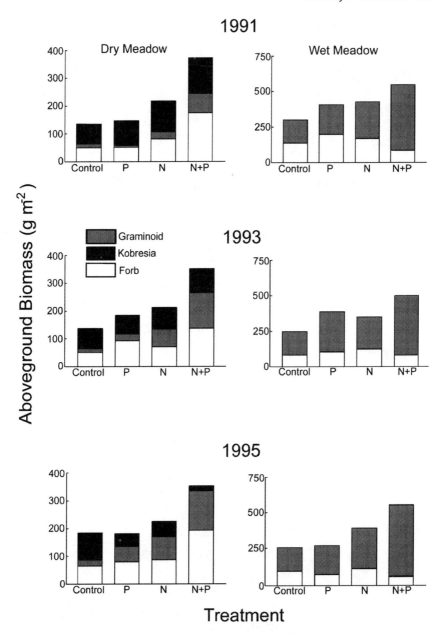

Figure 9.2. Temporal trend in aboveground production in dry and wet meadow communities under N, P, N + P fertilization treatments and control conditions. Fertilization of the plots was initiated in 1990 (Bowman et al. 1993). Note the decreasing biomass of *Kobresia* in the dry meadow and forbs in the wet meadow and the increasing biomass of graminoids in both communities under fertilized conditions.

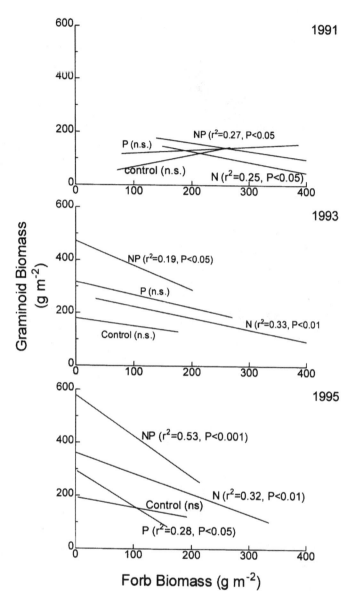

Figure 9.3. Relationship between forb and graminoid biomass in wet meadow plots under N, P, N + P fertilization treatments and control conditions. Fertilization of the plots was initiated in 1990 (Bowman et al. 1993). The increasing statistical significance and slope of the negative correlation between these two growth forms through time indicates competitive exclusion is occurring under fertilized conditions.

with the timing of snowmelt, soil moisture, and internal control (Billings and Bliss 1959; Holway and Ward 1965; May and Webber 1982; Walker et al. 1995). Vegetative emergence, and thus the initiation of photosynthesis, is linked with the disappearance of snow and thawing of the soil. The potential for photosynthetic compensation due to delayed emergence of leaves has not been examined, but Walker et al. (1995) found *Bistorta* plants produced larger leaves during high snow years when growth was delayed, potentially compensating for the shortened growing season.

Controls on plant senescence are poorly understood in alpine tundra. While day length, air temperatures, and soil moisture and nutrients all probably play roles in initiating leaf senescence, internal developmental controls are also important. *Saxifraga rhomboidea* leaves begin to senesce in mid-July, and *Acomastylis rossii* leaves in the fellfield and dry meadow take on a bright red hue in late July, prior to the onset of the first hard frosts and independent of drought conditions (Spomer and Salisbury 1968). May and Webber (1982) suggested that late phenological events are less temporally and spatially variable than is vegetative emergence, and thus biotic control may be a stronger constraint on the end of the growing season than climatic cues are.

## Physiological and Developmental Control in Plants

As discussed above, alpine plants vary in their capacity to respond to variation in the availability of resources and thus to climatic variation controlling resource supply. The biotic controls over plant production responses fall generally into the categories of physiological processes and developmental constraints.

Variation in field photosynthetic rates among alpine plants does not correlate with their abundance, but differential increases in photosynthesis rates following fertilization corresponds with the change in biomass of different growth forms (Bowman et al. 1993, 1995). Leaf gas exchange in moist meadow species appears to be more responsive to fertilization than that in dry meadow species, possibly as a result of more-conservative stomatal control in dry meadow plants. Increases in photosynthetic rates following fertilization were related more to increases in stomatal conductance than to increases in leaf N (Bowman and Conant 1994; Bowman et al. 1995). In general, however, increases in whole-plant photosynthesis, resulting from an increase in leaf area, is a more important response to increases in nutrient availability than is the increase in photosynthesis rates per unit area of leaf.

The capacity to increase growth following uptake of nutrients, or nutrient-use efficiency, varies among alpine plant species, with graminoids generally exhibiting greater N-use efficiency and P-use efficiency than forbs do (Bowman 1994; Bowman et al. 1995; see chapter 10 for a more thorough discussion). Forbs generally are thought to have more-conservative patterns of vegetative development than graminoids have, and thus cannot translate increased resource uptake into growth as readily as graminoids can. Forbs rely heavily on storage of resources to meet growth demands (Mooney and Billings 1960; Jaeger and Monson 1992) and exhibit a substantial amount of luxury consumption following fertilization (Lipson et al. 1996).

Biomass and nutrient allocation to different plant organs vary significantly among alpine species (see Theodose et al. 1996; chapter 10). The ability to alter

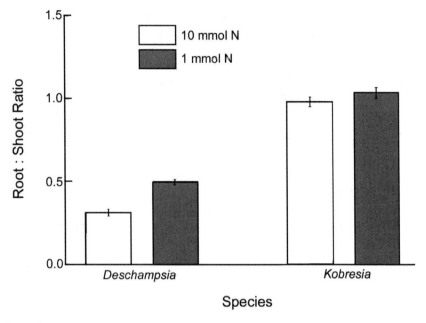

Figure 9.4. Root:shoot ratios under low and high N treatments (1 mmol and 10 mmol N, applied as $NH_4NO_3$ in a growth chamber experiment), indicating the relative plasticity of *Deschampsia* and the inflexibility of *Kobresia* to alter biomass allocation as a result of site fertility (data from Theodose 1995).

these allocation patterns in response to changes in resource availability also varies among species. This variation is exemplified by two dominant graminoids, *Kobresia myosuroides* from dry meadows and *Deschampsia caespitosa* from moist meadows (Theodose 1995). *Deschampsia* allocates more biomass to shoots than to roots and will increase belowground allocation as N and water availability decrease, whereas *Kobresia* maintains a higher and invariant root:shoot ratio (Fig. 9.4). The capacity to alter biomass allocation is an important component of response to changes in resource availability, potentially enhancing growth and reproduction (Poorter 1989), but it may not be advantageous in alpine environments, where changes in resource availability are unpredictable during the growing season (e.g., linked with summer precipitation events), and chronic resource shortage may be common.

Many arctic and alpine plants preform vegetative buds at least one year prior to emergence (Sørensen 1941; Diggle 1997; Aydelotte and Diggle 1997). Preformation has been described most extensively in forb species, although it may also occur in some sedge species, such as *Kobresia* (Bell and Bliss 1979). The impact of this preformation on plant growth is a dampening of the plant response to resource variation. For example, alpine willows will produce more leaves per shoot during a second year of fertilization, but not during the first year (Bowman and Conant 1994). *Polygonum* (*Bistorta*) *viviparum* requires 4 years for full leaf and flower development (Diggle 1997), and *Caltha leptosepala* has a 2–3 year lag between leaf initiation and maturation (Aydelotte and Diggle 1997). This predetermination of leaf number may help explain the lag between climate and production described in Walker et al. (1994). However, alpine plants may compensate for this delay in leaf development by producing larger leaves in years with higher resource availability (Bowman and Conant 1994; Walker et al. 1995). Not all alpine plants exhibit leaf and flower preformation, as sedges and grasses appear to have the capacity to form new vegetative meristems within a growing season and are more responsive to variation in resource supply.

In addition to bud preformation, reduced cell numbers per organ appears to be a common growth constraint in alpine plants. This explains in part the small size of alpine species relative to low elevation congeners (Körner and Pelaez Menendez-Riedl 1989).

Species that exhibit greater capacities to alter patterns of biomass and nutrient allocation to roots and shoots, and lower degrees of preformation of leaves, should be more responsive to variation in resource availabilities than those with conservative allocational and developmental patterns. This can be seen in the responses of different growth forms to fertilization, which indicate that grasses are more responsive than sedges and forbs are (Theodose and Bowman 1997). These differential responses to changes in resource availability have important implications for the response of tundra to increases in anthropogenic N deposition and climate change. Changes in the abundance of species can have important feedbacks to production, nutrient cycling, water use, and energy balance in alpine tundra (Bowman 2000; chapter 16).

Disturbance of alpine tundra plants and soils has both direct and indirect effects on primary production. While the direct effects have not been well quantified, removal of plant tissues by herbivores (chapter 14), burial of plants by gopher mounds (Thorn 1982; Cortinas and Seastedt 1996; chapter 8), and damage of plant tissues by freezing or ice needle formation would all detrimentally influence plant production. Indirect effects mediated by stimulation of soil microbial activity would include soil warming following removal of the plant canopy, mixing of soil horizons by pocket gophers, and increasing litter quality (lower C:N ratios) when clipped vegetation or feces is deposited on the soil surface. Soil moisture and nutrient retention decreases as a result of changes in soil texture and density following gopher disturbance (Litaor et al. 1996). The long-term influences of disturbance on alpine tundra are poorly understood, but are likely to be important in explaining spatial variation in production and plant species composition within and among alpine communities (Chambers et al. 1990).

## Summary and Conclusions

While rates of primary production are low relative to other ecosystems, the variation in rates among sites and communities and diversity of environments and biota in alpine tundra make it a dynamic system for study. Important controls on primary production include physical environmental factors, such as water and nutrient availability and growing season length, and species-level biological controls, such as physiological and developmental capacities to respond to resource variation. These factors work as mosaics across the alpine landscape, influencing the observed temporal variation in production associated with variation in climate. Plant species exhibit different responses to environmental factors, including availability of specific nutrients, which, along with differential sensitivity to water stress among species, likely contribute to species diversity in alpine tundra. Many questions related to biotic control of production and other ecosystem functions are beginning to be addressed in the alpine and are facilitated by the variation in plant dominants within similar microclimates. The outcome of such inquiries will provide a better understanding of biotic feedbacks to environmental variation and their subsequent short- and long-term impacts on primary production.

References

Aber, J. D., and J. M. Melillo. 1992. *Terrestrial ecosystems.* Philadelphia: Saunders.

Aydelotte, A. R., and P. K. Diggle. 1997. Analysis of developmental preformation in the alpine herb, *Caltha leptosepala. American Journal of Botany* 84:1646–1657.

Bell, K. L., and L. C. Bliss. 1979. Autecology of *Kobresia belardii:* Why winter snow accumulation limits local distribution. *Ecological Monographs* 49:377–402.

Benning, T. L., and T. R. Seastedt. 1995. Landscape-level interactions between topoedaphic features and nitrogen limitation in tallgrass prairie. *Landscape Ecology* 10:337–348.

Billings, W. D. 1973. Arctic and alpine vegetations: Similarities, differences, and susceptibility to disturbances. *BioScience* 23:697–704.

Billings, W. D., and L. C. Bliss. 1959. An alpine snowbank environment and its effect on vegetation, plant development and productivity. *Ecology* 40:389–397.

Bliss, L. C. 1966. Plant productivity in alpine microenvironments on Mount Washington, New Hampshire. *Ecological Monographs* 36:125–155.

———. 1985. Alpine. In *Physiological ecology of North American plant communities,* edited by B. F. Chabot and W. D. Billings. New York: Chapman and Hall.

Bowman, W. D. 1994. Accumulation and use of nitrogen and phosphorus following fertilization in two alpine tundra communities. *Oikos* 70:261–270.

Bowman, W. D. 2000. Biotic controls over ecosystem response to environmental change in alpine tundra of the Rocky Mountains. *Ambio* 29:396–400.

Bowman, W. D., and R. T. Conant. 1994. Shoot growth dynamics and photosynthetic response to increased nitrogen availability in the alpine willow *Salix glauca. Oecologia* 97:93–99.

Bowman, W. D., T. A. Theodose, and M. C. Fisk. 1995. Physiological and production responses of plant growth forms to increases in limiting resources in alpine tundra: Implications for differential community response to environmental change. *Oecologia* 101:217–227.

Bowman, W. D., T. A. Theodose, J. C. Schardt, and R. T. Conant. 1993. Constraints of nutrient availability on primary production in two alpine communities. *Ecology* 74:2085–2098.

Caldwell, M. M. 1968. Solar ultraviolet radiation as an ecological factor for alpine plants. *Ecological Monographs* 38:243–268.

Caldwell, M. M., R. Robberecht, R. S. Nowak, and W. D. Billings. 1982. Differential photosynthetic inhibition by ultraviolet radiation in species from the arctic-alpine life zone. *Arctic and Alpine Research* 14:195–202.

Caldwell, M. M., A. H. Teramura, and M. Tevini. 1989. The changing ultraviolet climate and the ecological consequences for higher plants. *Trends in Ecology and Evolution* 4:363–367.

Chambers, J. C., J. A. MacMahon, and R. W. Brown. 1990. Alpine seedling establishment: the influence of disturbance type. *Ecology* 71:1323–1341.

Chapman, S. B. 1986. Production ecology and nutrient budgets. In *Methods in Plant Ecology*, edited by P. D. Moore and S. B. Chapman. Oxford, England: Alden Press.

Cortinas, M. R., and T. R. Seastedt. 1996. Short- and long-term effects of gophers (*Thomomys talpoides*) on soil organic matter dynamics of alpine tundra. *Pedobiologia* 40:162–170.

Diemer, M., and Ch. Körner. 1996. Lifetime carbon balances of herbaceous perennial plants from low and high altitudes in the central Alps. *Functional Ecology* 10:33–43.

Diggle, P. K. 1997. Extreme preformation in alpine *Polygonum viviparum:* An architectural and developmental analysis. *American Journal of Botany* 84:154–169.

Ehleringer, J., and P. C. Miller. 1975. Water relations of selected plant species in the alpine tundra, Colorado. *Ecology* 56:370–380.

Enquist, B. J., and J. J. Ebersole. 1994. Effects of added water on photosynthesis of *Bistorta vivipara:* The importance of water relations and leaf nitrogen in two alpine communities, Pikes Peak, Colorado, USA. *Arctic and Alpine Research* 26:29–34.

Fahey, T. J., and J. W. Hughes. 1994. Fine root dynamics in a northern hardwood forest ecosystem, Hubbard Brook Experimental Forest, NH. *Journal of Ecology* 82:533–548.

Fisk, M. C. 1995. Nitrogen dynamics in an alpine landscape. Ph.D. diss., University of Colorado, Boulder.

Fisk, M. C., and S. K. Schmidt. 1995. Nitrogen mineralization and microbial biomass N dynamics in three alpine tundra communities. *Soil Science Society of America Journal* 59:1036–1043.

Fisk, M. C., S. K. Schmidt, and T. R. Seastedt. 1998. Topographic patterns of above- and belowground production and nitrogen cycling in alpine tundra. *Ecology* 79:2253–2266.

Giblin, A. E., K. J. Nadelhoffer, G. R. Shaver, J. A. Laundre, and A. J. McKerrow. 1991. Biogeochemical diversity along a riverside toposequence in arctic Alaska. *Ecological Monographs* 61:415–436.

Greenland, D., N. Caine, and O. Pollak. 1984. The summer water budget and its importance in the alpine tundra of Colorado. *Physical Geography* 5:221–239.

Hendrick, R. L., and K. S. Pregitzer. 1992. The demography of fine roots in a northern hardwood forest. *Ecology* 73:1094–1104.

Holway, J. G., and R. T. Ward. 1965. Phenology of alpine plants in northern Colorado. *Ecology* 46:73–83.

Jackson, L. E., and L. C. Bliss. 1984. Phenology and water relations of three plant life-forms in a dry treeline meadow. *Ecology* 65:1302–1314.

Jaeger, C. H., and R. K. Monson. 1992. Adaptive significance of nitrogen storage in *Bistorta bistortoides,* an alpine herb. *Oecologia* 92:578–585.

Johnson, D. A., and M. M. Caldwell. 1975. Gas exchange of four arctic and alpine tundra plant species in relation to atmospheric and soil moisture stress. *Oecologia* 21:93–108.

Keigley, R. B. 1987. Effect of experimental treatments on *Kobresia myosuroides* with implications for the potential effect of acid deposition. Ph.D. diss., University of Colorado, Boulder.

Klikoff, L. G. 1965. Photosynthetic response to temperature and moisture stress of three timberline meadow species. *Ecology* 46:516–517.

Komárková, V. 1979. Alpine vegetation of the Indian Peaks area, Front Range, Colorado Rocky Mountains. *Flora et Vegetatio Mundi* Bd VII. Vaduz: Liechtenstein J. Cramer.

Körner, Ch. 1982. $CO_2$ exchange in the alpine sedge Carex curvula as influenced by canopy structure, light, and temperature. *Oecologia* 53:98–104.

———. 1989. The nutritional status of plants from high altitudes. *Oecologia* 81:379–391.
———. 1999. *Alpine plant live, functional plant ecology of high mountain ecosystems.* Berlin: Springer.
Körner, Ch., and S. Pelaez Menendez-Riedl. 1989. The significance of developmental aspects in plant growth analysis. In *Variation in growth rate and productivity of higher plant,* edited by H. Lambers, M. L. Cambridge, H. Konings, and T. L. Pons. The Hague: SPB Academic Publishing.
Kuramoto, R. T., and L. C. Bliss. 1970. Ecology of subalpine meadows in the Olympic Mountains, Washington. *Ecological Monographs* 40:317–347.
Lauenroth, W. K. 1979. Grassland primary productivity. In *North American grasslands in perspective,* edited by N. R. French. Ecological Studies 32. New York: Springer-Verlag.
Lieth, H., and R. H. Whittaker. 1975. *Primary production of the biosphere.* New York: Springer-Verlag.
Lipson, D., R. K. Monson, and W. D. Bowman. 1996. Luxury uptake and storage of nitrogen in the rhizomatous alpine herb, *Bistorta bistortoides. Ecology* 77:1277–1285.
Litaor, M. I., R. Mancinelli, and J. C. Halfpenny. 1996. The influence of pocket gophers on the status of nutrients in alpine soils. *Geoderma* 70:31–48.
May, D. E., and P. J. Webber. 1982. Spatial and temporal variation of vegetation and its productivity on Niwot Ridge, Colorado. In *Ecological studies in the Colorado alpine: A festschrift for John W. Marr,* edited by J. Halfpenny. Occasional Paper 37, Institute of Arctic and Alpine Research, University of Colorado, Boulder.
Mooney, H. A., and W. D. Billings. 1960. The annual carbohydrate cycle of alpine plants as related to growth. *American Journal of Botany* 47:594–598.
Mooney, H. A., R. D. Hillier, and W. D. Billings. 1965. Transpiration rates of alpine plants of the Sierra Nevada of California. *American Midland Naturalist* 74:374–386.
Nadelhoffer, K. J., A. E. Giblin, G. R. Shaver, and J. A. Laundre. 1991. Effects of temperature and substrate quality on element mineralization in six arctic soil. *Ecology* 72:242–253.
Oberbauer, S., and W. D. Billings. 1981. Drought tolerance and water use by plants along an alpine topographic gradient. *Oecologia* 50:325–331.
Poorter, H. 1989. Interspecific variation in relative growth rate: on ecological causes and physiological consequences. In *Variation in growth rate and productivity of higher plant,* edited by H. Lambers, M. L. Cambridge, H. Konings, and T. L. Pons. The Hague: SPB Academic Publishing.
Raab, T. K., D. A. Lipson, and R. K. Monson. 1996. Non-mycorrhizal uptake of amino acids by roots of the alpine sedge *Kobresia myosuroides:* Implications for the alpine nitrogen cycle. *Oecologia* 108:488–494.
Ram, J., J. S. Singh, and S. P. Singh. 1989. Plant biomass, species diversity, and net primary production in a central Himalayan high altitude grassland. *Journal of Ecology* 77:456–468.
Rau, W., and H. Hofmann. 1996. Sensitivity to UV-B of plants growing in different altitudes in the Alps. *Journal of Plant Physiology* 148:21–25.
Rehder, H. 1976. Nutrient turnover in alpine ecosystems. I. Phytomass and nutrient relations in four mat communities in the northern calcareous alps. *Oecologia* 22:411–423.
Rehder, H., and A. Schäfer. 1978. Nutrient studies in alpine ecosystems IV. Communities of the Central Alps and comparative survey. *Oecologia* 34:309–327.
Robberecht, R., M. M. Caldwell, and W. D. Billings. 1980. Leaf ultraviolet optical properties along a latitudinal gradient in the arctic-alpine life zone. *Ecology* 61:612–619.
Sala, O. E., W. J. Parton, L. A. Joyce, and W. K. Lauenroth. 1988. Primary production of the central grassland region of the United States. *Ecology* 69:40–45.

Schimel, D. S., T. G. F. Kittel, A. K. Knapp, T. R. Seastedt, W. J. Parton, and V. B. Brown. 1991. Physiological interactions along resource gradients in a tallgrass prairie. *Ecology* 72:672–684.
Scott, D., and W. D. Billings. 1964. Standing crop and productivity of an alpine tundra. *Ecological Monographs* 34:243–270.
Shaver, G. R., and F. S. Chapin III. 1991. Production: Biomass relationships and element cycling in contrasting arctic vegetation types. *Ecological Monographs* 61:1–31.
Smith, W. K., and G. N. Geller. 1980. Plant transpiration at high elevations: theory, field measurements, and comparisons with desert plants. *Oecologia* 41:109–122.
Sørensen, T. 1941. Temperature relations and phenology of the northeast Greenland flowering plants. *Meddelelser øm Gronland* 125:1–305.
Spomer, G. G., and F. B. Salisbury. 1968. Eco-physiology of *Geum turbinatum* and implications concerning alpine environments. *Botanical Gazette* 129:33–49.
Sullivan, J. H., A. H. Termura, and L. H. Ziska. 1992. Variation in UV-B sensitivity in plants from a 3,000-m elevational gradient in Hawaii. *American Journal of Botany* 79:737–743.
Taylor, R. V., and T. R. Seastedt. 1994. Short- and long-term patterns of soil moisture in alpine tundra. *Arctic and Alpine Research* 26:29–34.
Theodose, T. A. 1995. Interspecific plant competition in alpine tundra. Ph.D. diss., University of Colorado, Boulder.
Theodose, T. A., and W. D. Bowman. 1997. Nutrient availability, plant abundance, and species diversity in two alpine tundra communities. *Ecology* 78:1861–1872.
Theodose, T. A., C. H. Jaeger, W. D. Bowman, and J. C. Schardt. 1996. Uptake and allocation of $^{15}N$ by alpine tundra plants: Implications for the role of competition in community structure in a stressful environment. *Oikos* 75:59–66.
Thorn, C. E. 1982. Gopher disturbance: Its variability by Braun-Blanquet vegetation units in the Niwot Ridge alpine tundra zone, Colorado Front Range, USA. *Arctic and Alpine Research* 14:45–51.
Vitousek, P. M., and R. W. Howarth. 1991. Nitrogen limitation on land and in the sea: How can it occur? *Biogeochemistry* 13:87–115.
Vogt, K. A., and H. Persson. 1991. Measuring growth and development of roots. In *Techniques and approaches in forest tree ecophysiology*, edited by J. P. Lassoie and T. M. Hinkley. Boca Raton, FL: CRC Press.
Walker, D. A., J. C. Halfpenny, M. D. Walker, and C. A. Wessman. 1993. Long-term studies of snow-vegetation interactions. *BioScience* 43:287–301.
Walker, M. D., R. C. Ingersoll, and P. J. Webber. 1995. Effects of interannual climate variation on phenology and growth of two alpine forbs. *Ecology* 76:1067–1083.
Walker, M. D., P. J. Webber, E. H. Arnold, and D. E. May. 1994. Effects of interannual climate variation on aboveground phytomass in alpine vegetation. *Ecology* 75:393–408.
Webber, P. J. 1978. Spatial and temporal variation of the vegetation and its productivity. In *Vegetation and production ecology of an Alaskan arctic tundra*, edited by L. L. Tieszen. New York: Springer-Verlag.
Wielgolaski, F. E. 1975. Primary productivity of alpine meadow communities. In *Fennoscandian tundra ecosystems*. Pt. 1. *Plants and microorganisms*, edited by F. E. Wielgolaski. Berlin: Springer-Verlag.
Wielgolaski, F. E., L. C. Bliss, J. Svoboda, and G. Doyle. 1981. Primary production of tundra. In *Tundra ecosystems: A comparitive analysis*, edited by L. C. Bliss, O. W. Heal, and J. J. Moore. Cambirdge: Cambridge University Press.
Zak, D. R., D. Tilman, R. R. Parmenter, C. W. Rice, F. M. Fisher, J. Vose, D. Milchunas, and C. W. Martin. 1994. Plant production and soil microorganisms in late-successional ecosystems: A continental-scale study. *Ecology* 75:2333–2347.

# 10

# Plant Nutrient Relations

Russell K. Monson
Renée Mullen
William D. Bowman

**Introduction**

Alpine soils do not generally exhibit high levels of inorganic fertility, which is the result of inadequate mineralization of organic litter, a consequence of the cool, short alpine growing season (Rehder and Schäfer 1978; Gokceoglu and Rehder 1977; Rehder 1976a, 1976b; Fisk and Schmidt 1995; chapters 11, 12). Slow mineralization rates, in turn, result in a soil that is high in organic humus, and more likely than the soil of other ecosystems, to sequester and bind inorganic nutrients, especially N and P. Accordingly, alpine plants are exposed to a difficult situation in their efforts to obtain the inorganic ions required to support growth and reproduction.

In accommodating the relative infertility of alpine soils, plants rely on a number of different traits, some of which are ubiquitous and some of which are more restricted in their distribution. Biomass allocation patterns favor high root:shoot ratios, increasing the potential for nutrient absorption by the roots relative to nutrient utilization by the shoot. Nutrient-use efficiencies (biomass produced per mass of senescent nutrient) tend to be high in alpine plants due to efficient resorption prior to leaf senescence. In several alpine growth forms, strict internal controls over seasonal phenology and growth (e.g., preformed buds and strongly enforced dormancy patterns) bring growth demands for nutrients more into balance with the limited supply provided by the soil. Luxury uptake and long-term storage during pulses of high nutrient availability provide plants with a means of bridging the gap between incongruent periods of high nutrient supply and high nutrient demand. Association of fungi with the roots of some alpine plants has the potential to enhance N and P acquisition. Finally, some alpine species can overcome the limitations imposed by scarce inorganic nutrient supplies through high rates of organic nutrient assimilation. It is the aim of this chapter to further consider each of these traits, with particular emphasis on their relationship to N and P acquisition. Topics concerning soil processes and their role in controlling nutrient availability have been covered elsewhere (chapter 8) and will not be repeated. Rather, this review focuses on nutrient relations from the plant's perspective.

## Alpine Plant Allocation Patterns in Relation to Nutrient Acquisition

As a group, alpine plants tend to allocate a high fraction of their biomass to shallow root systems. Measured belowground:aboveground live biomass ratios for alpine communities range from 4:1 to 25:1 on Niwot Ridge, depending on community-type (Webber and May 1977; Bowman et al. 1993; chapter 9). These are among the highest values reported for plants, which are typically less than 2:1 but can be as high as 45:1 in plants from arctic tundra (Dennis and Johnson 1970). The cause of higher belowground allocation in alpine plants has been the subject of debate. Dennis and Johnson (1970) have invoked the principle of allocational tradeoff by suggesting that the absence of strong aboveground carbon sinks causes an increase in allocation to belowground sinks. This hypothesis gains support from the fact that many alpine species exhibit heritable aboveground dwarfism, a growth habit that is typically characterized by increased relative belowground allocation (Larcher 1977, cited in Körner and Renhardt 1987). Wielgolaski (1972) has argued that the ratio of turnover rates of belowground structures relative to aboveground structures is lower in alpine and arctic plants compared with that of temperate ecosystems, resulting in a greater relative accumulation of belowground biomass. Webber and May (1977) point out that, compared with many lower elevation ecosystems, alpine growth forms are more likely to possess belowground rhizomes, corms, and rootstocks—structures that, in addition to playing a role in vegetative reproduction, are specialized for nutrient storage. These structures also tend to reflect high tissue biomass densities (i.e., high biomass per unit of surface area), compared with living aboveground structures or the fine roots typically associated with nutrient absorption. All of these hypotheses share the implication that the high belowground fraction on Niwot Ridge is the result of factors unrelated to selection for enhanced nutrient acquisition and should not necessarily be viewed as an adaptation to low fertility.

More recently, Körner and Renhardt (1987) have argued that the high belowground fraction of alpine plants provides advantages toward nutrient acquisition. In a comparative study of related taxa from low and high elevation sites in the Austrian Alps, these workers found that the fractional allocation of biomass to fine roots increased at the expense of allocation to aboveground stem and flower tissues as elevation increased (Fig. 10.1). By allocating more biomass to fine roots and less biomass to stems, the alpine species have effectively increased nutrient supply while decreasing nutrient demand. Concomitant with higher fine root fractions and lower stem fractions, the alpine species exhibited higher tissue N concentrations in roots, stems, and leaves (Fig. 10.1). This study provides evidence of the beneficial effects

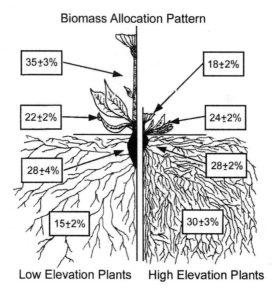
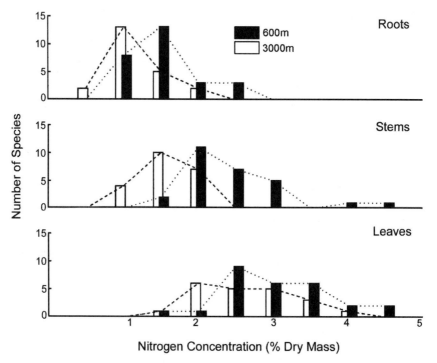

Figure 10.1. Biomass and N allocation patterns summarized for 22 lowland and 27 alpine species in the Austrian Alps. In the upper figure, average percentage allocation values are presented for each biomass compartment. It is obvious that high-elevation plants exhibit higher allocation to fine roots relative to aboveground stems and flowers. In the lower figures, the frequency distribution of N concentration in each of the main biomass compartments is presented. It is obvious that there are more high-elevation species with higher tissue N concentrations (from Körner and Renhardt 1987; Körner 1989).

of a higher allocation to the tissues of nutrient acquisition, when coupled with a lower allocation to the tissues of nutrient utilization, in alpine plants.

The evidence that can be drawn from the Körner and Renhardt (1987) studies to support an advantageous role for an increased belowground fraction in alpine plants, has to be characterized as "indirect" at best. There are still significant uncertainties about the function of alpine root systems. Principle among these is the question of whether belowground biomass in low- and high-elevation plants reflects allocation across comparable time scales. If the allocation pattern of alpine plants is the result of accumulation across several years (due to the slower turnover of alpine roots), but the allocation in low-elevation plants is the result of accumulation across fewer years (due to the faster turnover of lowland roots), then standing crop comparisons can be misleading. Additionally, there are uncertainties about the efficiency with which old roots can acquire nutrients. The slower turnover rate of belowground biomass in alpine species means that much of the belowground standing biomass is composed of old, often suberized, roots. It is likely that age and suberization decrease the efficacy of root nutrient uptake (e.g., Kramer and Bullock 1966; Robards et al. 1979).

The high belowground biomass fraction in alpine plants is typically concentrated in the most shallow soil layers. In a comparison of conspecifics from foothill and alpine habitats near Niwot Ridge, Holch et al. (1941) found shallower rooting depths for the alpine species (Fig. 10.2). In general, alpine species are rooted within the top 30 cm of the soil (Fig. 10.2). This places their roots within the warmest microenvironment of the soil profile, enhancing nutrient uptake rates. Slow decomposition of belowground litter within this zone results in development of a rich humic layer, a process that enhances the capacity for tight cation binding to the organic structure, and forces the roots of alpine plants to live almost entirely in a world dominated by organic molecules. In two community-types from Niwot Ridge, root standing crop biomass (live plus dead roots) was concentrated within the upper 5–10 cm of the soil profile, and humus was concentrated in the upper 40 cm (Fig. 10.3).

## Nutrient-Use Efficiency of Alpine Plants

The capacity to translate soil nutrient uptake into growth is an important ecological character in nutrient-limited environments. Nutrient-use efficiency is defined as the amount of growth per unit of nutrient taken up from the environment (Vitousek 1982; Pastor and Bockheim 1984; Shaver and Melillo 1984; Berendse and Aerts 1987). Although alpine plants generally exhibit slow growth rates and some degree of internal control over growth, they have relatively high nutrient-use efficiencies as a result of effective resorption and storage.

Nutrient-use efficiency consists of three components: (1) photosynthetic nutrient-use efficiency (photosynthesis rate/foliar nutrient concentration), (2) the ability to recycle nutrients (usually expressed as resorption efficiency), and (3) leaf longevity (Berendse and Aerts 1987). Nearly all of the plants of the southern Rocky Mountain alpine flora are herbaceous perennials that die back to ground level after the growing season, and have leaf life spans of 2–2.5 months. Thus, the first two

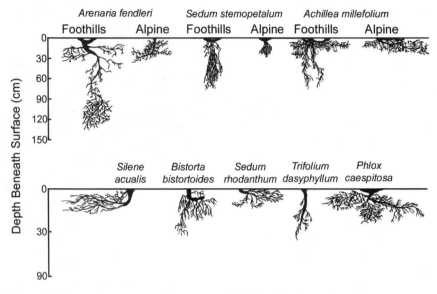

Figure 10.2. Rooting patterns for low- and high-elevation plants in the Colorado Front Range. In the upper panel, rooting depths and patterns are presented for different populations of the same species. In every case, the alpine population exhibited shallower roots. In the lower panel, additional alpine species are shown to illustrate the general shallow nature of these root systems.

components are the most important for determining nutrient-use efficiency of tundra plants on Niwot Ridge.

Compared with lowland plants, concentrations of foliar N and P in alpine plants are relatively high (Körner 1989; Bowman et al. 1993), but photosynthetic rates are comparable (Körner and Diemer 1987). Thus, photosynthetic N- and P-use efficiencies of alpine species are relatively low. Maintenance of higher leaf N and P concentrations in alpine plants may be due to the need for higher photosynthetic enzyme activities, given the lower $CO_2$ partial pressures at high elevation (Körner and Diemer 1987) and/or increased diffusive resistances to $CO_2$ (Terashima et al. 1995).

Alpine plants exhibit relatively high nutrient resorption efficiencies. Between 40 and 75% of the N and approximately 50% of the P in the current season's foliage is potentially available to support growth during the following year (Bowman 1994; Bowman et al. 1995). Generally, the capacity to resorb nutrients is related to leaf longevity and source-sink relations (Chapin and Kedrowski 1983; del Arco et al. 1991; Chapin and Moilanen 1991). A high nutrient storage capacity in many alpine species enhances potential sink strength and facilitates resorption from senescing nutrient sources (i.e., senescing leaves, flowers, and fine roots).

Variation among alpine species on Niwot Ridge in nutrient-use efficiencies can, to a large extent, be explained by growth form and the plant community in which they are found. Forbs tend to exhibit lower photosynthetic N-use efficiency and N-production efficiencies (biomass production/peak season nutrient concentration)

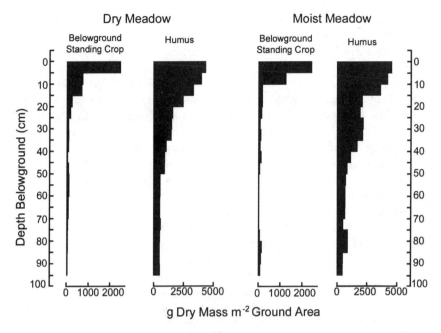

Figure 10.3. Belowground patterns of standing crop and humus in two communities on Niwot Ridge. The data illustrate the shallow nature of belowground biomass distribution and its influence on high organic accumulation within the upper 50 cm of the soil profile (from Webber and May 1977).

compared to graminoids. This results from higher foliar N concentrations and tighter internal controls over growth in forbs (Bowman et al. 1995). However, there is little difference between graminoids and forbs in N-use efficiency based on external supply requirements, as both have relatively high N resorption efficiencies. At the community level, there is higher nutrient-use efficiency for the specific nutrient that limits production. For example, wet meadow species have higher P-use efficiencies than dry meadow species have, whereas dry meadow species have higher N-use efficiencies than wet meadow species have. This pattern corresponds to a greater N limitation of production in the dry meadow and a greater P limitation of production in the wet meadow (Bowman 1994). While such a response might be expected, since increased growth would occur as a result of uptake of a limiting element, both communities also exhibited greater resorption efficiencies of the most limiting nutrient.

## Internal Controls over Growth in Relation to Alpine Nutrient Limitations

Limitations to growth can occur when demand for nutrients exceeds supply. Balance between demand and supply in alpine plants is often achieved through strict control over meristematic activity (Körner and Larcher 1989; Körner and Palaez Menendez-Riedl 1990; Diggle 1997). This reduces the frequency of short-term imbalance between demand and supply and ensures long-term persistence, a trait that is presumably linked to fitness in alpine environments. It has been suggested that selection for reduced growth rate is the underlying cause of constrained meristematic activity in alpine plants, though not necessarily selection aimed at slowing resource demand and establishing greater balance with resource supply (Körner and Pelaez Menendez-Riedl 1990). Other reasons for selection of reduced growth rate might include the advantages conferred through dwarfism, such as occurs in alpine cushion plants. The cushion growth form flourishes in the warm microclimate next to the ground surface and produces a tightly woven web of branches and leaves that effectively traps litter from its own production or that which is blown in from other plants (Körner and Pelaez Menendez-Riedl 1990).

Whatever the ultimate cause of internal growth controls, the proximal effect is to produce a limited meristematic potential, which has several ramifications in terms of nutrient use. (1) *Growth does not respond quickly to stochastic pulses of soil nutrients.* One corollary to limited meristematic potential is the evolution of "preformation." Many alpine plants have acquired the ability to form vegetative and floral buds up to several years in advance, allowing for rapid expansion each spring (Diggle 1997; Aydelotte and Diggle 1997; chapter 9). Preformation comes at a cost in terms of adjusting growth patterns in response to stochastic resource pulses. Evidence for this cost has been seen recently in studies on Niwot Ridge, including the two-year delayed growth response to fertilizaton in *Salix glauca* (Bowman and Conant 1994) and lack of a growth response to fertilization for over 4 years in *Bistorta bistortoides* (Lipson et al. 1996). (2) *Surplus nutrients acquired through "luxury consumption" may be used inefficiently.* The acquisition of nutrients when supply exceeds demand is termed luxury consumption, or resource accumulation (Millard 1988; Chapin et al. 1990). To capitalize on accumulated resources, at some point demand has to exceed supply. This requirement can be met in one of two ways: Either supply is diminished or demand is enhanced. The internal processes that control meristematic activity would have no control over supply-side dynamics. However, they would have considerable control over demand-side dynamics, principally by controlling plasticity in growth and its response to nutrient surplus. Plasticity should be constrained by internal controls over meristematic activity, including preformation. Thus, by evolving strong controls over meristematic activity, alpine plants may lose some potential for plasticity and utilization of nutrient pulses that are stored following luxury consumption. (3) *Temporal displacement of growth and nutrient availability may decrease nutrient-use efficiency.* Storage of nutrients and later utilization cannot be as efficient as immediate utilization. Storage carries a cost in terms of constructing storage reservoirs, converting resources to chemical forms that are compatible with storage, and transporting compounds to and from the site of storage. Nutrient use would be most efficient if nutrient acquisition and growth occurred simultaneously. Internal controls over meristematic activity can cause growth, and nutrient acquisition to be out of phase, thus reducing nutrient-use efficiency. This may occur in alpine plants, which have high growth rates during the early summer, but because of a need for storage and preformation are reduced later in the season.

In contrast, nutrient availability may be highest during the mid- to late summer when soil temperatures and microbial activity increase. Such patterns may cause plants to rely on storage to support the next summer's early season growth spurt (Jaeger et al. 1999).

Debate exists as to the degree of control exerted by inherent developmental growth constraints in alpine plants. Körner (1989) found evidence in a worldwide survey of alpine plants that internal constraints dominate controls over growth. The evidence to support this view includes high tissue nutrient concentrations and lack of growth response to fertilization. Leaf N content for alpine herbs growing across a range of northern latitudes averaged 3.2% of dry mass in the subarctic zone and 1.3–2.1% of dry mass in the tropical zone, but values as high as 5% have been observed (Körner 1989). The higher N concentration of subarctic, alpine plants was attributed to their shorter growing season. Foliar N concentration in alpine forbs on Niwot Ridge average 3.1% of dry mass, but values as high as 4.2% have been observed (Bowman et al. 1993). Similarly, high concentrations of tissue P can be found in alpine species (Körner 1989; Bowman 1994). These foliar N and P concentrations lie at the extreme of those reported for vascular plants and suggest a lack of nutrient limitation to photosynthesis and growth. Limited responses of growth in alpine species to artificial fertilization has also been offered as evidence that internal growth constraints limit the growth of alpine plants more than external nutrient constraints do (Körner 1989). More recently, fertilization experiments on Niwot Ridge have demonstrated that both growth and foliar N concentration can increase when supplied with additional N (Bowman et al. 1993). These increases, however, tend to lag by one or two growing seasons behind fertilization and are not sustained in the long term. In fertilization experiments in the Austrian Alps and on Niwot Ridge, species replacement is the long-term community-level response to fertilization, not sustained growth of dominant species (Körner 1989; Theodose and Bowman 1997b; chapter 9). When taken together, the results from these experiments demonstrate that although there is evidence of some limitation to growth by external nutrient supply to alpine plants, internal meristematic constraints probably dominate and underlie the high observed tissue nutrient concentrations.

Species with conservative growth patterns may maintain their dominance in communities through a combination of adaptation to low soil resource supplies and manipulation of nutrient cycling. The former is related to an ability to establish persistant populations at low resource supplies (sometimes referred to as R*, the minimum level of resource supply required for population establishment (Tilman 1982)) and to competition for resources despite low relative growth rates (Theodose and Bowman 1997a). Plants may also manipulate soil nutrient cycling to maintain low nutrient supply, thereby limiting competitive exclusion (Tilman 1988). This is exemplified in alpine moist meadow communities, characterized by relatively high soil resource supply, by the dominance of *Acomastylis rossii* (Fig. 10.4) over *Deschampsia caespitosa*. *Acomastylis* has a low relative growth rate, preforms its buds 2–3 years in advance (C. Meloche, unpublished data) and produces high concentrations of phenolics (20% in litter; Steltzer and Bowman 1998), whereas *Deschampsia* has high relative growth rates, high fine root turnover, and responds more than most alpine plants to N fertilization. The phenolics produced by *Acomastylis* stimulate

Figure 10.4. Alpine avens (*Acomastylis rossii*) is characterized by high concentrations of low molecular-weight phenolics, which correspond with low rates of net N mineralization in moist meadow soils (Steltzer and Bowman 1998) (Photo by William D. Bowman).

microbial growth and immobilization of N and are associated with an order of magnitude lower net N mineralization under canopies of *Acomastylis* relative to *Deschampsia*. The growth rate of *Deschampsia* plants grown in pots with *Acomastylis* litter are lower than those grown in pots with *Deschampsia* litter (Bowman and Steltzer, unpublished data), consistent with the suggestion that *Acomastylis* is manipulating soil N supplies to lower competitive exclusion.

## Nutrient Storage in Alpine Plants

Resource storage is defined as the formation of reserves for the purpose of bridging a gap between resource supply and demand (Chapin et al. 1990). Many alpine plants rely on storage, laying down carbohydrate and nutrient reserves in the fall and utilizing those reserves to drive growth the following spring and early summer. Resource storage in alpine plants can occur in belowground organs (roots, rhizomes, or corms) or perennating shoot buds that overwinter near the surface. Autumnal replenishment of stored reserves probably occurs through acquisition of nutrients directly from the soil as well as resorption of nutrients from senescing tissues.

Although the proximate pattern of storage has been well described in alpine plants (e.g., Mooney and Billings 1960), the ultimate reason(s) for the evolution of storage is not as clear. In studies on Niwot Ridge, two hypotheses have been tested with respect to the possible adaptive significance of storage (Jaeger and Monson 1992; Jaeger et al. 1999). One hypothesis was that resource storage in alpine plants enables them to initiate growth during the early spring when soils are cold, thus extending the short alpine growing season. A second hypothesis was that soil microorganisms are such effective competitors for N during the early weeks of the growing season, that alpine plants have evolved the capacity to drive early-season growth with stored reserves, thus avoiding competition. Late in the growing season, when the capacity for microbial immobilization of N is reduced by episodic freeze-thaw events or depletion of labile soil C, plants might replenish their stores of soil nutrients.

Through observations of seasonal N allocation patterns in the alpine herb, *Bistorta bistortoides*, it was shown that plants store appreciable quantities of N in their belowground rhizomes and in preformed shoot buds (up to 30–35 mg N). Each season, 5–15 mg of this N is translocated from the rhizome to the expanding shoot to support growth (Fig. 10.5). By careful accounting of all N pools, it was estimated that stored reserves can account for approximately 60% of the N used to drive seasonal growth (Jaeger and Monson 1992) (Table 10.1). Approximately half of the mobilized reserve came from rhizome storage and half came from the shoot primordia (Table 10.1). In addressing the first hypothesis, the question was asked whether this use of stores extends the growing season of *B. bistortoides* into the cold, early-spring months. The answer was no. Growth *and* use of stored N were initiated during the early summer, after soils began to warm, new roots had formed, and soil N slowly became available. There seemed to be no obvious use of stores during the early, cold part of the growing season. It was concluded that stored N in this species is used to alleviate the internal competition between vegetative and reproductive sinks, both of which reach a maximum during the early summer.

In addressing the second hypothesis, the question was asked whether the use of stored N allows plants to avoid early-spring competition with soil microbes. Once again, the answer was no. The results revealed a consistent pattern in which plants accumulate N early in the growing season, and microbes accumulate N late in the growing season (Jaeger et al. 1999). In fact, soil microbial biomass tends to be low in this ecosystem compared with other ecosystems (chapter 12), and in competition experiments using added $^{15}NO_3^-/^{15}NH_4^+$ at different times during the growing season, microbes consistently immobilized N at only 4–7% the rate of plant uptake (Jaeger et al. 1999). Thus, plants had ample opportunity to obtain soil N resources, without the pressure of microbial competition. Once again, it was concluded that the principal function of stored N was to support the high demands of simultaneous vegetative and reproductive growth during the early portion of the growing season.

Given the reliance of seasonal growth on stored N, it was hypothesized that storage pools limit the growth of *B. bistortoides*. This was studied through field observations of N storage and growth in plants that had been fertilized for 2 years (Lipson et al. 1996). In response to fertilization, rhizome N concentration increased but annual aboveground production did not (first two columns of Table 10.2). Thus, the hypothesis was not supported. The results demonstrated that the additional N assimilated by fertilized plants represented "luxury" N uptake. It was concluded that additional growth was constrained by meristematic preformation, limitations by other nutrients (e.g., P), and/or the prohibitive carbon costs of storing the excess N.

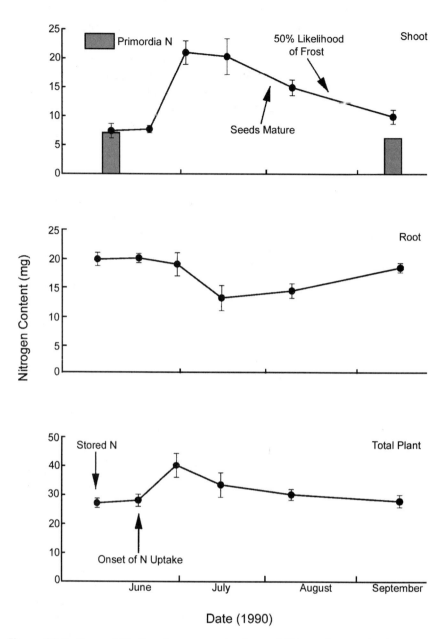

Figure 10.5. Seasonal N allocation patterns in plants of *Bistorta bistortoides* from a dry meadow population on Niwot Ridge. In the upper panel, it is clear that primordia N reserves at the beginning of the season provide a considerable fraction of the N required for seasonal production of aboveground biomass. Key phenological phases are presented to demonstrate the short period in which plants must complete reproductive development. In the middle panel, early season decreases in rhizome N content are obvious as N is translocated from belowground storage reserves to the expanding shoot. Late-season replenishment of the belowground reserve is evident. In the bottom panel, total plant N dynamics are presented, showing the large fraction of stored N used relative to N actually taken up from the soil (from Jaeger and Monson 1992).

Table 10.1. Reliance on Stored N Reserves for Aboveground Seasonal Growth in *Bistorta bistortoides* from Two Communities on Niwot Ridge

|  | Community-Type | |
| --- | --- | --- |
|  | Dry Meadow | Snowbed |
| Total N, from stored reserves | 63 | 60 |
| Mobilized N reserve, from shoot primordia (i.e., aboveground stores) | 33 | 24 |
| Mobilized N reserve, from belowground rhizome (i.e., belowground stores) | 30 | 36 |

*Source:* From Jaeger and Monson (1992).

*Note:* Values are in %.

There is evidence that the uptake and storage of N in excess of that required for growth exerts strong negative feedback on further N acquisition in *B. bistortoides*. The change in N content and N concentration in rhizomes and shoots was assessed between the beginning of the growing season and the time of peak biomass (last four columns of Table 10.2). It is apparent that following luxury uptake and storage, there is an increased reliance on stored N to drive shoot growth at the expense of N taken up during the current season; that is, all of the N required for shoot growth can be accounted for by the decrease in total rhizome N. In the fertilized plants, net N uptake ceased. This is evidenced in the data of Table 10.2 by the good match between decreases in total rhizome N and increases in total shoot N; that is, if net N uptake had continued it should have reduced the decrease in rhizome N below the increase in shoot N. Thus, in response to high levels of stored N a "switch" was activated, such that growth was driven entirely by stored, excess N, rather than by a combination of stored N and N assimilated during the current season.

Studies of the biochemistry and cellular properties in *B. bistortoides* have provided insight into the carbon costs of nitrogen storage (Lipson et al. 1996). It was shown that in this species, N is stored as amino acids, with arginine and the nonprotein amino acid, δ-acetylornithine, representing the principal chemical species (Fig. 10.6). Storage of excess N through luxury uptake did not alter the types of storage compounds used. In studying N storage at the cellular level, two questions were addressed: (1) Can the carbon costs of N storage be observed through reductions in rhizome carbon pools? and (2) when N stores are increased through luxury uptake, are new cells produced or is the increased storage accommodated by increasing the N concentration of existing cells? In answering the first question, the costs of storage *were* reflected in depleted pools of rhizome sucrose, presumably to accommodate the synthesis of stored amino acids (last column in Table 10.3). In answering the second question, all stored N, even that obtained through luxury consumption, was stored in existing cortical and pith cells—new cells were not produced to accommodate additional N storage. It was reasoned that if new cells were produced, an increase in rhizome biomass or a decrease in cell size should have been observed

Table 10.2. Rhizome N Dynamics and Plant Growth in *Bistorta bistortoides* and Results of ANOVA

| Treatment | Rhizome N Concentration (mg N/g dry mass) | Total Annual Aboveground Plant Biomass (g) | Change in Rhizome Dry Mass (g) | Change in Rhizome N Concentration (mg/g) | Change in Total Rhizome N (mg) | Change in Total Shoot N (mg) |
|---|---|---|---|---|---|---|
| Unfertilized | 15.2 ± 1.0 | 0.52 ± 0.05 | 0.00 ± 0.24 | 3.23 ± 1.27 | 5.18 ± 3.56 | 9.21 ± 2.69 |
| Fertilized | 21.5 ± 1.5 | 0.45 ± 0.05 | 0.00 ± 0.36 | 11.68 ± 2.45 | 14.45 ± 8.43 | 13.35 ± 2.53 |
| F-values | 51.11 | 0.07 | 0.92 | 9.83 | 0.29 | 0.09 |
| P | 0.0001 | 0.79 | 0.35 | 0.0057 | 0.59 | 0.30 |

*Source*: From Lipson et al. (1996).

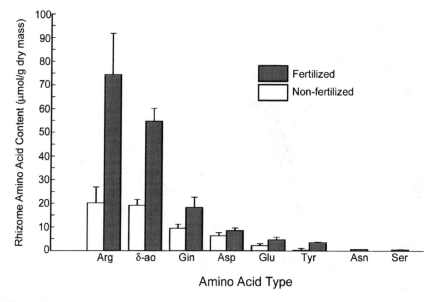

Figure 10.6. The types of amino acids stored in the rhizome of *Bistorta bistortoides* and the response of their concentrations to N fertilization. Arg = arginine; δ-ao = δ-acetylornithine; Gln = glutamine; Asp = aspartate; Glu = glutamate; Tyr = tyrosine; Asn = asparagine; Ser = serine (from Lipson et al. 1996).

as more cells were packed into existing rhizome biomass, which was not found (first three columns of Table 10.3).

Storage and utilization of P has been studied in the Niwot Ridge snowbed plant, *Ranunculus adoneus* (Mullen and Schmidt 1993). This species is one of the earliest to emerge from snowbanks during the spring, often adding substantial amounts of aboveground biomass while still covered with snow. Analyses of root and shoot P contents have shown that concentrations are high during the early spring, declining significantly by midsummer, and undergoing replenishment during the autumn (Fig. 10.7). Autumnal replenishment has been linked to mycorrhizal associations with fungi (see next section).

Table 10.3. Storage Dynamics at the Cellular Level for Rhizomes of *Bistorta bistortoides* and Results of ANOVA

| Treatment | Mass per Cell (ng/cell) | Volume per Cell (pL/cell) | N per Cell (pg N/cell) | Starch per Cell (ng/cell) | Sucrose per Cell (ng/cell) |
|---|---|---|---|---|---|
| Unfertilized | 9.4 ± 0.9 | 22.8 ± 2.8 | 77.3 ± 7.1 | 2.1 ± 0.3 | 0.35 ± 0.04 |
| Fertilized | 7.8 ± 0.8 | 21.0 ± 2.0 | 126.5 ± 17.0 | 1.6 ± 0.2 | 0.19 ± 0.02 |
| F-values | 1.25 | 0.21 | 6.54 | 1.56 | 11.12 |
| P | 0.29 | 0.66 | 0.03 | 0.28 | 0.018 |

*Source:* From Lipson et al. (1996).

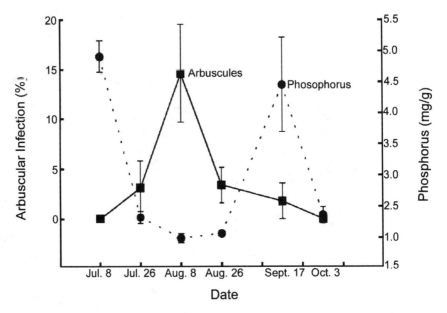

Figure 10.7. The development of arbuscles of the coarse endophyte and shoot P concentrations in *Ranunculus adoneus* growing in the snowbed habitat-type at Niwot Ridge. Note the late season acquisition of P at approximately the same time that storage pools would be replenished (from Mullen and Schmidt, 1993).

## Fungal Associations and the Nutrient Status of Alpine Plants

Although a comprehensive picture of mycorrhizal associations in alpine plants and their significance in alpine systems is far from complete, studies show that most alpine plants are colonized by at least one type of root fungal endophyte (Allen et al. 1987; Gardes and Dahlberg 1996; Haselwandter 1987; Haselwandter and Read 1980, 1982; Lesica and Antibus 1986; Trappe and Luoma 1992; Trappe 1988). The main fungal components of these symbioses belong to one of three groups: (1) arbuscular mycorrhizal (AM) fungi, (2) ectomycorrhizal (ECM) fungi, or (3) dark or dematiaceous septate (DS) fungi (Allen et al. 1987; Haselwandter 1987; Haselwandter and Read 1980 1982; Lesica and Antibus 1986; Trappe 1988).

The most common of mycorrhizal fungi, the AM fungi, have been found throughout alpine habitats, although their presence often varies with community-type, elevation, and season (Allen et al. 1987; Gardes and Dahlberg 1996; Mullen and Schmidt 1993; Read and Haselwandter 1981). AM fungi are well known for their contributions to the phosphorus (P) nutrition of greenhouse-grown plants, although this role cannot necessarily be extrapolated to alpine plants in the field. Thus far, it has not been possible to accurately determine the specific identities of AM fungi or to assess their abundance due to the lack of reproductive structures and the difficulty in associating resting spores in soil with colonizing structures within plant roots. However, molecular methods using specific primers for AM fungi are currently being developed and soon will allow researchers to amplify AM fungal DNA from within plant roots and thus to identify colonizing symbionts.

Ectomycorrhizal fungi are most commonly found colonizing trees and shrubs and thus are less frequent in the alpine ecosystem compared to AM fungi (Gardes and Dahlberg 1996). The dominant ectomycorrhizal fungus found in alpine tundra is *Cenococcum geophilum,* an ascomycete that colonizes alpine shrubs of the genera *Salix* and *Arctostaphylos* as well as the herbs *Kobresia myosuroides, Bistorta viviparum,* and *Dryas octopetala.* The identification of *C. geophilum* as a common alpine associate, however, may be due to its prominent, jet black hyphae and characteristic soil sclerotia, which makes identification especially easy. Using Restriction Fragment Length Polymorphism analysis to examine ECM fungi in *K. myosuroides,* Gardes and Dahlberg (1996) found that roots were colonized by two yet-to-be-identified basidiomycetes in addition to *C. geophilum.* There is much work still to be done in identifying ECM fungi associated with the roots of alpine plants.

A third group of fungi common to alpine root associations is characterized by dark, septate hyphae (hereafter referred to as dark septate, or DS fungi) (Bissett and Parkinson 1979; Currah and Van Dyke 1986; Read and Haselwandter 1982; Stoyke and Currah 1991; Trappe 1988). In alpine ecosystems of the European Alps, DS fungi represent the dominant root endophyte (Haselwandter and Read 1981). In a study of DS colonization in low-elevation and high-elevation plants, Currah and Van Dyke (1986) found that 6% of low-elevation plants and 79% of high-elevation plants were colonized by these fungi. DS fungi have also been found to inhabit the roots of many plant species of the alpine tundra in the Colorado Front Range (Luoma and Trappe, unpublished data). The taxonomic identity of DS fungi is variable and at this time remains a mystery. In a study of isolates taken from subalpine roots, Stoyke et al. (1992) found that 66% were genetically related to the ascomycete *Phialocephala fortinii.* Other more recent studies indicate that the DS fungi may belong to many genera (Harney 1995).

Despite the numerous observations of DS fungi in alpine ecosystems, a clear understanding of their ecology and their effects on the plants they inhabit continues to elude researchers (Read 1993). Results of controlled environment reinoculation experiments have been mixed. In one case, pathogenic effects were attributed to DS fungal infection (Wilcox and Wang 1987), but in the alpine sedge, *Carex firma,* infection increased plant biomass and plant phosphorus uptake (Haselwandter and Read 1982). In another alpine sedge, *C. sempervirens,* infection by the DS fungus produced no growth effect, and the authors speculated that a nutrient other than P limited growth (Haselwandter and Read 1982). As will be discussed, increased levels of DS infection in *R. adoneus* corresponded with increased plant N uptake, suggesting that these fungi may be contributing in some way to plant N nutrition.

*Ranunculus adoneus,* an alpine herb with the somewhat uncommon ability to emerge through snow in full bloom, is common in late-melting snowbeds in the Rocky Mountains. To begin to understand the role of these fungi and their importance to alpine systems, development of AM and other fungal endophytes were monitored seasonally in natural populations of *R. adoneus.* Plant phenology, phosphorus, and nitrogen levels were then related to fungal development. The appearance

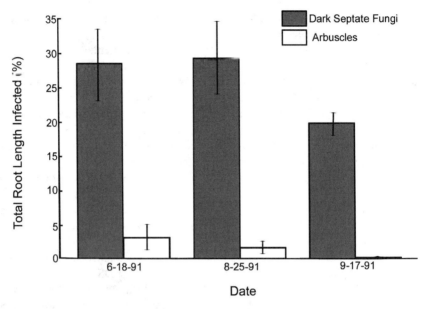

Figure 10.8. Infection levels of dark septate (DS) fungi and VAM arbuscles in roots that had overwintered at least 1 year (percentage of total root length infected ± SE). Note the high abundance of DS fungi and the low abundance of arbuscles during this part of the growing season when N uptake is maximal.

of arbuscles, structures that mediate nutrient transfer and thus indicate an active symbiosis between plant and fungus, corresponded with increased plant P levels late in the growing season (Fig. 10.7; Mullen and Schmidt 1993). P accumulation occurred well after plant reproduction and most plant growth, but approximately at the time of bud preformation and the establishment of storage pools for the next season's growth. Although these data do not unequivocally indicate a causal relationship between fungal colonization, P uptake, and subsequent bud formation, they provide evidence of the probable role of mycorrhizal fungi to these alpine plants.

Nitrogen levels in *R. adoneus* increased dramatically during the early season when soils were still cold and new roots had not formed. As previously described, arbuscular peaks occurred late in the growing season (August), and most N uptake occurred in June. Thus, N uptake does not appear to be related to AM colonization in *R. adoneus*. During this time of maximal N uptake, however, infection by DS fungi was high (Fig. 10.8). Although this could be merely coincidental, DS fungi subsequently isolated from the roots of *R. adoneus*, were able to take up and grow on both inorganic and organic forms of N at temperatures as low as 2°C. These results indicate that DS fungi are able to access the various forms of N that would be available during the flush of N that occurs during snowmelt (Brooks et al. 1996). DS fungal hyphae and sclerotia are found within *R. adoneus* cortical cells, and the plants do not appear to be affected pathologically by the association. It is likely that some type of nutrient transfer occurs between plant and fungus, whether the plant can digest fungal material within cortical cells or the fungus is a weak pathogen; however, the relationship between the two has not yet been elucidated.

## The Uptake of Inorganic versus Organic Nitrogen

To date, there is little information on the forms of soil N absorbed by alpine plants. Plants from most temperate ecosystems preferentially utilize inorganic N ions (i.e., $NO_3^-$ or $NH_4^+$) that are freed from organic matter and secreted as excess nitrogen by carbon-starved microbes. This causes a tight coupling between microbial mineralization and plant N uptake. In alpine ecosystems, however, slow rates of mineralization constrain plant N uptake. In this case, plants may rely on the uptake of organic N to satisfy their N demands.

The uptake of soil organic N is known for a number of plant systems (e.g., Stribley and Read 1980; Abuzinadah and Read 1986a; Abuzinadah et al. 1986b; Chapin et al. 1993; Kielland 1994). Most examples of organic N uptake involve mycorrhizae and include the uptake of small proteins (e.g., Bajwa and Read 1986; Abuzinadah and Read 1986a; Abuzinadah et al. 1986b; Leake and Read 1990; Finlay et al. 1992) and/or amino acids (Stribley and Read 1980; Bajwa and Read 1986). In some cases, the mycorrhizal uptake of amino acids occurs at greater rates than the uptake of inorganic N (Stribley and Read 1980). Most of these studies have dealt with documenting the phenomenon of organic-N uptake by mycorrhizae, the role of fungal proteases in facilitating uptake, and the role of organic-N uptake in plant nutrition under controlled growth conditions.

Recently, it was shown that some non-mycorrhizal plant roots also have the capacity to take up amino acids. Studies of the arctic sedge *Eriophorum vaginatum* concluded that as much as 60% of the N required for annual growth could be taken up as glycine (Chapin et al. 1993). Kielland (1994) described amino acid uptake by excised roots of six arctic species, including deciduous shrubs and three sedge species. (The non-mycorrhizal status of the roots was only certain for the sedge species.) It was estimated that amino acid uptake in these species can supply between 10 and 82% of their annual N requirement.

In studies with the Niwot Ridge alpine sedge, *Kobresia myosuroides,* non-mycorrhizal plants were capable of considerable amino acid uptake (Raab et al. 1996). These experiments were conducted with growth-chamber grown plants. Plants that had been fertilized with glycine exhibited glycine uptake rates that exceeded $NO_3^-$ uptake rates for plants that had been fertilized with $NO_3^-$ (Fig. 10.9). Roots of *K. myosuroides* were not capable of $NH_4^+$ uptake. This is the first known report of a plant whose roots cannot take up $NH_4^+$.

Plants of *K. myosuroides* grown with the amino acid glycine as their sole N source grew faster than did plants grown with the inorganic ions $NO_3^-$ and $NH_4^+$. One could argue that glycine added to the pots was mineralized to $NO_3^-$ and $NH_4^+$ before uptake. However, measurements using stable isotope composition of the plant biomass support the conclusion that intact glycine was taken up. Plants were grown with glycine that carries a distinctive, natural $^{13}C/^{12}C$ signature ($\delta^{13}C$ value of $-35$ ‰). After four months of growth, the $\delta^{13}C$ signature of plants grown with

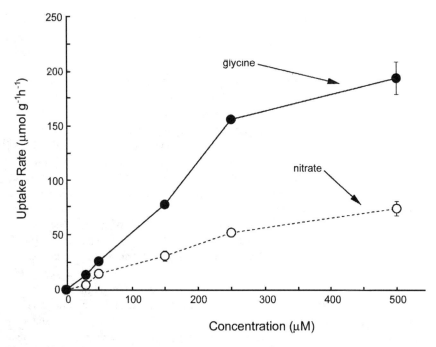

Figure 10.9. The rate of glycine uptake by plants of *Kobresia myosuroides* grown with glycine as the sole N source or the rate of nitrate uptake by plants of *K. myosuroides* grown with $NH_4^+/NO_3^-$ as the sole N source as a function of $NO_3^-$ or glycine concentration in the rooting medium. The results illustrate the higher rates of glycine uptake at all concentrations (from Raab et al. 1996).

the labeled glycine was $-30.7 \pm 0.2‰$, compared to $-29.8 \pm 0.1‰$ for the control plants grown with $NH_4NO_3$. Using the two-member mixing model (Treseder et al. 1995), it was estimated that the uptake of intact glycine contributed 16% of the carbon assimilated into biomass over this 4-month period. (This should be an underestimate of the amount of glycine actually absorbed since some might have been lost through respiration.)

The results presented here only have ecological meaning if amino acids are a principal component of alpine soils. Measurements of $NO_3^-$, $NH_4^+$, and amino acids were made over two growing seasons at two dry-meadow, *Kobresia*-dominated sites using microlysimeters (Raab et al. 1996). Results demonstrated that at one site with shallow soils characterized by average soil organic contents, free amino acid concentrations in the soil pore water were low (generally less than 20 μM). At the second site, which was characterized by high soil organic content, amino acid concentrations in the soil pore water were between 20 and 100 μM throughout the season, with values generally higher than those for $NO_3^-$ and $NH_4^+$. For comparison, free amino acid concentrations in agricultural soils range from 10 to 20 μM (Schobert et al. 1988) and in arctic tundra soils from 8 to 80 μM (Chapin et al. 1993; Kielland 1994). The most abundant amino acids in the alpine dry meadows were glycine (10–

100 μM), glutamate (5–70 μM), and late in the season, cysteine (5–15 μM). From the results described, it is possible that in some sites, especially those characterized by high soil organic matter contents, plants utilize organic N sources to satisfy their growth N demands.

Two issues that emerge from the observations of root organic N uptake and soil organic N availability are whether mycorrhizal fungi contribute to a plant's ability to acquire organic N and why glycine is so abundant in alpine soils. With respect to fungal/plant interactions, it was noted in the previous section that certain fungi do indeed utilize organic N as a primary N source. It has also been hypothesized that infection of roots by some of these fungi exhibits a phenology that is consistent with the early-season N uptake of Niwot Ridge alpine species (Mullen et al. 1998). Studies using root boxes constructed with two compartments such that only hyphae are present on one side and the roots of *K. myosuroides* plus hyphae are present on the other side have shown that, indeed, intact molecules of glycine are transferred through the hyphae to the roots, and the N from the glycine is transferred to the plant shoot tissues (Lipson et al. 1999a). Thus, mycorrhizal relationships between alpine plants and fungi may have an important role in the utilization of organic N as a primary nutrient source. With respect to the question of why glycine is the most abundant amino acid in alpine soils, it has been shown recently that glycine is the amino acid most slowly mineralized by microbes native to Niwot Ridge soils (Lipson et al. 1999b). This may be related to the low C/N ratio of glycine and its suitability as a C source to these soil microbes. In effect, there appears to be an interesting pattern of resource partitioning between *K. myosuroides* and associated soil microbes, whereby glycine is preferentially ignored by mineralizing microbes but preferentially absorbed by plant roots and associated fungi.

## Conclusions and Generalizations

The alpine environment places unique constraints on the availability of soil nutrients to plants. Cold soil temperatures through all but the shallowest soil depths and short growing seasons cause mineralization rates to be low and the time for nutrient acquisition to be short. Strong vertical stratification of microclimate through space and seasonal changes in microclimate exist in this ecosystem, forcing the biota to converge on the nearest tens of centimeters above and below the ground surface and the tens of days during the short growing season—a narrow zone within which temperatures can support the metabolic rates needed to utilize atmospheric and soil nutrients. Plants have responded to these limitations through shallow, abundant roots, efficient use of those nutrients they acquire, nutrient storage patterns, and the evolution of a capacity to utilize organic nutrient forms. This is a combination of environmental constraints and evolved functional responses shared by relatively few ecosystems and lying at the extreme of those nutrient-use patterns reported for ecosystems of less-harsh climates. As such, studies of alpine plant nutrient relations should continue to broaden our understanding of plant/soil interactions, especially with respect to new selective forces and the limits of evolved responses to soil infertility.

## References

Abuzinadah, R. A., R. D. Finlay, and D. J. Read. 1986a. The role of proteins in nitrogen nutrition of ectomycorrhizal plants. II. Utilization of protein by mycorrhizal plants of *Pinus contorta. New Phytologist* 103:495–506.

Abuzinadah, R. A., and D. J. Read. 1986b. The role of proteins in the nitrogen nutrition of ectomycorrhizal plants I. Utilization of peptides and proteins by ectomycorrhizal fungi. *New Phytologist* 103:481–493.

Allen, E. B., J. C. Chambers, K. F. Connor, M. F. Allen, and R. W. Brown. 1987. Natural reestablishment of mycorrhizae in disturbed alpine ecosystems. *Arctic and Alpine Research* 19:11–20.

Arco, J. M. del, A. Escudero, and M. V. Garrido. 1991. Effects of site characteristics on nitrogen retranslocation from senescing leaves. *Ecology* 72:701–708.

Aydelotte, A. R., and P. K. Diggle. 1997. Analysis of developmental preformation in the alpine herb, *Caltha leptosepala*. *American Journal of Botany* 84:1646–1657.

Bajwa, R., and D. J. Read. 1986. Utilization of mineral and amino-nitrogen sources by the ericoid mycorrhizal endophyte Hymenoscyphus ericae and by mycorrhizal and non-mycorrhizal seedlings of Vaccinium. *Transactions of the British Mycology Society* 87:269–277.

Berendse, F., R. and Aerts. 1987. Nitrogen-use efficiency: A biologically meaningful definition? *Functional Ecology* 1:293–296.

Bissett, J., and D. Parkinson. 1979. The distribution of fungi in some alpine soils. *Canadian Journal of Botany* 57:1609–1629.

Bowman, W. D. 1994. Accumulation and use of nitrogen and phosphorus following fertilization in two alpine tundra communities. *Oikos* 70:261–270.

Bowman, W. D., and R. T. Conant. 1994. Shoot growth dynamics and photosynthetic response to increased nitrogen availability in the alpine willow *Salix glauca*. *Oecologia* 97:93–99.

Bowman, W. D., T. A. Theodose, and M. C. Fisk. 1995. Physiological and production responses of plant growth forms to increases in limiting resources in alpine tundra: Implications for differential community response to environmental change. *Oecologia* 101:217–227.

Bowman, W. D., T. A. Theodose, J. C. Schardt, and R. T. Conant. 1993. Constraints of nutrient availability on primary production in two alpine tundra communities. *Ecology* 74:2085–2097.

Brooks, P. D., M. W. Williams, and S. K. Schmidt. 1996. Microbial activity under alpine snowpacks. *Biogeochemistry* 32:93–113.

Chapin, F. S., III, and R. A. Kedrowski. 1983. Seasonal changes in nitrogen and phosphorus fractions and autumn retranslocation in evergreen and deciduous taiga trees. *Ecology* 64:376–391.

Chapin, F. S., III, and L. Moilanen. 1991. Nutritional controls over nitrogen and phosphorus resorption from Alaskan birch leaves. *Ecology* 72:709–715.

Chapin, F. S., L. Moilanen, and K. Kielland. 1993. Preferential use of organic nitrogen for growth by a non-mycorrhizal arctic sedge. *Nature* 361:150–153.

Chapin, F. S., E.-D. Schulze, and H. A. Mooney. 1990. The ecology and economics of storage in plants. *Annual Review of Ecology and Systematics* 21:423–447.

Currah, R. S., and M. Van Dyke. 1986. A survey of some perennial vascular plant species native to Alberta for occurrence of mycorrhizal fungi. *Canadian Field Naturalist* 100:330–342.

Dennis, J. G., and P. L. Johnson. 1970. Shoot and rhizome root standing crops of tundra vegetation at Barrow, Alaska. *Arctic and Alpine Research* 2:253–266.

Diggle, P. K. 1997. Extreme preformation in alpine *Polygonum viviparum:* An architectural and developmental analysis. *American Journal of Botany* 84:154–169.
Finlay, R. D., A. Frostefard, and A. M. Sonnerfeldt. 1992. Utilization of organic and inorganic nitrogen sources by ectomycorrhizal fungi in pure culture and in symbiosis with *Pinus contorta* Dougl. ex Loud. *New Phytologist* 120:105–115.
Fisk, M. C., and S. K. Schmidt. 1995. Nitrogen mineralization and microbial biomass N dynamics in three alpine tundra communities. *Soil Science Society of America Journal* 59:1036–1043.
Gardes, M., and A. Dahlberg. 1996. Mycorrhizal diversity in arctic and alpine tundra: An open question. *New Phytologist* 133:147–157.
Gokceoglu, M., and H. Rehder. 1977. Nutrient turnover studies in alpine ecosystems. III. Communities of lower altitudes dominated by *Carex sempervirens* Vill. and *Carex ferruginea* Scop. *Oecologia* 28:317–331.
Harney, S. 1995. Dematiaceous endophytes from plant roots: Molecular characterization and interactions. Ph.D. diss., State University of New York, College of Environmental Science and Forestry, Syracuse.
Haselwandter, K. 1987. Mycorrhizal infection and its possible significance in climatically and nutritionally stressed alpine plant communities. *Angewandte Botanik* 61:107–114.
Haselwandter, K., and D. J. Read. 1980. Fungal associations of roots of dominant and subdominant plants in high-alpine vegetation systems with special reference to mycorrhiza. *Oecologia* 45:57–62.
———. 1982. The significance of a root-fungus association in two Carex species of high alpine communities. *Oecologia* 53:352–354.
Holch, A. E., E. W. Hertel, W. O. Oakes, and H. H. Whitwell. 1941. Root habits of certain plants in the foothill and alpine belts of Rocky Mountain National Park. *Ecological Monographs* 11:329–345.
Jaeger, C. H., and R. K. Monson. 1992. The adaptive significance of nitrogen storage in *Bistorta bistortoides,* an alpine herb. *Oecologia* 92:578–585.
Jaeger, C. H., R. K. Monson, M. Fisk, and S. K. Schmidt. 1999. Seasonal partitioning of nitrogen by plants and soil microorganisms in an alpine ecosystem. *Ecology* 80:1883–1891.
Kielland, K. 1994. Amino acid absorption by arctic plants: Implications for plant nutrition and nitrogen cycling. *Ecology* 75:2373–2383.
Körner, Ch. 1989. The nutritional status of plants from high altitudes. A worldwide comparison. *Oecologia* 81:379–391.
Körner, Ch., and M. Diemer. 1987. *In situ* photosynthetic responses to light, temperature, and carbon dioxide in herbaceous plants from low and high altitude. *Functional Ecology* 1:179–194.
Körner, Ch., and W. Larcher. 1989. Plant life in cold climates. In *Plant and temperature,* edited by S. P. Long and F. I. Woodward. Symposium of Society of Experimental Biology 42. Cambridge, England: The Company of Biologists.
Körner, Ch., and S. Pelaez Menendez-Riedl. 1990. The significance of developmental aspects in plant growth analysis. In *Variation in growth rate and productivity,* edited by H. Lambers, M. L. Cambridge, H. Konings, and T. L. Pons. The Hague, Netherlands: SPB Academic Publishing bv.
Körner, Ch., and U. Rehnhardt. 1987. Dry matter partitioning and root length/leaf area ratios in herbaceous perennial plants with diverse altitudinal distribution. *Oecologia* 74:411–418.
Kramer, P. J., and H. C. Bullock. 1966. Seasonal variations in the proportions of suberized and unsuberized roots of trees in relation to the absorption of water. *American Journal of Botany* 53:200–204.

Larcher, W. 1977. Ergebnisse des IBP-Projekts "Zwergstrauchheide Patscherkofel." *Sitzungsber Österreichische Akademie Wissenschafter, Mathematisch-Naturwissenschaftliche Klasse, Abt I* 186:301–371.

Leake, J. F., and D. J. Read. 1990. Proteinase activity in mycorrhizal fungi. II. The effects of mineral and organic nitrogen sources on induction of extracellular proteinase in *Hymenoscyphus ericae* (Read) Korf and Kernan. *New Phytologist* 116:123–127.

Lesica, P., and R. K. Antibus. 1986. Mycorrhizae of alpine fell-field communities on soils derived from crystalline and calcareous parent materials. *Canadian Journal of Botany* 64:1691–1697.

Lipson, D., W. D. Bowman, and R. K. Monson. 1996. Luxury consumption and storage of nitrogen in the alpine herb, *Bistorta bistortoides*. *Ecology* 77:1277–1285.

Lipson, D. A., C. W. Schade, S. K. Schmidt, and R. K. Monson. 1999a. Ectomycorryzal transfer of amino acid nitrogen to the alpine sedge, *Kobresia myosuroides*. *New Phytologist* 142:163–167.

Lipson, D. A., T. K. Raab, S. K. Schmidt, and R. K. Monson. 1999b. Variation in competitive abilities of plants and microbes for specific amino acids. *Biology and Fertility of Soils* 29:257–261.

Millard, P. 1988. The accumulation and storage of nitrogen by herbaceous plants. *Plant, Cell and Environment* 11:1–8.

Mooney, H. A., and W. D. Billings. 1960. The annual carbohydrate cycle of alpine plants as related to growth. *American Journal of Botany* 47:594–598.

Mullen, R. B., and S. K. Schmidt. 1993. Mycorrhizal infection, phosphorus uptake, and phenology in *Ranunculus adoneus:* Implications for the functioning of mycorrhizae in alpine systems. *Oecologia* 94:229–234.

Mullen, R. B., S. K. Schmidt, and C. H. Jaeger. 1998. Nitrogen uptake during snow melt by the snow buttercup, *Ranunculus adoneus*. *Arctic and Alpine Research* 30:121–125.

Pastor, J., and Bockheim, J. G. 1984. Distribution and cycling of nutrients in an aspen-mixed-hardwood-spodosol ecosystem in northern Wisconsin. *Ecology* 65:339–353.

Raab, T. K., D. A. Lipson, and R. K. Monson. 1996. Non-mycorrhizal uptake of amino acids by roots of the alpine sedge, *Kobresia myosuroides:* Implications for the alpine nitrogen cycle. *Oecologia* 108:488–494.

Read, D. J. 1993. Mycorrhiza in plant communities. *Advanced Plant Pathology* 9:1–31.

Read, D. J., and K. Haselwandter. 1981. Observations on the mycorrhizal status of some alpine plant communities. *New Phytologist* 88:341–352.

Rehder, H. 1976a. Nutrient turnover in alpine ecosystems. Pt. I. Phytomass and nutrient relations in four mat communities in the northern calcareous Alps. *Oecologia* 22:411–423.

———. 1976b. Nutrient turnover in alpine ecosystems. Pt. II. Phytomass and nutrient relations in the Caricetum firmae. *Oecologia* 23:49–62.

Rehder, H., and A. Schäfer. 1978. Nutrient turnover studies in alpine ecosystems. Pt. IV. Communities of the central Alps and comparative survey. *Oecologia* 34:309–327.

Robards, A. W., D. T. Clarkson, and J. Sanderson. 1979. Structure and permeability of the epidermal/hypodermal layers of the sand sedge (*Carex arenaria* L.). *Protoplasma* 101:331–347.

Schobert, C., W. Köckenberger, and E. Komor. 1988. Uptake of amino acids by plants from the soil: A comparative study with castor bean seedlings grown under natural and axenic soil conditions. *Plant Soil* 109:181–188.

Shaver, G. R., and J. M. Melillo. 1984. Nutrient budgets of marsh plants: Efficiency concepts and relation to availability. *Ecology* 65:1491–1510.

Steltzer H., and W. D. Bowman. 1998. Differential influence of plant species on soil nitrogen transformations within moist meadow Alpine tundra. *Ecosystems* 1:464–474.

Stoyke, G., and R. S. Currah. 1991. Endophytic fungi from the mycorrhizae of alpine ericoid plants. *Canadian Journal of Botany* 69:347–352.

Stoyke G., K. N. Egger, and R. S. Currah. 1992. Characterization of sterile endophytic fungi from the mycorrhizae of subalpine plants. *Canadian Journal of Botany* 70:2009–2016.

Stribley, D. P., and D. J. Read. 1980. The biology of mycorrhiza in the Ericaceae. Pt. VII. The relationship between mycorrhizal infection and the capacity to utilize simple and complex organic nitrogen sources. *New Phytologist* 86:365–371.

Terashima, I., T. Masuzawa, H. Ohba, and Y. Yokoi. 1995. Is photosynthesis suppressed at higher elevations due to low $CO_2$ pressure? *Ecology* 76:2663–2668.

Theodose, T. A., and W. D. Bowman. 1997a. The influence of interspecific competition on the distribution of an Alpine graminoid: Evidence for the importance of plant competition in an extreme environment. *Oikos* 79:101–114.

———. 1997b. Nutrient availability, plant abundance, and species diversity in two alpine tundra communities. *Ecology* 78:1861–1872.

Tilman, D. 1982. *Resource competition and community structure.* Princeton, NJ: Princeton University Press.

———. 1988. Plant strategies and the dynamics and structure of plant communities. Princeton, NJ: Princeton University Press.

Trappe, J. M. 1988. Lessons from alpine fungi. *Mycologia* 80:1–10.

Trappe, J. M., and D. L. Luoma. 1992. The ties that bind: Fungi in ecosystems. In *The fungal community: Its organization and role in the ecosystem,* edited by G. C. Caroll, and D. T. Wicklow. New York: Marcel Dekker.

Treseder, K. K., D. W. Davidson, and J. R. Ehleringer. 1995. Absorption of ant-provided carbon dioxide and nitrogen by a tropical epiphyte. *Nature* 375:137–139.

Vitousek, P. M. 1982. Nutrient cycling and nutrient use efficiency. *American Naturalist* 119:553–572.

Webber, P. J., and D. E. May. 1977. The magnitude and distribution of belowground plant structures in the alpine tundra of Niwot Ridge, Colorado. *Arctic and Alpine Research* 9:157–174.

Wielgolaski, F. E. 1972. Vegetation types and plant biomass in tundra. *Arctic and Alpine Research* 4:291–306.

Wilcox H. E., and C. J. K. Wang. 1987. Mycorrhizal and pathological associations of dematiaceous fungi in roots of 7-month-old tree seedlings. *Canadian Journal of Forest Research* 17:884–899.

# 11

# Controls on Decomposition Processes in Alpine Tundra

Timothy R. Seastedt
Marilyn D. Walker
David M. Bryant

## Introduction

The snowpack gradient in the alpine generates a temperature and moisture gradient that largely controls organic matter decomposition. While low temperatures constrain decomposition and mineralization (chapter 12), moisture appears to be the strongest source of landscape variation in the alpine, with surface decay rates of plant materials highest in moist and wet meadow habitats. Despite a longer snow-free season and higher surface temperatures in dry meadows, decay in these areas is substantially lower than in moist meadows. Studies of decay rates of roots within the soil indicate that decay is uniformly low in all habitats and is limited by low temperatures and perhaps by the absence of certain groups of decomposer invertebrates. As in other ecosystems, substrate quality indices such as nitrogen and lignin content can be shown to be important factors influencing the rate of decay of specific substrates.

Alpine ecosystems were overlooked during the flurry of activity associated with the extensive ecosystem science programs of the 1960s and 1970s. With the few exceptions to be discussed here, decomposition studies in cold regions were conducted in arctic tundra or northern temperate and boreal forests. The need for this information in conjunction with efforts to understand carbon cycling in the alpine stimulated a substantial research effort in the 1990s. Studies have included both the effects of landscape location on decay (O'Lear and Seastedt 1994; Bryant et al. 1998), information on the importance of substrate chemistry on decomposition processes (Bryant et al. 1998), and preliminary information on some of the decomposer organisms (O'Lear and Seastedt 1994; Addington and Seastedt 1999).

Niwot Ridge researchers also participated in the Long-term Intersite Decomposition Experiment Team (LIDET) study, which involved placement of a dozen different litter types in the alpine and in 27 other sites from the tropics to the arctic tundra (Harmon 1995). All but one of the plant species used in the LIDET experiments were exotic to the alpine. Collectively these studies have provided sufficient information to represent the alpine in global decomposition modeling efforts. While providing excellent information on effects of substrate quality and global climate on decay, LIDET studies were replicated only within a single habitat type in the alpine. Thus, the work does not provide information on within-ecosystem sources of variation on decay.

Decomposition, as discussed here, is the physical and chemical conversion of organic matter to inorganic components. Most organic material is composed of CHO compounds, hence the final potential end points of about 90% of decomposition are $CO_2$ and $H_2O$. During this process, the other nutrients contained in the initial mass of organic materials are released into the soil solution as inorganic nutrients (mineralization). A fraction of the CHO compounds form recalcitrant materials known as humus, and this material comprises a large fraction of the soil organic matter (SOM) of the soil.

Unfortunately, methods employed to measure decomposition often confuse "disappearance" with decay, respiration, and mineralization. In fact, litter can disappear from the alpine surface but not decompose. Wind transport, fragmentation, and surface and ground water movements undoubtedly contribute to litter disappearance in the alpine. Moreover, the quantities lost to wind and water likely vary across the landscape, depending on topographic position and vegetation type. In particular, losses of surface plant litter in the dry meadows and fellfields may be largely due to transport rather than decay but, except for anecdotal observations, data to quantify such losses are lacking.

Decomposition is controlled by three variables: (1) substrate quality, (2) composition and abundance of the decomposer microflora and fauna, and (3) the microclimate of the decomposing substrate (Swift et al. 1979). The relative significance of these controls varies with respect to the landscape gradient of the alpine tundra. As discussed in chapter 12, the landscape position can control much of the soil microclimate, which affects plant species composition, which in turn affects substrate quality, which then will affect the composition of decomposer organisms that feed on these materials. Therefore, all three of the variables responsible for controlling decomposition rates interact to some degree.

## The General Pattern of Decomposition

At senescence, plants resorb and transport to perennial organs a fraction of the labile components from the senescing tissues. Fisk (1995) estimated resorption of amino acid N in foliage of alpine plants to average about 50%. Phosphorus often is resorbed at similar rates (e.g., Bowman, 1994), but other elements that are structurally bound in the tissue are not internally recycled to the living portions of the plant. This process results in creation of a litter substrate that is generally of reduced nitrogen and phosphorus content (i.e., lower quality for decomposers) compared to the living plant material.

Labile compounds not resorbed back into living plant tissues, including carbohydrates and water soluble materials such as phenols, often disappear very rapidly. Microbes may consume a portion of these materials in situ, but water-soluble substrates will also leach into the soil. Cellulose, often a large fraction of new plant detritus, can decompose rapidly, leaving behind a suite of increasingly complex and recalcitrant materials such as lignin and lignin-like materials that can be formed by the combination of phenolic materials with proteins. Humus formation occurs during the decomposition process, and the half-life of humus is believed to be several thousands of years or longer in cold climates such as the alpine (Parton et al. 1987). Humus increases soil water holding capacity and cation exchange capacity; hence, its presence often is associated with fertile soils, including those of the alpine (Burns, 1980; chapter 8). At the same time, however, incomplete and/or slow rates of decomposition results in an accumulation of essential plant nutrients in detritus, and these organic materials have the potential to become the "bottleneck" for nutrients, thereby limiting plant productivity (Vitousek and Howarth 1991). Work by Bowman and colleagues (Bowman et al. 1993; chapter 9) has demonstrated that nitrogen or nitrogen and phosphorus availability in the soil strongly influences plant productivity. This occurs despite the internal growth constraints exhibited by many alpine plant species (chapter 10). Thus, with the exception of very labile materials such as sodium and perhaps potassium, controls on decomposition processes concurrently represent the major controls on most nutrient fluxes as well.

Litter decay studies at Niwot Ridge have included *Acomastylis rossii* foliage (Webber et al. 1976; O'Lear and Seastedt 1994), surface and buried *Kobresia myosuroides* foliage (Cortinas and Seastedt 1996), mixed sedge foliage (Bryant 1996; Bryant et al. 1998), and buried mixed sedge roots (Bryant 1996; Table 11.1). The litter used in the LIDET study included a wide variety of substrates exotic to the alpine, and these alien substrates appeared to decay, on average, more slowly than the native litters (Table 11.2). Surface foliage decay of the common native plant species averaged 47% mass loss in the first year compared with an average of only about 15% mass loss for the nonnative litter. Root decomposition, however, was similar between native and nonnative species.

Decomposition within the soil appears to exhibit a different pattern of decay than litter decomposing on the surface of the tundra exhibits. Initial decay in the soil is equal to or more rapid than surface rates, while subsequent decay is slower than that observed on the surface (Bryant 1996; Bryant et al. 1998; Cortinas and Seastedt 1996). Since most primary production is belowground (Webber and Ebert-May 1977, Fisk et al. 1998; chapter 9), this observation explains the large amounts of organic materials found in most alpine tundra soils. While productivity in the alpine is generally less than or equal to that observed in temperate grasslands, decomposition is even slower, thereby generating the large storage component of soil carbon. This reinforces the idea that decomposition processes in the alpine can have important feedbacks to plant productivity and species composition. Plant species adapted to low nutrient conditions would be favored by mechanisms that enhance these nu-

Table 11.1. Decomposition Rates for Foliage and Root Litter Used in the LIDET Study at Niwot Ridge

| Plant Species | Type/Location of Litter[a] | Percent of Initial Mass Remaining | |
|---|---|---|---|
| | | Year 1 | Year 2 |
| Sugar maple | Foliage | 78.3 (3.03) | 65.0 (3.42) |
| Big bluestem | Foliage | 89.6 (0.69) | — |
| | Foliage[b] | 80.3 (1.94) | 78.7 (1.57) |
| | Roots | 82.0 (4.05) | 77.5 (0.84) |
| Drypetes | Foliage | 76.2 (1.47) | 60.9 (1.17) |
| Slash pine | Roots | 82.0 (1.01) | 72.5 (1.02) |
| Red pine | Needles | 93.9 (0.31) | 92.5 (2.54) |
| | Roots | 87.3 (1.09) | 81.2 (0.07) |
| Chestnut oak | Foliage | 78.9 (0.82) | 72.2 (0.19) |
| Red cedar | Needles | 91.2 (1.47) | 84.5 (0.91) |
| Wheat | Foliage | 90.1 (1.44) | 75.8 (1.92) |
| Average | Foliage | 85.4 (1.40) | 74.7 (2.08) |
| Average | Roots | 83.8 (1.50) | 77.1 (1.17) |

Note: Values are means (s.e.) of 3–4 samples per collection date.

[a]Foliage or needles were surface litter; roots were buried litter.
[b]Separate study by Seastedt (unpublished results).

Table 11.2. Rates of Decomposition of Native Litter on Niwot Ridge

| Plant Species | Type/Location of Litter | Percent of Initial Mass Remaining | | Reference |
|---|---|---|---|---|
| | | Year 1 | Year 2 | |
| Alpine avens[a] | Foliage | 48.0 (54–44) | 38.0 (42–33) | 1 |
| Alpine avens | Foliage | 58.4 (3.20) | 47.2 (1.18) | 2 |
| Alpine avens | Foliage | 59.3 (1.86) | 49.7 (1.86) | 3 |
| Kobresia[b] | Foliage (surface) | 59.2 (1.03) | 49.8 (1.56) | 4 |
| Kobresia[b] | Foliage (buried) | 50.7 (2.02) | 43.2 (2.41) | 4 |
| Mixed sedges[c] | Foliage | 71.8 (85–66) | — | 5 |
| Mixed sedges[c] | Roots | 85.0 (88–78) | — | 5 |

Note: Values are means with standard errors or ranges in parentheses.

1. Webber et al. (1976).
2. O'Lear and Seastedt (1994).
3. Seastedt (unpublished LTER results).
4. Cortinas and Seastedt (1996).
5. Bryant (1996).

[a]This study used green litter. Values in parentheses are ranges of values observed over the landscape.
[b]Litter harvested prior to senescence.
[c]This study used a mixture of litter harvested prior to senescence and senesced materials. Values in parentheses are ranges of values observed over the landscape.

trient limitations (Wedin and Tilman 1996). The slow decay rates of colder regions does just this by sequestering essential nutrients in these organic substrates.

## Effects of Climate and Microclimate on Decomposition Rates

Temperature and moisture control biotic activities, and accordingly, these variables can be expected to show very large effects on decomposition rates (e.g., Olson 1963; Meentemeyer 1978). Arctic researchers (e.g., Flanagan and Veum 1974; Heal and French 1974; Heal et al. 1981) summarized the pattern of decomposition with respect to soil temperature and moisture (Fig. 11.1). Biotic activities cease in frozen soils; hence, temperatures clearly restrict decomposition activities to that fraction of the year where $H_2O$ exists in liquid phase. At levels of soil moisture approaching or above field capacity, decomposition is also constrained by oxygen limitations. Flanagan and Veum (1974) indicated that microbial respiration is maximized at levels of about 60–80% of saturation. If these values hold for alpine soils, then soil moisture in wet meadows (e.g., Taylor and Seastedt 1994; Walker et al. 1994) is sufficient to slow decomposition within the soil. The lack of water during the growing season can also reduce microbial activity. We know this occurs in surface litter, but moisture values observed in the soil (which seldom range below 30% and are usually higher; Taylor and Seastedt 1994) appear sufficient to maintain microbial activity.

Tundra microbial populations have been reported to exhibit high sensitivities to temperature changes, with $Q_{10}$ values of 3.5–4.0 reported for temperature ranges of $-5°-+10°C$ (Heal et al. 1981). Microbial and invertebrate activities continue at temperatures below 0°C (e.g., Brooks 1995; Brooks et al. 1996; Addington and Seastedt 1999), with the lowest limit of activity probably set by the freezing point of the soil solution. Thus, soil temperature data can be used to estimate the period of decomposition activities.

The physical processes associated with freeze-thaw events also have significant effects, both on the physical breakdown of litter and indirectly via effects on microbial populations (Killham 1994). At Niwot Ridge, freeze-thaw does not appear to severely disrupt microbial populations (Lipson 1998). A large pool of microbial biomass often develops during the winter (Brooks 1995; Lipson 1998). This growth may be related to the availability of labile substrates from new litter produced in autumn, and this biomass may also be responsible for the relatively large rates of decomposition observed under snow. Soil invertebrates can be mobile as well beneath snow (Addington and Seastedt 1999). These animals can supercool and survive extreme temperatures, but food particles in their guts may cause lethal ice crystal formation. Hence, feeding by these animals is restricted during winter, and this lack of feeding may also explain the large increase in their main food source, the microbial biomass.

Organic matter decay in the alpine shows a correlation with the amount and duration of snowpack (Webber et al. 1976; O'Lear and Seastedt 1994; Bryant 1996). This correlation is assumed to be due to abiotic as well as biotic factors. Snowpack

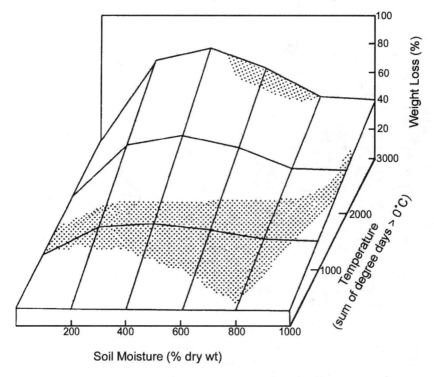

Figure 11.1. Relationship between temperatures and soil moisture on surface litter decay in arctic tundra (adapted from Heal and French 1974).

affects both soil temperatures and soil moisture and as such should be extremely important in explaining landscape variations in surface litter decomposition (Fig. 11.2). As mentioned above, microbial activity can continue under snowpack as long as temperatures remain at or near freezing. Thus, despite the short growing season, the period of decomposition is potentially extended under snowpack. This may be particularly true of early snows in autumn that can slow or even prevent the freezing of alpine soils. The relationship between decay and moisture is not linear. As suggested in both Figs. 11.1 and 11.2, an optimum rate of decomposition is found at intermediate moisture regimes. Sites too wet become anaerobic and exhibit reduced microbial and invertebrate activity, and sites with prolonged cold temperatures (i.e., in permanent snowbanks or areas entombed in ice) will have rates of decomposition limited by temperature. Thus, a curvilinear rate of decay in response to a snow gradient is predicted, and this pattern has been observed in alpine tundra litterbag studies (Webber et al. 1976; O'Lear and Seastedt 1994). The negative effect of excessive snowpack on surface litter decomposition appears to diminish somewhat over time (Fig. 11.2). More recent litterbag results that have evaluated a wider range of alpine communities suggest that this decline due to excessive moisture may be limited to very few communities within the alpine tundra, and that wet sites

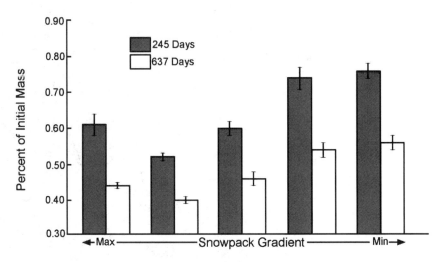

Figure 11.2. Alpine avens (*Acomastylis rossii*) litter weight loss after 245 and 637 days in the field. Mass loss was measured at five sites across a snowpack gradient (adapted from O'Lear and Seastedt 1994).

under substantial snowpack do exhibit high rates of decay, at least on the surface (Bryant et al. 1998; Seastedt, unpublished results).

Decay within the soil appears to lack the landscape-associated variance observed for surface litter (Bryant et al. 1998). Since both temperature and moisture control decay rates, root detritus at warm, dry sites may decay at a similar rate to litter at a wet, cool site. Fisk (1995) used this interpretation to explain the similarity of a number of soil characteristics across the alpine landscape. Buried litter has faster rates of decomposition during the summer growing season than surface litter has (Webber et al. 1976; Cortinas and Seastedt 1996), probably due to the removal of moisture limitations on the decomposition process. During the nongrowing season and during the second and subsequent years of decay, however, belowground litter appears to slow in its rate of decay relative to similar litter on the surface (e.g., Table 11.1). This observation may in part be an artifact of the litterbag method; fragmentation may result in "losses" in surface litter (losses out of the bags) that do not occur belowground.

## Substrate Quality

Quality tends to be defined after the fact. The adage is that substrates that decay rapidly are of high quality; substrates that decay slowly are of low quality. The environment of the decaying substrate needs to be qualified for this assertion. Sugar is a high-quality substrate, provided it is decomposed in an environment that can provide nutrients to the organisms consuming this material. Also, the amount of surface area of a given substrate also affects its rate of decay. Sawdust decays much faster than a equivalent weight of solid wood decays. Given these caveats, lignin (Van Cleve 1974) or lignin:nitrogen ratios (Melillo et al. 1982) tend to be standard indices of substrate quality. The LIDET study has provided a wealth of different substrate qualities to illustrate the range of decomposition rates that can be observed in the alpine with respect to lignin and nitrogen values and ratios (Fig. 11.3; Bryant et al. 1998). Lignin, nitrogen, and the lignin:nitrogen ratio all produced significant relationships with either the percentage mass remaining or the decay constant (k), but none of the relationships was particularly robust or superior to the others (Fig. 11.3). Surprisingly, initial nitrogen content of these litters was not correlated with initial lignin content; hence, one could argue that N content does affect decomposition independent of its (usual) inverse relationship with lignin. Bryant (1996) was not able to demonstrate that this relationship extended to litter augmented with nitrogen fertilization supplements. These potentially conflicting findings may be reconciled by noting that the N content of some of the LIDET litter was well below that measured for the native litters used in Bryant's experiments.

Studies using native litter suggest that rates of decay for these materials are faster than that of the average LIDET litter. There are possible biases affecting these results. First, there has been a tendency by Niwot investigators to use plant materials that have not fully senesced in their litterbag studies. Since green litter is generally a higher quality substrate, it does decay faster, particularly during the first year (Dearing 1997). However, decomposition of senescent litter of alpine avens, the most common plant species at Niwot Ridge, also appears to be fairly rapid relative to other species. This plant has a high concentration of low-molecular-weight phenolics, and while phenols may initially protect the material from both microbial and animal consumers, these materials are lost during the first year of decay (Dearing 1997). Decay of alpine avens after the first year of decay appears to match rates observed by the LIDET litter.

The significance of surface area (a variable resulting from substrate characteristics, freeze-thaw processes, and invertebrate grazing activities) has not been directly tested in the alpine. However, the degree to which a substrate is vulnerable to physical and biological decomposition processes clearly is related to the surface fraction of the substrate exposed to these conditions. Work by Caine (1979) demonstrated that even rock, when crushed and exposed to water, can exhibit measurable rates of "decomposition" (weathering). Physical factors affecting decay, including leaching and freeze-thaw fragmentation, clearly are important in the alpine. Photooxidation may also be significant.

The N content of the environment does not appear to affect rates of litter decay in the alpine and may even suppress longer termed decay rates (Bryant et al. 1998). This finding has relevance in terms of current trends of increasing nitrogen deposition in rainfall and dryfall, indicating that the decomposer system will not respond with increased decay rates. These results do not consider a nitrogen deposition-induced change in the substrate quality of the litter that would, however, affect decomposition rates (e.g., Fig. 11.3). Such changes could occur if N content of alpine plant species is enhanced overall but also would be the consequence of shifts in species composition to species with higher N contents.

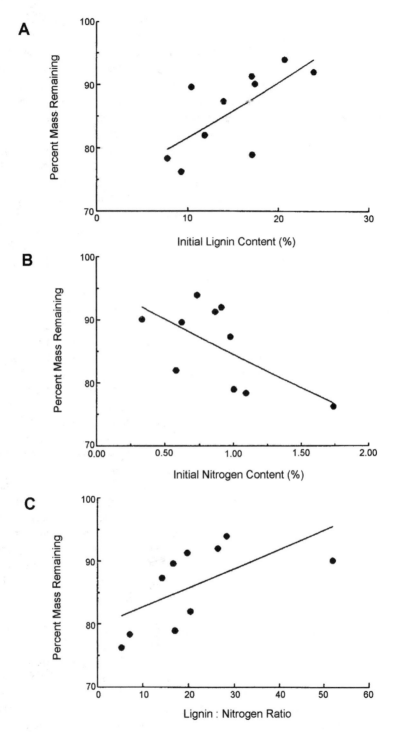

Figure 11.3. Effects of (a) initial lignin, (b) initial nitrogen concentrations, and (c) the initial lignin:nitrogen ratio of LIDET litters placed on Niwot Ridge. Results are shown for the first year of decomposition.

## Biotic Processes

Substrate quality and microclimate characteristics of a site can provide precise predictions of the rates of decomposition (e.g., Meentemeyer 1978; Parton et al. 1987). A predictive model of decay can therefore be created without even acknowledging the direct causal agents of decomposition, the microbes and associated decomposer invertebrates. This fact tends to frustrate biologists, who find the analysis conducive to arguments that biota or biotic diversity are not important in ecosystem processes. Nonetheless, the activities of the biota are so adequately predicted by the other two variables that their effects become embedded into the substrate-microclimate relationship.

Biota should not be taken for granted, particularly under nonequilibrium conditions where the abiotic surrogates of biotic activity may not correlate with the composition or activities of the biota. The loss or removal of species that have important functions in the decomposition process will change rates of decay, as demonstrated in a study by O'Lear and Seastedt (1994). When an arthropod repellant (naphthalene) was added to mushroom litter, decay at moist sites was significantly reduced compared with untreated controls. However, the repellant had no effect at dry sites. The authors demonstrated that important fauna (in this case, Diptera larvae) were not present at the dry site, but were abundant at the moist site. Hence, the repellant had no effect where there were no fauna, and the strong moisture effect demonstrated for decomposition was in fact related to the presence of a significant detritivore at the wetter sites. This study illustrates that biota are important, but their effects can sometimes be correlated or predicted from microsite characteristics.

As discussed above, the biota are the active causal agents of most decomposition processes. If biotic effects were perfectly predictable from temperature, moisture, and substrate quality, then there would exist no reason to discuss them as an additional variable affecting decomposition. We know that the composition of the biota is not constant over the alpine landscape, but definitive studies linking patterns of decomposer species to those of decomposition have yet to be conducted. Shulls and Mancinelli (1982), who studied relatively dry sites, suggested that changes in the enzymatic characteristics of microbes varied with altitude. They thought that while fungi and bacteria both were active in decomposition processes in the alpine, bacteria might dominate. O'Lear and Seastedt (1994) suggested that drier sites of the tundra were dominated by bacteria and bacterivores and functioned much like temperate grasslands. A group of arthropods known as the prostigmatid mites dominate these habitats. In contrast, wetter tundra sites as well as sites with abundant willows were dominated by oribatid mites, many of which are associated with a fungivorious diet, suggesting that these areas might have decomposition processes dominated by fungi.

Soil invertebrate densities in the litter and upper soil horizons are generally positively correlated with amounts of litter and organic matter (Swift et al. 1979; Seastedt 1984; Blair et al. 1994). Hence, substantial variation in invertebrate densities could be expected and densities may mimic the alpine landscape pattern seen for soil organic matter. These densities are inversely related to rates of decomposition when viewed at a global scale (Seastedt 1984), and this same inverse pattern can be seen in the alpine. Seastedt (unpublished results) found the highest densities of microarthropods within tree islands, where litter is more abundant and decay is slower than in the surrounding tundra. This finding illustrates that invertebrate activities—not invertebrate densities—are the critical characteristic of the fauna with respect to rates of decomposition.

O'Lear and Seastedt (1994) reviewed the limited information available on the decomposer fauna of the alpine. The smaller arthropods ("microarthropods"), mostly mites and collembolans, predominate in the comminution (fragmentation) and decomposition of decaying plant litter in these ecosystems (Douce and Crossley 1977, 1982; Tolbert et al. 1977). Missing from the results in the alpine are inventories and information on nematodes and enchytraeid worms, two groups likely to also be very significant in the decomposition process. Collectively, the soil decomposer fauna is responsible for most of the animal diversity of the alpine, and these fauna are suggested to be essential, not as primary decomposers, but as agents that maintain the composition and activities of the bacteria and fungi (Coleman and Crossley 1996). The literature is fairly consistent in demonstrating that consumption of bacteria and fungi can stimulate the overall activity of the microflora, and this observation likely holds for the alpine as well.

## Alpine and Arctic Decomposition: How Do Patterns Differ?

Several differences are suggested in the relative importance of climate and substrate quality as controls of alpine and arctic decomposition processes. First, snowpack amount and duration clearly affects decomposition in the alpine (O'Lear and Seastedt 1994), but snowpack differences do not appear to significantly affect decay rates in the arctic tundra (Seastedt, unpublished results). Two reasons are hypothesized for this difference. First, snowpack can cover thawed soils in the autumn of the alpine, thereby extending the nonfrozen period and increasing overall rates of decomposition. This mechanism, if it occurs at all in the arctic, appears less important. Snow cover in the arctic sites appears to be more uniform over the landscape, and the insulation effect of snowpack may therefore be more uniform than that observed in the alpine. The often wet, flat plains that comprise much of the arctic tundra offer less opportunity to develop a large landscape level temperature and moisture gradient. In contrast to the arctic, a large portion of the alpine landscape that is represented by the dry meadows and fellfields fails to receive sufficient, sustained snow cover to provide a thermal insulation effect. Thus the combination of wind and topography of the alpine likely produces a larger gradient in snowpack.

Secondly, substrate quality has been suggested to be a major factor affecting decomposition and soil organic matter across the arctic landscape (Nadelhoffer et al. 1991; Hobbie 1996). In contrast, the range in substrate quality of the herbaceous vegetation in the alpine does not appear to be as significant a factor in generating landscape differences (Fisk et al. 1998).

Temperature is emphasized as a controlling variable in arctic studies (Heal et al. 1981; Nadelhoffer et al. 1991; Hobbie 1996). However, as previously mentioned, the warmer meadows in the alpine are also drier (Greenland 1991) and have the lowest rates of surface litter decomposition. The large variances observed in landscape gradients of temperature and moisture in the alpine are not reported or emphasized in the studies of the arctic tundra. Hence, one might describe the dominant climatic control as temperature in the arctic, but as a snowpack-generated, temperature, and moisture interaction in the alpine.

Finally, patterns in the composition of the soil biota also show a landscape pattern in the alpine that has not been documented at arctic sites (O'Lear and Seastedt 1994). Collectively, the major difference between the alpine and arctic tundra in controls on decomposition processes appears to be the relatively large within-ecosystem variation generated in the alpine by the topography-climate interaction. This interaction is largely absent or at a much coarser spatial scale in the arctic tundra.

## Future Research

We still know very litter about the diversity of soil organisms and the functional attributes of the soil biota. In addition, there are a number of relevant areas that need to be addressed to understand patterns of decomposition and soil organic matter abundance in the alpine. First, lateral movements and redeposition of litter should be measured across the alpine landscape because wind in the alpine is hypothesized to be a major agent of litter comminution and redistribution. This work might be done in conjunction with efforts designed to monitor deposition of loess from outside the immediate alpine ecosystem. We know that the alpine acts as a "source ecosystem" for a number of materials supplied to the subalpine and montane systems, but the magnitude of these transfers have not been quantified. Organic matter may, in fact, be one of these materials transported in significant amounts by wind and water to lower elevation systems.

Additonal work is needed to quantify the relative sensitivity of temperature and moisture in decomposer processes. Current ecosystem models such as Century (Bryant 1996; Hartz 1997; Conley et al. 2000) seem to only poorly represent the decomposition process across the alpine landscape. The Century model does, however, emphasize the importance of this process in regulating nitrogen availability. Current data (e.g., O'Lear and Seastedt 1994; Fisk 1995; Bryant 1996) suggest that temperatures are limiting decomposition in moist habitats, while moisture constrains the decomposition process in dry habitats. Plant species characteristics may control both microclimate as well as substrate quality and contribute large sources of variation in addition to general landscape controls (Steltzer and Bowman 1998). At present, we hypothesize that the highest amounts of soil carbon in aerobic sites are not necessarily found only at the most productive areas of the alpine (the water-subsidized moist meadows) (Bowman et al. 1993; Walker et al. 1994), but also at the relatively drier *Kobresia* meadows (Fisk et al. 1998). The plants themselves may be creating a microclimate favorable for productivity but not conducive to decomposition by removing soil water prior to the period of maximum warming in late summer.

Finally, the role of decomposition as a mechanism for constraining nutrient availability and thereby affecting the composition of alpine vegetation needs further study (Fisk 1995; Hartz 1997; Conley et al. 2000). Immobilization of nutrients, associated with potentially long-term storage in humus and/or export of organics in soil water, may provide the buffering mechanism whereby the present vegetation may persist despite enhanced N inputs. If so, then once this buffering capacity is reached, a threshold response is anticipated. Plant species composition and associated impacts on litter production and decomposition may then respond much more quickly to increased N inputs from atmospheric deposition.

*Acknowledgments* This chapter substantially benefited from critical reviews by John Blair, Susan Sherrod, William Bowman, and Diane McKnight.

References

Addington, R. N., and T. R. Seastedt. 1999. Activity of soil microarthropods beneath snowpack in alpine tundra and subalpine forest. *Pedobiologia* 43:47–53.

Blair, J. M., R. W. Parmelee, and R. L. Wyman. 1994. A comparison of forest floor invertebrate communities of four forest types in the northeastern US. *Pedobiologia* 38:138–145.

Bowman, W. D. 1994. Accumulation and use of nitrogen and phosphorus following fertilization in two alpine tundra communities. *Oikos* 70:261–270.

Bowman, W. D., T. A. Theodose, J. C. Schardt, and R. T. Conant. 1993. Constraints of nutrient availability on primary production in two alpine tundra communities. *Ecology* 74:2085–2097.

Brooks, P. D. 1995. Microbial activity and nitrogen cycling under seasonal snowpacks, Niwot Ridge, Colorado. Ph.D. diss., University of Colorado.

Brooks, P. D., M. W. Williams, and S. K. Schmidt. 1996. Microbial activity under alpine snowpacks, Niwot Ridge, Colorado. *Biogeochemistry* 32:93–113.

Bryant, D. M. 1996. Litter decomposition in an alpine tundra. Master's thesis, University of Colorado.

Bryant, D. M., E. A. Holland, T. R. Seastedt, and M. D. Walker. 1998. Analysis of decomposition in alpine tundra. *Canadian Journal of Botany* 76:1295–1304.

Burns, S. F. 1980. Alpine soil distribution and development, Indian Peaks, Colorado Front Range. Ph.D. diss., University of Colorado, Boulder.

Caine, N. 1979. Rock weathering rates at the soil surface in an alpine environment. *Catena* 6:131–144.

Coleman, D. C., and D. A. Crossley Jr. 1996. *Fundamentals of soil ecology*. New York: Academic Press.

Conley, A. H., E. A. Holland, T. R. Seastedt, and W. J. Parton. 2000. Simulation of carbon and nitrogen cycling in an alpine tundra. *Arctic, Antarctic, and Alpine Research* 32:147–154.

Cortinas, M. R., and T. R. Seastedt. 1996. Short- and long-term effects of gophers (*Thomomys talpoides*) on soil organic matter dynamics in alpine tundra. *Pedobiologia* 40:162–170.

Dearing, D. 1997. The manipulation of plant toxins by a food hoarding herbivore, the North American pika, Ochotona princeps. *Ecology* 78:774–781.

Douce, K. G., and D. A. Crossley Jr. 1977. Acarina abundance and community structure in an arctic coastal tundra. *Pedobiologia* 17:32–42.

———. 1982. The effect of soil fauna on litter mass loss and nutrient loss dynamics in arctic tundra at Barrow, Alaska. *Ecology* 63:523–537.
Fisk, M. C. 1995. Nitrogen dynamics in an alpine landscape. Ph.D. diss., University of Colorado, Boulder.
Fisk, M. C., S. K. Schmidt, and T. R. Seastedt. 1998. Topographic patterns of above- and belowground production and nitrogen cycling in alpine tundra. *Ecology* 79:2253–2266.
Flanagan, P. W., and A. K. Veum. 1974. Relationships between respiration, weight loss, temperature and moisture in organic residues on tundra. In *Soil organisms and decomposition in tundra,* edited by A. J. Holding, O. W. Heal, S. F. MacLean Jr., and P. W. Flanagan. Stockholm, Sweden: Tundra Biome Steering Committee.
Greenland, D. 1991. Surface energy budgets over alpine tundra in summer. *Mountain Research and Development* 11:339–351.
Harmon, M. 1995. *Long-term intersite decomposition experiment team (LIDET).* Publication 19. Seattle: LTER Network Office.
Hartz, A. A. 1997. Simulation of carbon and nitrogen cycling in an alpine tundra ecosystem. Master's thesis, University of Colorado, Boulder.
Heal, O. W., P. W. Flanagan, D. D. French, and S. F. MacLean Jr. 1981. Decomposition and accumulation of organic matter. In *Tundra ecosystems: A comparative analysis,* edited by L. C. Bliss, O. W. Heal, and J. J. Moore. New York: Cambridge University Press.
Heal, O. W., and D. D. French. 1974. Decomposition of organic matter in tundra. In *Soil organisms and decomposition in tundra,* edited by A. J. Holding, O. W. Heal, S. F. MacLean Jr., and P. W. Flanagan. Stockholm, Sweden: Tundra Biome Steering Committee.
Hobbie, S. E. 1996. Temperature and plant species control over litter decomposition in Alaska tundra. *Ecological Monographs* 66:503–522.
Killham, K. 1994. *Soil ecology.* Cambridge: Cambridge University Press.
Lipson, D. A. 1998. Plant-microbe interactions and organic nitrogen availability in an alpine dry meadow. Ph.D. diss., University of Colorado, Boulder.
Meentemeyer, V. 1978. Macroclimate and lignin control of decomposition. *Ecology* 59:465–472.
Melillo, J. M., J. D. Aber, and J. F. Muratore. 1982. Nitrogen and lignin control of hardwood leaf litter decomposition dynamics. *Ecology* 63:621–626.
Nadelhoffer, K. J., A. E. Giblin, G. R. Shaver, and J. A. Laundre. 1991. Effects of temperature and substrate quality on element mineralization in six arctic soils. *Ecology* 72:242–253.
O'Lear, H. A., and T. R. Seastedt. 1994. Landscape patterns of litter decomposition in alpine tundra. *Oecologia* 99:95–101.
Olson, J. 1963. Energy storage and the balance of producers and decomposers in ecological systems. *Ecology* 44:322–331.
Parton, W. J., D. S. Schimel, C. V. Cole, and D. S. Ojima. 1987. Analysis of factors controlling soil organic matter levels in great plains grasslands. *Soil Science Society of America Journal* 51:1173–1179.
Seastedt, T. R. 1984. The role of microarthropods in decomposition and mineralization processes. *Annual Review of Entomology* 29:25–46.
Shulls, W. A., and R. L. Mancelli. 1982. Comparative study of metabolic activities of bacteria in three ecoregions of the Colorado Front Range, U.S.A. *Arctic and Alpine Research* 14:53–58.
Steltzer H., and W. D. Bowman. 1998. Differential influence of plant species on soil nitrogen transformations within moist meadow Alpine tundra. *Ecosystems* 1:464–474.

Swift, M. J., O. W. Heal, and J. M. Anderson. 1979. *Decomposition in terrestrial ecosystems*. Oxford: Blackwell Scientific.

Taylor R. V., and T. R. Seastedt. 1994. Short- and long-term patterns of soil moisture in alpine tundra. *Arctic and Alpine Research* 26:14 20.

Tolbert, W. W., V. R. Tolbert, and R. E. Ambrose. 1977. Distribution, abundance, and biomass of Colorado alpine tundra arthropods. *Arctic and Alpine Research* 9:221–234.

Van Cleve, K. L. 1974. Organic matter quality in relation to decomposition. In *Soil organisms and decomposition in tundra*, edited by A. J. Holding, O. W. Heal, S. F. MacLean Jr., and P. W. Flanagan. Stockholm, Sweden: Tundra Biome Steering Committee.

Vitousek, P. M., and R. W. Howarth. 1991. Nitrogen limitation on land and in the seas: How can it occur? *Biogeochemistry* 13:87–115.

Walker, M. D., P. J. Webber, E. H. Arnold, and D. Ebert-May. 1994. Effects of interannual climate variation on aboveground phytomass in alpine vegetation. *Ecology* 75:393–408.

Webber, P. J., and D. Ebert-May. 1977. The magnitude and distribution of belowground plant structures in the alpine tundra of Niwot Ridge, Colorado. *Arctic and Alpine Research* 9:157–174.

Webber, P. J., J. C. Emerick, D. Ebert-May, and V. Komarkova. 1976. The impact of increased snowfall on alpine vegetation. In *Ecological impacts of snowpack augmentation in the San Juan Mountains of Colorado*, edited by H. W. Steinhoff and J. D. Ives. Final Report, San Juan Ecology Project prepared for USDI Bureau of Reclamation (NTIS PB 255012) Fort Collins, CO: Colorado State University. CSU-FNR-7052–1.

Wedin, D., and D. Tilman. 1996. Influence of nitrogen loading and species composition on the carbon balance of grasslands. *Science* 27:1720–1723.

# 12

## Nitrogen Cycling

Melany C. Fisk
Paul D. Brooks
Steven K. Schmidt

### Introduction

In this chapter, we discuss the current understanding of internal N cycling, or the flow of N through plant and soil components, in the Niwot Ridge alpine ecosystem. We consider the internal N cycle largely as the opposing processes of uptake and incorporation of N into organic form and mineralization of N from organic to inorganic form. We will outline the major organic pools in which N is stored and discuss the transfers of N into and from those pools. With a synthesis of information regarding the various N pools and relative turnover of N through them, we hope to provide greater understanding of the relative function of different components of the alpine N cycle.

Because of the short growing season, cold temperatures, and water regimes tending either toward very dry or very wet extremes, the alpine tundra is not a favorable ecosystem for either production or decomposition. Water availability, temperature, and nutrient availability (N in particular) all can limit alpine plant growth (chapter 9). Cold soils also inhibit decomposition so that N remains bound in organic matter and is unavailable for plant uptake (chapter 11). Consequently, N cycling in the alpine often is presumed to be slow and conservative (Rehder 1976a, 1976b; Holzmann and Haselwandter 1988). Nonetheless, studies reveal large spatial variation in primary production and N cycling in alpine tundra across gradients of snowpack accumulation, growing season water availability, and plant species composition (May and Webber, 1982, Walker et al., 1994, Bowman, 1994, Fisk et al. 1998; chapter 9). Furthermore, evidence for relatively large N transformations under seasonal snowcover (Brooks et al., 1995a, 1998) and maintenance of high microbial biomass in frozen soils (Lipson et al. 1999a) provide a complex temporal component of N cycling on Niwot Ridge. Our discussion of N cycling on Niwot Ridge will focus on two main points: first, the spatial variation in N turnover in relation to snowpack regimes and plant community distributions; and second, the temporal variability of N transformations during both snow-free and snow-covered time periods.

## Nitrogen Pools

Total ecosystem N and the distribution of N among plant and soil components has been documented for topographic gradients of dry, moist, and wet meadow plant communities on Niwot Ridge. Fisk et al. (1998) found from 700 g N m$^{-2}$ in dry meadows to 800 g N m$^{-2}$ in wet meadows, to a depth of 15 cm (Table 12.1). The majority of this N occurs in soil organic matter, accounting for 85–95% of the ecosystem totals (Table 12.1). Of the remaining 5–15% of organic N quantified by Fisk et al. (1998), most was in nonliving plant biomass (5–10%), followed by living plant biomass (1–2%), and finally microbial biomass (0.5–1%; Table 12.1).

Nitrogen distributions in soil and biomass at Niwot Ridge are similar to those reported from tundra ecosystems in the Alps (Rehder 1976a, 1976b; Rehder and Shäfer 1978). Total ecosystem N varied from approximately 600 to 900 g m$^{-2}$ in high-elevation tundra communities of the Central Alps, but lower quantities of N (approximately 300 g m$^{-2}$) were found in high-elevation tundra of the Northern Alps (Rehder and Schäfer 1978). The wider range of N in these studies compared with Niwot Ridge may reflect the broader geographic scale encompassed in the Alps, including differences in parent materials, soil ages, and wider ranges of elevation. Nevertheless, the distribution of N among plant and soil pools was similar to that found on Niwot Ridge. Greater than 90% of total N in the European alpine was found in soil organic matter, and less than 3% of the total was found in aboveground pools (Rehder 1976a, 1976b; Rehder and Schäfer 1978).

Total quantities of N in plant and microbial biomass vary with topographic position, tending to be smallest in dry meadow and greatest in wet meadow plant communities on Niwot Ridge. Total plant matter (both living and nonliving) contained 40 g N m$^{-2}$ in dry meadows, 60 g N m$^{-2}$ in moist meadows, and 100 g N m$^{-2}$ in

Table 12.1. Average N Content (g N m$^{-2}$) of Organic Pools, Measured in the First Week of August 1992 and 1993

| Plant Community | Soil | Plant Biomass N | | | | Microbial N |
| | | Aboveground | | Belowground | | |
| | | Living | Dead | Living | Dead | |
| --- | --- | --- | --- | --- | --- | --- |
| Dry meadow | 657 | 2.4 | 6.6 | 5.3 | 26.8 | 3.1 |
| Moist meadow | 696 | 5.0 | 8.9 | 12.0 | 31.9 | 4.0 |
| Wet meadow | 690 | 6.4 | 8.4 | 10.5 | 74.0 | 7.8 |

*Sources:* Data are from Fisk (1995) and Fisk et al. (1998).

*Note:* All values are to 15 cm depth.

wet meadows (Table 12.1). Fisk et al. (1998) found N in living plant biomass (above- plus belowground) at the time of peak aboveground biomass (late July–early August) to vary between 8 g m$^{-2}$ in dry meadows and 17 g m$^{-2}$ in wet meadows. Bowman (1994) sampled to greater soil depths and documented higher values of 25 g m$^{-2}$ in dry meadows and 42 g m$^{-2}$ in wet meadows. May and Webber (1982) conducted a more-extensive study of fellfields, shrub tundra, and snowbed plant communities in addition to dry, moist, and wet meadows on Niwot Ridge. They documented the highest plant biomass in shrub tundra and the lowest in snowbeds, and it is likely that these communities represent extremes in plant biomass N content as well as in biomass.

Microbial N is the smallest of the organic N pools (Table 12.1). Microbial N can vary considerably both within communities and over time (Brooks et al. 1995a, 1998; Jaeger et al. 1999; Fisk and Schmidt 1995), yet it appears on average to increase over gradients from shallow to deep snow cover (Brooks et al. 1998) and from dry to wet meadows (Fisk et al. 1998; Table 12.1).

Production and decomposition processes that contribute to large-scale patterns of N accumulation in the alpine are regulated by climate associated with topography and snowpack regime. Both net primary production and decomposition increase across topographic moisture gradients from dry to wet meadows on Niwot Ridge (O'Lear and Seastedt 1994; Bryant et al. 1998; see following section; chapter 11). Walker et al. (1993) propose that soil water regimes are a major factor determining plant biomass, and other studies illustrate that total plant N content increases over soil moisture gradients from dry to wet meadows (Fisk et al. 1998; Bowman et al. 1993; Bowman 1994). As growing season decomposition is more temperature- than moisture limited, it is predicted to increase to a lesser extent than production over these gradients (chapter 11), thus favoring the greatest soil N storage in wet meadows.

Despite this expectation, Fisk et al. (1998; Table 12.1) observed similar quantities of soil N among communities, and it is likely that wintertime processes need to be taken into account in predicting patterns of soil N storage in the alpine. Brooks et al. (1997, 1998) have shown that soil C and N pools are responsive to variability in soil freezing controlled by winter snowpack accumulation, and dry meadow soils can have high microbial biomass during the winter (Lipson et al. 1999a). Winter decomposition can mineralize organic matter quantities equivalent to 2–25% of the annual aboveground primary production (Brooks et al. 1997), with variation in rates depending on both temporal and spatial severity of soil freezing and in the duration of soil thaw under snow. Overwinter decomposition is favored in soils that experience a hard freeze before consistent snow cover develops, whereas moderate decomposition is expected in consistently snow-covered soils, and low decomposition results in soils with shallow or sporadic snow cover. Comparable results have been found for alpine sites in Wyoming (Sommerfeld et al. 1993, 1996). Similarly, in situ measurements using buried litter bags indicate significant differences in overwinter C loss according to snowpack regime on Niwot Ridge (Bryant 1996), while long-term manipulations of snow cover in Montana have decreased soil C and N pools by 20% (Weaver and Welker, unpublished data). Overall these results suggest that the decomposition pattern in winter will be greater in wet meadows and lower in dry meadows, in contrast to growing-season patterns of decomposition.

At a smaller spatial scale, soil disturbance probably contributes to variation in N storage (Pauker and Seastedt 1996). For example, both Burns (1980) and Fisk and Schmidt (1995) found significant spatial variability in soil N within individual plant communities. Soils on Niwot Ridge are high in organic C, varying from 10 to 20% and with C:N varying from 12 to 15 (Pauker and Seastedt 1996; Fisk and Schmidt 1995; Burns 1980). Presumably most N in soil organic matter is in humus of extremely long-term stability. Although residence times in these organic pools are on average extremely long, small-scale soil disturbances that have the potential to alter these stores of N are common in the alpine. Gopher activity is the most commonly noted disruption of tundra soil. Pocket gophers remove vegetation and mix soil, allowing organic matter loss via erosion (Litaor et al. 1996), and frequent disturbance by gophers can deplete soil organic matter and N (Cortinas and Seastedt 1996). A second biotic alteration of soil organic matter and hence organic N occurs with the movement of krummholz tree islands (Pauker and Seastedt 1996). This movement, over time scales of 100–1000 years, leaves behind soil low in organic matter and N compared with adjacent tundra soil.

## Plant N Uptake and Turnover

### Snow-Free Season

Patterns of growing-season plant N uptake for production parallel those of N in biomass in the different plant communities on Niwot Ridge. The most complete current estimates of whole-plant N uptake for production, summing aboveground and fine roots, vary from 4.5 g N m$^{-2}$ yr$^{-1}$ in dry meadows to 7.2 g N m$^{-2}$ yr$^{-1}$ in wet meadows (Fisk et al. 1998; Table 12.2). Annual plant uptake of N for aboveground production was quantified using the N content of the current year's senescent biomass, and N uptake for root production was measured using a root ingrowth technique (chapter 9).

Table 12.2. Net N Fluxes (g N m$^{-2}$) Through Plant and Soil Pools During the Snow-Free (June–October) and Snow-Covered (November–May) Seasons

| Time Period | Plant Community | Net N Mineralization | Microbial N | Plant Biomass | | |
|---|---|---|---|---|---|---|
| | | | | Aboveground | Belowground | Total |
| Snow-free | Dry meadow | 0.9 | 2.2 | 1.4 | 3.1 | 4.5 |
| | Moist meadow | 1.1 | 4.7 | 2.1 | 3.5 | 5.6 |
| | Wet meadow | 1.0 | 7.7 | 2.4 | 4.8 | 7.2 |
| Snow-covered | Dry meadow | 2.0 | 2.0 | | | |
| | Moist meadow | 6.0 | 6.0 | | | |
| | Wet meadow | ND | ND | | | |

*Sources:* Data are from Fisk and Schmidt (1995) and Brooks et al. (1998).

*Note:* Values for microbial N represent the difference between maximum and minimum biomass N.

Spatial patterns of plant N uptake were correlated with an environmental gradient in soil water availability (Fisk et al. 1998) and also appear related to an interaction between water availability and temperature. The optimum combination of high water availability and high temperatures probably occurs in wet meadows, whereas soils usually have dried substantially in dry and moist meadows by the time maximum temperatures are reached (Taylor and Seastedt 1994; Fisk and Schmidt 1995; Fisk et al. 1998). From fertilization experiments, however, it is obvious that plant N cycling is dependent on the plant species present as well as the particular point along an environmental gradient. Nitrogen use by the dry meadow dominant *Kobresia myosuroides* is probably constrained within a narrow range. Bowman et al. (1993, 1995) found that *K. myosuroides* increased aboveground production following N fertilization, but little change in tissue N content was noted. Most of the aboveground production response was due to the increased abundance of other species, primarily non-mycorrhizal forbs and other graminoids. In contrast, *Deschampsia caespitosa*, the dominant moist meadow graminoid, increased growth in response to N (Bowman et al. 1995). Wet meadow species responded to N additions with higher tissue N content; however, aboveground production did not change with N additions alone (Bowman et al., 1993). Given these varied responses, Bowman et al. (1995) proposed that changes in plant species and growth form distributions would accompany any changes in N cycling that might result from N deposition or from regional changes in climate.

Interannual variation in N cycling processes is likely due to variation in production and/or climate. Fisk et al. (1998) found differences in plant biomass N and aboveground N use for production between years. Furthermore, within-season patterns of plant N accumulation differed from year to year (Fisk 1995; Fig. 12.1). We have no estimate of year-to-year variability in root production and N use, but based on the differences in root biomass between years, we expect that N uptake for root production varies over time as well. The one year in which root production was measured was probably a year of low production because both aboveground and belowground biomass were relatively low (Fisk et al. 1998).

Because of the cold, often dry, and nutrient-poor environment in alpine tundra, we may expect much of the plant biomass and N to be belowground (Scott and Billings 1964; May and Webber 1982). For dry, moist, and wet meadows, there were some small differences in the partitioning of N between above and belowground plant biomass, but the majority of N was found belowground (Table 12.1). Nitrogen used for root production was 70% of the total in moist and wet meadows and 80% in dry meadows (Fisk et al. 1998). The biomass data of May and Webber (1982; chapter 9) suggest that proportionately more N also would be found belowground in fellfields and snowbeds and to a lesser extent shrub tundra because of the woody aboveground biomass in that community. Likewise, the majority of N uptake for plant production was allocated belowground.

We can estimate cycling of N through living plant biomass by comparing annual N uptake to average peak-season plant N pools. Fisk et al. (1998) found little difference in the proportion of plant biomass N turned over annually among communities, either aboveground or belowground. More N is present in peak-season

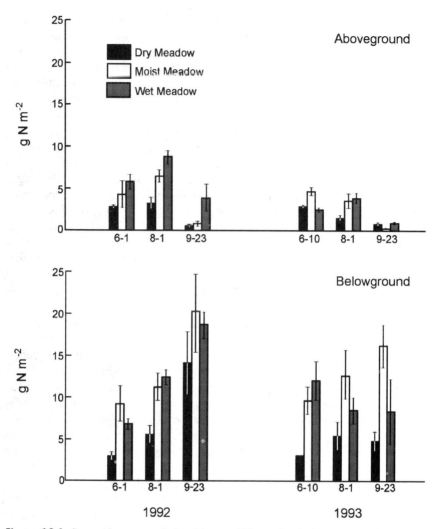

Figure 12.1. Seasonal patterns of plant biomass N immediately following snowmelt (6-1-92 and 6-10-93), at mid-season (8-1), and following aboveground senescence (9-23) in three tundra communities on Niwot Ridge, CO. Data are from Fisk (1995).

aboveground biomass compared to senescent litter (Tables 12.1 and 12.2), and so a considerable amount of the N used for aboveground production must be recycled internally. N storage can be important to plant growth given the short growing season and low availability (Rehder 1976a, 1976b; Jaeger and Monson 1992), and resorption efficiency from foliage can vary from 45 to 75% (Bowman et al. 1995; chapter 10). For total aboveground biomass, including aboveground stems and flowering structures, Fisk et al. (1998) found that 65% of the peak biomass aboveground N content was resorbed prior to senscence (Tables 12.1 and 12.2). The remaining 35% was lost as litter each year (Fisk et al. 1998).

A greater proportion of N used for belowground production probably turns over annually compared to aboveground production. However, it is more difficult to adequately assess N turnover through living plant roots because of the uncertainty surrounding resorption of N from roots. Assuming that no N is recycled internally for root production, Fisk et al. (1998) estimated that 80–90% of root biomass N turns over annually. Using peak biomass N data of Bowman et al. (1995) and Bowman (1994) again shows little variation among communities, but considerably lower turnover through the root N pool is suggested (15%). Substantial year-to-year variation in root biomass and N (Webber and Ebert-May 1977; May and Webber 1982; Fisk et al. 1998) is likely to contribute to some interannual varation in N cycling through alpine plant roots. Quantities of root N also vary between studies. Bowman (1994) found more than two times the quantity of N in root biomass than Fisk et al. (1998) did. While differences in sampling methods likely contributed to this difference, the patterns also suggest significant temporal as well as spatial variability in root N and root turnover.

Most of the work on N cycling processes on Niwot Ridge has been conducted in dry, moist, and wet tundra communities, which together account for approximately 50% of the areal extent of the Niwot Ridge study area (May and Webber 1982). We do not know the extent to which we can extend our understanding of N cycling to the less-productive alpine environments such as the fellfield and snowbed communities or the more-productive shrub tundra. While we can estimate relative quantities of N in plant matter based on the biomass data of May and Webber (1982), we can be less sure of whether annual uptake of N for plant production can be extrapolated. Fellfield soils are drier and rockier than dry meadows are, and snowbeds have the shortest growing seasons. While we find little variation in biomass:production relationships over the more simple environmental gradient encompassing dry, moist, and wet meadows, it is more likely that different relationships between biomass and production would be found in the more-extreme environments in fellfields and snowbeds and as a result of the growth-form differences, with greater woody tissues in shrub tundra and fellfields. Extending N cycling research into these communities that are more environmentally extreme and that have the most distinctly different vegetation (May and Webber 1982; Walker et al. 1994) will be an important step toward understanding the biotic and climatic interactions that influence productivity and nutrient cycles in the alpine.

## Snow-Covered Season

A common assumption in cold-dominated ecosystems is that biological activity, and thus N cycling processes, do not occur to any measurable degree under the alpine snowpack or in the exposed, frozen soils. However, there is increasing evidence that wintertime processes are an important part of annual N budgets on Niwot Ridge. Under consistent snowpack, soils thaw enough to allow biological activity (Cline 1995; Brooks et al. 1995b, 1996), and microbial biomass is highest during the winter in dry meadow soils (Lipson et al. 1999a). Lewis and Grant (1980) found far greater losses of $NO_3^-$ from alpine watersheds in years of low compared with years of high snowpack, and they suggested that biota prevent losses by sequestering nutrients under the deeper snowpack. By the time the snowpack starts to melt, it appears that plants do take up available N. Seasonal patterns of plant biomass N suggest that many of the plants take up N very early in the season (Fig. 12.1). Mullen et al. (1998) demonstrated N accumulation in *Ranunculus adoneus* while under the melting snowpack. Bilbrough and Bowman (unpublished) recovered 30% of $^{15}$N added to snowpack in five dominant plant species of moist and wet meadow communities before the snow had fully melted. It is clear that periods of snow cover, which we generally consider the nongrowing season, are an important part of annual plant nutrient budgets in the alpine and may also create differences in biota and nutrient cycling between low and high snowpack areas.

## Microbial N Transformations

Microbial mineralization of N from plant litter or other organic form determines the availability of inorganic N for biotic uptake or for loss. The microbial N pool is a major portion of actively cycling N, despite its relatively small size. Both the magnitude and timing of microbial mineralization and immobilization processes thus are critical to our understanding of patterns of plant-available N and the interactions between plants and soil processes during the short alpine growing season. Moreover, microbial immobilization and turnover of N are key factors regulating availability of N for loss during the snowcovered season when we would expect that plants are a smaller sink for N. Although nongrowing season processes are not often considered in studies of N cycling, soil microorganisms are active during the winter in soils at Niwot Ridge and other tundra sites (Sommerfeld et al. 1991, 1993, 1996; Brooks et al. 1995b, 1996; Lipson et al. 1999a). As previously mentioned for plants, this subnivian activity is an important component of annual cycling of N in these ecosystems.

### Snow-Free Season

Fisk et al. (1998) measured gross N mineralization and microbial N immobilization in short-term laboratory incubations of dry, moist, and wet meadow soils. Gross N mineralization and N immobilization were closely related, and both were several times greater in wet than in moist or dry meadow soils at several times during the snow-free season on Niwot Ridge. Across the topographic gradient from dry to wet meadows, both processes also were highly correlated with soil moisture (Fig. 12.2), suggesting that spatial patterns of soil N turnover are determined, directly or indirectly, by snowpack and associated soil water availability (Fisk et al. 1998). The spatial relationship between gross N mineralization and soil water availability also agrees with aboveground decomposition studies on Niwot Ridge. O'Lear and Seastedt (1994) found an increase in litter decomposition rates across a topographic snowpack gradient from dry to snowmelt-subsidized meadows (chapter 11). However, the decay of roots in the soil did not exhibit a clear topographic trend (Bryant 1996). This inconsistency in the patterns of root decay and mineralization processes is unexpected given the importance of belowground biomass and production in this

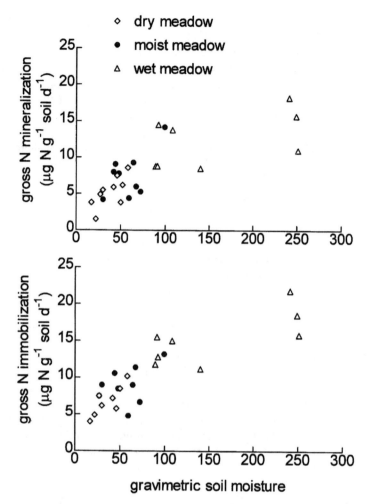

Figure 12.2. Gross N mineralization and immobilization patterns according to soil water content in dry, moist, and wet meadow tundra at three times during the snow-free season (after snowmelt, midseason, following senescence; see Fig. 12.1). Rates were determined in laboratory incubations at 15°C. Data are from Fisk (1995).

ecosystem, emphasizing the need for a more-complete understanding of the N dynamics of decaying roots in alpine tundra.

Because plant species composition, soil moisture, and soil temperature all vary spatially at Niwot Ridge, there could be multiple factors interacting to regulate decomposition and nutrient supply. Microbial activity was strongly related to soil moisture (Fisk et al. 1998), but we cannot conclude that moisture is a direct or indirect control or a covariate with other controls of microbial activity. For instance, litter quality is likely to differ among plant communities on Niwot Ridge, but it might covary with the gradient in soil moisture as a result of environmental controls over plant species composition. Plant species effects on N transformations have been documented within communities that vary in plant species dominance. Steltzer and Bowman (1998) found a tenfold difference in net N mineralization between soils under canopies of *Acomastylis rossii* and *Deschampsia caespitosa,* both dominants of the moist meadow. These within-community effects would have been averaged together in studies addressing community-level processes (Fisk and Schmidt 1995; Fisk et al. 1998), suggesting that studies of intercommunity comparisons may be innapropriate to understand how biota interact with the environment to affect nutrient cycles.

Rates of gross N immobilization and patterns of net N increases in microbial biomass both suggest that N flow through microbial biomass is a dominant component of alpine N cycling. In the absence of net gains or losses of N from microbial biomass, the immobilization potentials measured by Fisk et al. (1998) suggest that, in each plant community, about 5% of the microbial N pool turns over daily. Turnover is probably overestimated given differences in soil temperatures between laboratory incubations (15°C) and field conditions. Nevertheless, it is apparent that N cycles through microbial biomass more than once during the growing season. Furthermore, large changes in the microbial biomass N pool do occur over time on Niwot Ridge (Fig. 12.3). At the end of the growing season, Jaeger et al. (1999) documented an increase in microbial N of approximately 5 g m$^{-2}$, which may be the initiation of the winter soil microbial community (Lipson et al. 1999a). During the growing season, Fisk and Schmidt (1995) found changes in the microbial N pool of up to 8 g N m$^{-2}$ over monthly intervals. This was not noted in a subsequent year (Fisk et al. 1998), and it is unclear to what extent this may be due to sampling intervals (Fisk 1995) or to year-to-year variation in the soil environment.

Net N mineralization represents the balance between gross N mineralization and N immobilization and generally is thought to be the best index of plant-available inorganic N. High immobilization rates resulted in no net N mineralization in the short-term laboratory incubations conducted by Fisk et al. (1998). However, in

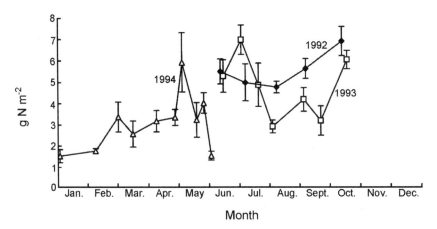

Figure 12.3. Temporal variation in microbial biomass N measured in snow-free (1992 and 1993) and snow-covered (1994) conditions in moist meadow soils on Niwot Ridge. Data are from Fisk and Schmidt (1995), Fisk (1995), and Brooks et al. (1998).

longer-term, in situ measurements on Niwot Ridge, net N mineralization was reported to average 1.0 and 1.2 g N m$^{-2}$ yr$^{-1}$ in two consecutive years (Fisk and Schmidt 1995). These estimates are unexpectedly low relative to soil inorganic N concentrations, which vary from 0.5 to 1.5 g N m$^{-2}$ throughout the snow-free season (Fisk 1995). They are also fairly low relative to most temperate ecosystems but fall within the range of net mineralization in the European alpine (0.14–5.0 g N m$^{-2}$ yr$^{-1}$) (Rehder 1976a, 1976b; Rehder and Schäfer 1978; Gokceoglu and Rehder 1977) and are greater than mineralization in arctic tundra (0.1–0.5 g N m$^{-2}$ yr$^{-1}$) (Giblin et al. 1991).

Spatial variation in net N mineralization was high on Niwot Ridge, and estimates of net N mineralization from dry, moist, and wet meadow soil during the snow-free season ranged from 0.3 to 1.6 g m$^{-2}$ yr$^{-1}$ (Fisk and Schmidt 1995). Unlike studies in tundra elsewhere and in contrast to other N cycling processes on Niwot Ridge, this spatial variation apparently was not related to plant community or soil moisture regime. As mentioned already, high spatial variation may be associated more with individual plant species (Steltzer and Bowman 1998).

Although we would expect rates of N availability to coincide with quantities of plant N uptake, net mineralization estimates of Fisk and Schmidt (1995) underestimate N use for plant production measured by Fisk et al. (1998; Table 12.2). These results are not unique to Niwot Ridge tundra. Plant uptake of N exceeded net N mineralization in the arctic (Giblin et al. 1991; Shaver et al. 1991) and also in the Alps (Rehder 1976a, 1976b; Rehder and Shafer 1978). It is likely that in soils with high organic matter content or high microbial N immobilization rates, the net N mineralization assay is a poor indicator of plant available N. High microbial immobilization of N can result in underestimation of plant-available N using the in situ net N mineralization technique (Vitousek and Andariese 1986). But it also is evident that alpine plants have access to other sources of available N. These include organic N, overlooked so far in this discussion of internal N cycling, and external inputs of N through deposition and fixation.

The use of soluble organic N by plants or by their mycorrhizal associates has been proposed as a means to improve competitiveness under nutrient poor conditions (Northup et al. 1995; Kaye and Hart 1997). Temperature and sometimes moisture limitations of decomposition can lead to the accumulation of organic N in alpine soils (chapter 11), and so we may expect to find these mechanisms facilitating N uptake in alpine plants. Mullen (1995) has shown that fungal symbionts of the dominant dry meadow plant, *Kobresia myosuroides,* can take up a wide array of soluble organic N compounds at temperatures as low as 2°C (chapter 10). In addition, Raab et al. (1996) have demonstrated the uptake of simple forms of organic N by *K. myosuroides* in the absence of mycorrhizal infection. Direct uptake of organic N has also been found for *Carex* species in arctic tundra (Kielland 1994). Amino acids and simple organic acids are abundant in alpine tundra soil (Raab et al. 1996), and although other plant species found on Niwot Ridge have not yet been investigated, it seems likely that the capacity for organic N uptake is not restricted to *K. myosuroides* (chapter 10).

Nitrogen deposition on Niwot Ridge is likely supplementing what becomes available through decomposition and mineralization processes. The current deposition rate of 0.6 g N m$^{-2}$ yr$^{-1}$ (chapter 3) is high relative to estimated net N mineralization of 1.0 g N m$^{-2}$ yr$^{-1}$ (Fisk and Schmidt 1995) and is equal to about 10% of annual plant N uptake (Fisk et al. 1998). N added to alpine tundra is taken up into plant and microbial biomass (Bowman 1994; Fisk and Schmidt 1996) and appears to accumulate in soil organic matter (Fisk and Schmidt 1996), suggesting that N deposited during the growing season may be affecting the N balance of the tundra at Niwot Ridge.

Another source of N inputs to the internal cycle is via N$_2$ fixation. Fixation by free-living bacteria appears minimal in Niwot Ridge soils (Wojciechowski and Heimbrook 1984). Symbiotic N$_2$ fixation, associated with three species of *Trifolium*, varied from 0.1 g m$^{-2}$ yr$^{-1}$ in wet meadows to 0.8 g m$^{-2}$ yr$^{-1}$ in fellfields, for a spatially averaged value of 0.5 g m$^{-2}$ yr$^{-1}$ for Niwot Ridge (Bowman et al. 1996). Based on these estimates, N$_2$ fixation is the source of 1–13% of annual plant N uptake documented for dry, moist, and wet meadows.

## Snow-Covered Season

Of the annual N deposition to Niwot Ridge, 50–80% is stored in the seasonal snowpack and released during melt (Bowman 1992; Brooks et al. 1997). Thus, snowmelt is an important time for understanding inputs and exports of N in this ecosystem. Despite the large quantities of N inputs to snow, Brooks et al. (1997) demonstrated that internal transformations of N in the soil are even greater and dominate the N cycle before and during snowmelt. Net N mineralization exhibited a predictable spatial pattern under the seasonal snowpack that appears to be controlled by the timing of snow cover (Brooks et al. 1998; Table 12.2). Soil thaw and microbial activity under snow are heterogeneous, which results in a spatial pattern in microbial processes that is closely related to growing season soil moisture gradients. While soil covered continuously by early snow may remain unfrozen during most of the winter (Sommerfeld et al. 1993), soil under shallow or inconsistent snow cover often remains frozen until the initiation of spring snowmelt. When frozen soil first thawed, either under the seasonal snowpack or during spring snowmelt, soil inorganic N pools averaged 1.5 g N m$^{-2}$ (Brooks et al. 1996, 1998). In moist meadows under deeper snow cover, where soil remained unfrozen during the winter, soil inorganic N increased to 2.0–7.5 g N m$^{-2}$ at the initiation of snowmelt. The highest net N mineralization rates (greater than 6 g N m$^{-2}$) were found in soils that experienced an extreme freeze early in the winter before snow cover developed and then thawed during late winter and early spring (Brooks et al. 1996). Within the subnivian environment, net overwinter N mineralization may be significantly greater than are estimates of net growing season N mineralization, reaching up to 7.5 g N m$^{-2}$ yr$^{-1}$ (Brooks et al. 1995b, 1996). These high net N mineralization rates are consistent with the highest levels of overwinter CO$_2$ flux (Brooks et al. 1997) and apparently result from the release of labile C by freeze-thaw cycles (Schimel and Klein 1996).

Net changes in microbial biomass N under the snowpack also appear to equal or exceed those during the snow-free season (Brooks et al. 1996; Lipson et al. 1999a; Table 12.2; Fig. 12.3). Within sites with some degree of snow cover, net N immobilization in microbial biomass appears to be related to the length of time that soil remains snow covered and thawed during the winter. When soil first thaws under snow, microbial biomass N ranges from 1.0 to 1.8 g N m$^{-2}$. Under consistent snow cover, microbial N slowly increases before melt, and then increases rapidly at the initiation of snowmelt (Fig. 12.3). Under shallow or inconsistent snow cover, microbial biomass N follows a similar pattern during melt, but with a much lower magnitude than under deeper, consistent snow cover (Brooks et al. 1998). For example, during the spring of 1994, microbial biomass immobilized 4.7 ($\pm$1.4) g N m$^{-2}$ at deep snow sites but only 1.7 ($\pm$0.3) g N m$^{-2}$ at sites with shallow, inconsistent snow cover. The ability of soil microbial biomass to sequester N during the winter and to the greatest extent under deeper snowcover concurs with greater streamwater NO$_3^-$ losses from the Niwot Ridge area during a year of low snowpack compared with a year of high snowpack (Lewis and Grant 1980). Large increases in both inorganic and microbial N before and during melt indicate high gross and net mineralization rates under continuous snow cover, yet low N export indicated that these environments are better able to immobilize episodic increases in N availabilty when compared with shallow, discontinuous snow cover sites (Brooks et al. 1998).

The transition period between winter and summer is critical for N transfers between microbial, plant, and soil pools. As snowmelt progresses, microbial biomass decreases at both shallow and deep snowpack sites, reaching 1.5–2.0 g N m$^{-2}$ as sites become snow free (Brooks et al. 1997). At all sites, the decrease in microbial biomass N during the later stages of melt was approximately equal to the N immobilized earlier in the winter and occurred without a concurrent increase in soil inorganic N pools or significant N export in meltwater or gaseous forms (Brooks et al. 1996, 1997, 1998; Williams et al. 1997). The decrease in microbial biomass during snowmelt in dry meadow soils is accompanied by large increases in soil protein and amino acid concentrations and soil protease activity (Lipson et al. 1999a) and appears to be associated with a decrease in soil C availability (Lipson et al. 1999b). This period may be critical to plant N uptake (Jaeger et al. 1999), and evidence indicates that plants are actively taking N up even before the completion of snowmelt. In addition, a transition in the composition of microbial communities occurs during snowmelt, mediated by a change in soil temperatures (Lipson et al. 1999a). Interannual variability in climate during the snow-covered season thus has the potential to significantly impact both aquatic and terrestrial biogeochemical cycles during the growing season.

## Summary

The alpine tundra environment on Niwot Ridge is characterized by long, cold winters followed by short, cool growing seasons. Consequently, N cycling during the growing season is strongly constrained by the timing and quantity of seasonal snow cover and soil frost. Both water availability and soil temperature early in the growing season are affected by the timing and distribution of snowmelt runoff, which in turn exhibits significant control over both plant and microbial activity. As a result, there are some consistent patterns in both the N storage and transformations in soil, plant, and microbial pools. Plant biomass N, plant N use for production, gross mineralization by soil microorganisms, and flux of N through soil microorganisms all tend to be lowest in dry tundra meadows and greatest in wet tundra meadows.

The most striking result emerging from the N cycling studies at Niwot is the dynamic nature of the microbial biomass N pool. Large changes occur in the microbial N pool, and during spring it appears that plants begin immobilizing N at the same time as it is rapidly turned over through the microbial pool. The transfer of N through the microbial pool and from microbial to plant biomass is thus an important component of the alpine N cycle that deserves more detailed study. Furthermore, a significant portion of the annual N cycle in this snow-dominated system occurs before the traditionally defined growing season, hence magnifying the relative importance of N cycling through microbial biomass. Within this environment, the timing of snowpack accumulation has a major effect on N dynamics by controlling the timing of soil thaw and the size and composition of the active microbial community at the initiation of snow melt. With current interests in N saturation (Baron et al. 1994; Williams et al. 1997) and biotic sequestration of N, this may be one of the most important times of the year for constructing comprehensive N budgets in high-elevation areas.

References

Baron J. S., D. S. Ojima, E. A. Holland, and W. J. Parton. 1994. Analysis of nitrogen saturation potential in Rocky Mountain tundra and forest: Implications for aquatic systems. *Biogeochemistry* 27:61–82.

Bowman, W. D. 1992. Inputs and storage of nitrogen in winter snowpack in an alpine ecosystem. *Arctic and Alpine Research* 24:211–215.

———. 1994. Accumulation and use of nitrogen and phosphorus following fertilization in two alpine tundra communities. *Oikos* 70:261–270.

Bowman, W. D., J. C. Schardt, and S. K. Schmidt. 1996. Symbiotic $N_2$-fixation in alpine tundra: Ecosystem input and variation in fixation rates among communities. *Oecologia* 108:345–350.

Bowman, W. D., T. A. Theodose, and M. C. Fisk. 1995. Physiological and production responses of plant growth forms to increases in limiting resources in alpine tundra: Implications for differential community response to environmental change. *Oecologia* 101:217–227.

Bowman, W. D., T. A. Theodose, J. C. Schardt, and R. T. Conant. 1993. Constraints of nutrient availability on primary production in two alpine tundra communities. *Ecology* 74:2085–2097.

Brooks, P. D., S. K. Schmidt, D. A. Walker, and M. W. Williams. 1995a. The Niwot Ridge snow fence experiment: Biogeochemical responses to changes in the seasonal snowpack. In *Biogeochemistry of Snow-Covered Catchments,* edited by K. Tonnessen, M. Williams, and M. Tranter. Wallingford, England: International Association of Hydrological Sciences 228.

Brooks P. D., S. K. Schmidt, and M. W. Williams. 1995b. Snowpack controls on soil nitrogen dynamics in the Colorado alpine. In *Biogeochemistry of Snow-Covered Catchments,* edited by K. Tonnessen, M. Williams, and M. Tranter. Wallingford, England: International Association of Hydrological Sciences 228.

———. 1997. Winter production of $CO_2$ and $N_2O$ from alpine tundra; environmental controls and relationship to inter-system C and N fluxes. *Oecologia* 110:403–413.

Brooks, P. D., M. W. Williams, and S. K. Schmidt. 1996. Microbial activity under alpine snow packs, Niwot Ridge, CO. *Biogeochemistry* 32:93–113.
———. 1998. Inorganic N and Microbial biomass dynamics before and during spring snowmelt. *Biogeochemistry* 43:1–15.
Bryant, D. M. 1996. Litter decomposition in an alpine tundra. Master's thesis, University of Colorado.
Bryant, D. M., E. A. Holland, T. R. Seastedt, and M. D. Walker. 1998. Analysis of decomposition in alpine tundra. *Canadian Journal of Botany* 76:1295–1304.
Burns, S. F. 1980. Alpine soil distribution and development, Indian Peaks, Colorado Front Range. Ph.D. diss., University of Colorado.
Cline, D. 1995. Snow surface energy exchanges and snowmelt at a continental alpine site. In *Biogeochemistry of Snow-Covered Catchments,* edited by K. Tonnessen, M. Williams, and M. Tranter. Wallingford, England: International Association of Hydrological Sciences 228.
Cortinas M. R, and T. R. Seastedt. 1996. Short- and long-term effects of gophers (*Thomomys talpoides*) on soil organic matter dynamics in alpine tundra. *Pedobiologia* 40:162–170.
Fisk, M. C., and S. K. Schmidt. 1995. Nitrogen mineralization and microbial biomass N dynamics in three alpine tundra communities. *Soil Science Society of America Journal* 59:1036–1043.
Fisk, M. C., and S. K. Schmidt. 1996. Microbial responses to excess nitrogen in alpine tundra soils. *Soil Biology and Biochemistry* 28:751–755.
Fisk, M. C., S. K. Schmidt, and T. R. Seastedt. 1998. Topographic patterns of above- and belowground production and nitrogen cycling in alpine tundra. *Ecology* 79:2253–2266.
Fisk M. C. 1995. Nitrogen dynamics in an alpine landscape. Ph.D. diss., University of Colorado, Boulder.
Giblin, A. E., K. J. Nadelhoffer, G. R. Shaver, J. A. Laundre, and A. J. McKerrow. 1991. Biogeochemical diversity along a riverside toposequence in arctic Alaska. *Ecological Monographs* 61:415–435.
Gokceoglu, M., and H. Rehder. 1997. Nutrient turnover studies in alpine ecosystems. 3. Communities of lower altitudes dominated by *Carex sempervirens* vill. and *Carex feruginea. Oecologia* 28:317–331.
Holzmann, H. P., and K. Haselwandter. 1988. Contribution of nitrogen fixation to nitrogen nutrition in an alpine sedge community (*Caricetum curvulae*). *Oecologia* 76:298–302.
Jaeger, C. H., and R. K. Monson. 1992. Adaptive significance of nitrogen storage in *Bistorta bistortoides,* an alpine herb. *Oecologia* 92:578–585.
Jaeger, C. H., R. K. Monson, S. K. Schmidt, and M. C. Fisk. 1999. Seasonal partitioning of N between plants and soil microorganisms in an alpine ecosystem. *Ecology* 80:1883–1891.
Kaye, J. P., and S. C. Hart. 1997. Plant-microbe competition for nitrogen. *Trends in Ecology and Evolution* 12:139–143.
Kielland, K. 1994. Amino acid absorption by arctic plants: Implications for plant nutrition and nitrogen cycling. *Ecology* 75:2373–2383.
Lewis, W. M., and M. C. Grant. 1980. Relationships between snow cover and winter losses of dissolved substances from a mountain watershed. *Arctic and Alpine Research* 12:11–17.
Lipson, D. A., S. K. Schmidt, and R. K. Monson. 1999a. Links between microbial population dynamics and nitrogen availability in an alpine ecosystem. *Ecology* 80:1613–1631.
Lipson, D. A., T. K. Raab, S. K. Schmidt, and R. K. Monson. 1999b. Variation in competitive abilities of plants and microbes for specific amino acids. *Biology and Fertility of Soils* 29:257–261.

Litaor, M. I., R. Mancinelli, and J. C. Halfpenny. 1996. The influence of pocket gophers on the status of nutrient in alpine soils. *Geoderma* 70:37–48.

May, D. E., and P. J. Webber. 1982. Spatial and temporal variation of the vegetation and its productivity on Niwot Ridge, Colorado. Pages 35–62. In *Ecological studies of the Colorado alpine, a Festschrift for John W. Marr.* Occasional paper no. 37, Institute of Arctic and Alpine Research, University of Colorado, Boulder.

Mullen, R. B. 1995. Fungal endophytes in the alpine buttercup Ranunculus adoneus: Implications for nutrient cycling in alpine systems. Ph.D. diss., University of Colorado, Boulder.

Mullen, R. B., S. K. Schmidt, and C. H. Jaeger III. 1998. Nitrogen uptake during snow melt by the snow buttercup, *Ranunculus adoneus*. *Arctic and Alpine Research* 30:121–125.

Northup, R. R., Z. Yu, and K. A. Vogt. 1995. Polyphenol control of nitrogen release from pine litter. *Nature* 377:227–229.

O'Lear, H. A., and T. R. Seastedt. 1994. Landscape patterns of litter decomposition in alpine tundra. *Oecologia* 99:95–101.

Pauker, S. J., and T. R. Seastedt. 1996. Effects of mobile tree islands on soil carbon storage in tundra ecosystems. *Ecology* 77:2563–2567.

Raab, T. K., D. A. Lipson, and R. K. Monson. 1996. Non-mycorrhizal uptake of amino acids by roots of the alpine sedge *Kobresia myosuroides:* Implications for the alpine nitrogen cycle. *Oecologia* 108:488–494.

Rehder, H. 1976a. Nutrient turnover in alpine ecosystems. Pt. I. Phytomass and nutrient relations in four mat communities in the northern calcareous alps. *Oecologia* 22:411–423.

———. 1976b. Nutrient turnover in alpine ecosystems. Pt. II. Phytomass and nutrient relations in the *Caricetum firmae*. *Oecologia* 23:49–62

Rehder, H., and A. Schäfer. 1978. Nutrient studies in alpine ecosystems. Pt. IV. Communities of the Central Alps and comparative survey. *Oecologia* 34:309–327.

Schimel, J. P., and J. P. Klein. 1996. Microbial response to freeze-thaw cycles in tundra and taiga soils. *Soil Biology and Biochemistry* 28:1061–1066.

Scott, D., and W. D. Billings. 1964. Effects of environmental factors on standing crop and productivity of an alpine tundra. *Ecological Monographs* 34:243–270.

Shaver, G. R., K. J. Nadelhoffer, and A. E. Giblin. 1991. Biogeochemical diversity and element transport in a heterogeneous landscape, the North slope of Alaska. In *Quantitative methods in landscape ecology,* edited by M. G. Turner and R. H. Gardner. Ecological Studies Vol. 82. New York: Springer-Verlag.

Sommerfeld, R. A., W. J. Massman, and R. C. Musselman. 1996. Diffusional flux of $CO_2$ through snow: Spatial and temporal variability among alpine-subalpine sites. *Global Biogeochemical Cycles* 10:473–482.

Sommerfeld, R. A., A. R. Mosier, and R. C. Musselman. 1993. $CO_2$, $CH_4$, and $N_2O$ flux through a Wyoming snowpack. *Nature* 361:140–143.

Sommerfeld, R. A., R. C. Musselman, J. O. Ruess, and A. R. Mosier. 1991. Preliminary measurements of $CO_2$ in melting snow. *Geophysical Research Letters* 18:1225–1228.

Steltzer, H., and W. D. Bowman. 1998. Differential influence of plant species on soil nitrogen transformations within moist meadow alpine tundra. *Ecosystems* 1:464–474.

Taylor, R. V., and T. R. Seastedt. 1994. Short- and long-term patterns of soil moisture in alpine tundra. *Arctic and Alpine Research* 26:14–20.

Vitousek, P. M., and S. W. Anderiese. 1986. Microbial transformations of labelled nitrogen in a clearcut pine plantation. *Oecologia* 68:601–605.

Walker, D. A., J. C. Halfpenny, M. D. Walker, and C. A. Wessman. 1993. Long-term studies of snow-vegetation interactions *Bioscience* 43:287–301.

Walker, M. D., P. J. Webber, E. A. Arnold, and D. Ebert-May. 1994. Effects of interannual climate variation on aboveground phytomass in alpine vegetation. *Ecology* 75:393–408.

Webber, P. J., and D. Ebert-May. 1977. The magnitude and distribution of belowground plant structures in the alpine tundra of Niwot Ridge, Colorado. *Arctic and Alpine Research* 9:157–174.

Williams, M. W., P. D. Brooks, A. R. Mosier, and K. A. Tonnessen. 1996. Mineral N transformations in and under seasonal snow in a high-elevation catchment, Rocky Mountains, USA. *Water Resources Research* 32:3161–3171.

Wojciechowski, M. F., and M. E. Heimbrook. 1984. Dinitrogen fixation in alpine tundra, Niwot Ridge, Front Range, Colorado, USA. *Arctic and Alpine Research* 16:1–10.

# 13

# Soil-Atmosphere Gas Exchange

Steven K. Schmidt
Ann E. West
Paul D. Brooks
Lesley K. Smith
Charles H. Jaeger
Melany C. Fisk
Elisabeth A. Holland

## Introduction

The alpine, while not extensive in global area, has several advantages for trace gas research, particularly the spatial landscape heterogeneity in soil types and plant communities. This variation can be viewed as a "natural experiment," allowing field measurements under extremes of moisture and temperature. While the atmospheric carbon dioxide ($CO_2$) record at Niwot Ridge extends back to 1968 (chapter 3), and NOAA has done extensive measurements on atmospheric chemistry at the subalpine climate station (e.g., Conway et al. 1994), work on tundra soil-atmosphere interactions were not initiated until recently. In 1992, studies were begun on Niwot Ridge to gain a comprehensive understanding of trace gas fluxes from alpine soils. Our sampling regime was designed to capture the spatial and temporal patterns of trace gas fluxes in the alpine. In addition, we coupled our studies of trace gas fluxes with ongoing studies of nitrogen cycling on Niwot Ridge (Fisk and Schmidt 1995, 1996; Fisk et al. 1998; chapter 12).

Methane ($CH_4$), carbon dioxide ($CO_2$), and nitrous oxide ($N_2O$) were studied because of their role in global environmental change and because they could be easily monitored at our remote sites. On a per-molecule basis, $CH_4$ and $N_2O$ are much more potent as greenhouse gases than $CO_2$ is (Lashof and Ahuja 1990; Rodhe 1990). In addition, $N_2O$ plays a role in ozone depletion in the stratosphere. The global $CH_4$ and $N_2O$ budgets are still poorly understood and the relative importance of soils in these budgets is even less clear. For example, estimates of the global soil sink for

$CH_4$ range from 9.0 to 55.9 Tg per year (Dörr et al. 1993). This range is large compared with the approximately 30 Tg of excess $CH_4$ that is accumulating in the atmosphere every year.

To better assess the role of soil in trace gas budgets, our work focused on investigating landscape patterns of gas fluxes ($CH_4$, $N_2O$, and $CO_2$) and environmental controls on these fluxes. We used natural landscape gradients in soil temperature and soil moisture to provide information on these controls and the ranges of gas fluxes observed in the alpine. We also supplemented these observational studies with an extensive series of laboratory manipulations to further our understanding of control mechanisms. Presentation of experimental results is limited here to where these assist in interpretations of landscape patterns.

## Study Sites

All of the studies have been carried out in the Saddle research site at Niwot Ridge or on adjacent north- and south-facing slopes. Landscape studies emphasizing variation in soil moisture and plant species composition are the primary focus of our review, primarily in dry meadow communities dominated by *Kobresia myosuroides*, a tussock-forming sedge; moist meadow communities dominated by *Acomastylis rossii*, a forb; and wet meadow communities dominated by *Carex scopulorum* (see chapter 6 for more complete description of vegetation in these communities). Soil moisture was highest in wet meadows followed by moist and dry meadows (West et al. 1999; chapter 8). Moist meadows also had significantly higher percentage of water-filled pore space than did dry meadows. Soil temperatures followed the same spatial pattern as soil moisture. Wet meadow soils were the coldest at 15 cm depth, and moist meadows were cooler than dry meadows. Soil tended to warm and dry after snowmelt, but these trends were more pronounced in the moist and wet meadows where snow accumulates during the winter. Soil pH also varied significantly across the alpine landscape (Fisk and Schmidt 1995). Dry meadows had the highest pH (mean = 5.4) followed by wet meadows (mean = 5.2) and finally moist meadows (mean = 4.7) (Fisk and Schmidt 1995).

## Methane Fluxes

### Spatial Patterns of $CH_4$ Flux

During the snow-free season, methane fluxes showed distinct spatial patterns that corresponded to plant community-type and to a lesser extent to landscape position (West et al. 1999). Moist and dry communities were net consumers of methane, whereas wet meadows showed varying amounts of methane production (Fig. 13.1).

In the dry meadows, only $CH_4$ uptake was observed (mean flux = $-0.77$ mg $CH_4$ $m^{-2} d^{-1}$ integrated over the growing season). In moist meadow soils, $CH_4$ was predominately consumed (mean = $-0.43$ mg $CH_4$ $m^{-2} d^{-1}$), but some zero and even some very small positive net fluxes were measured (West et al. 1999). The south-

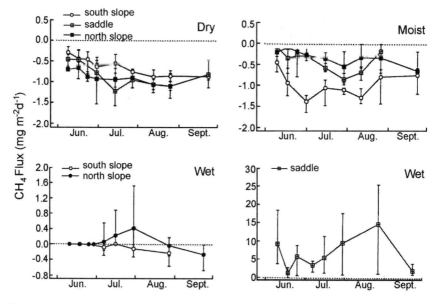

Figure 13.1. Spatial and temporal patterns of methane fluxes for dry, moist, and wet meadows on Niwot Ridge for 1992.

slope site usually had the highest rate of methane oxidation of the moist meadow sites compared to other slope exposures (Fig. 13.1).

Overall, the rates of $CH_4$ oxidation in moist and dry meadows were generally lower than methane oxidation rates observed in other dry tundra systems. For example, Adamsen and King (1993) reported methane oxidation rates of $-2.4--4.3$ mg $CH_4$ m$^{-2}$ d$^{-1}$ for subarctic tundra, and Whalen and Reeburgh (1990a) observed a rate of $-2.7$ mg $CH_4$ m$^{-2}$ d$^{-1}$ at a moist tundra site in the Aleutian Islands. One possible reason for the lower methane oxidation rates on Niwot Ridge relative to arctic tundra is higher rates of N deposition (Sievering et al. 1996), which has been shown to decrease $CH_4$ fluxes of soils (Hütsch et al. 1994; Neff et al. 1994).

In wet meadows, net $CH_4$ fluxes were usually positive or zero, but intersite variability was extremely high (Fig. 13.1). The north-slope site had zero net fluxes early in the summer followed by production of $CH_4$ in midsummer and consumption of $CH_4$ in late summer and fall. The south-slope site showed a similar pattern, except that the magnitude of the $CH_4$ production phase was lower. In fact, only the Saddle site showed a net positive methane flux for the growing season. The mean flux for this site was $+8.45$ mg $CH_4$ m$^{-2}$ d$^{-1}$ (West et al. 1999). In contrast, mean fluxes were very low, $-0.06$ and $+0.05$ mg $CH_4$ m$^{-2}$ d$^{-1}$, for the south- and north-slope sites, respectively. In addition, Neff et al. (1994) measured a relatively low emission rate of $+0.05$ mg $CH_4$ m$^{-2}$ d$^{-1}$ from a site very close to our south-slope site. The $CH_4$ emissions observed in our wet meadow plots are smaller than were most fluxes from wet meadows of Arctic tundra (Table 13.1). Our estimates are similar, however, to fluxes from moist sites in the Arctic. Most of the high positive fluxes

Table 13.1. A Comparison of Methane Fluxes from Methane-Producing Tundra of Arctic and Alpine Environments

| Site | Mean Daily Flux (mg $CH_4$ $m^{-2}$ $d^{-1}$) | Reference |
|---|---|---|
| Arctic | | |
|   Wet meadow[a] | 149.7 | Christensen (1993) |
|   Tussock[b] | 44.0 | Christensen (1993) |
|   Intertussock | 1.0 | Christensen (1993) |
|   Wet meadow[a] (1987–1989) | 19.9 | Whalen and Reeburgh (1992) |
|   Wet meadow[a] (1990) | 451.9 | Whalen and Reeburgh (1992) |
|   Tussock[b] (1987–1990) | 59.0 | Whalen and Reeburgh (1992) |
|   Intertussock (1987–1990) | 11.6 | Whalen and Reeburgh (1992) |
|   Wet meadow | 143.6 | Bartlett et al. (1992) |
|   Upland tundra | 2.3 | Bartlett et al. (1992) |
|   Polygonal ground (low center) | 46.1 | Morrissey and Livingston (1992) |
|   Polygonal ground (high center) | 4.9 | Morrissey and Livingston (1992) |
|   Alpine tundra | 0.6 | Whalen and Reeburgh (1990b) |
| Alpine | | |
|   Wet meadow[c] (mean of three sites) | 2.8 | West et al. (1999) |
|   Wet meadow[c] (fertilized) | 0.17 | Neff et al. (1994) |
|   Wet meadow[c] (control) | 0.05 | Neff et al. (1994) |
|   Wet meadow[a] (mean of two sites) | 7.7 | Smith and Lewis (1992) |

*Note:* Fluxes are mean values for the snow-free season unless indicated otherwise.

[a]Dominated by *Carex aquatilis*.
[b]Dominated by *Eriophorum vaginatum*.
[c]Dominated by *Carex scopulorum*.

shown in Table 13.1 are from sites that are saturated with water for most of the growing season.

## Controls of $CH_4$ Flux

In the dry and moist meadow communities, the highest methane oxidation rates were generally observed later in the growing season (Fig. 13.1) after soils had dried and warmed. In the dry meadows, $CH_4$ flux did not correlate with gravimetric soil moisture or soil temperature, but there was a weak correlation with percentage of water-filled pore space (West et al. 1999). West et al. (1999) did find, however, that methane oxidation correlated with constructed model variables that represented the days since the last precipitation event. In other words, $CH_4$ oxidation rates were higher if there had been a recent precipitation event. The effect of precipitation events were stronger in the two driest meadow plots (south and north slopes). To further investigate the effects of recent rainfall events on methane fluxes from dry meadows, West et al. (1999) intensively sampled the north-slope plot before and immediately following an 11-mm rainfall event that was supplemented with an additional 20 mm of deionized water. This experiment was conducted in the middle of a summer dry period. Watered plots showed the highest rate of methane oxidation ($-1.4$ mg $CH_4$ m$^{-2}$ d$^{-1}$) recorded for any dry meadow plot in 1992. These results indicate that during summer dry periods, methanotrophs are probably water limited in the dry meadows. Several studies in other systems have shown that water can stimulate $CH_4$ oxidation in dry soils (Striegl et al. 1992; Torn and Harte 1996).

In moist meadows, methane oxidation was correlated with extant soil moisture levels but not with recent rainfall events (West et al. 1999). Methane flux also correlated with soil temperature, especially at deeper soil depth (15 cm). Soil moisture and temperature taken together explained 63% of the variation in moist meadow fluxes, with soil moisture accounting for more variation than soil temperature did (West et al. 1999). In comparing methane fluxes in the dry and moist meadows, West et al. (1999) also noted that the maximum $CH_4$ oxidation occurred in the driest moist meadow plot but in the wettest two dry meadow plots. Likewise, Fisk and Schmidt (1995) found that seasonal patterns of water availability were important in controlling nitrogen mineralization in the dry meadow but not the moist meadow communities.

In the wet meadows, soil moisture (to 10 cm depth) did not correlate with $CH_4$ flux in any of the sites. Instead, $CH_4$ fluxes appeared to be sensitive only to soil temperature. In addition, there seems to be a threshold temperature below which no fluxes were observed. In the south- and north-slope sites, positive $CH_4$ fluxes were only observed when soil temperatures at a 15 cm depth were greater than 4°C (West et al. 1999).

In summary, $CH_4$ fluxes during the growing season are correlated with moisture and temperature regimes, which in turn are determined by winter snowpack distribution. As moisture availability increases from dry to moist to wet meadows, seasonally averaged $CH_4$ fluxes were more positive. However, no one factor correlated with $CH_4$ flux across all plant community types. Methane oxidation in the dry meadows was primarily limited by lack of water through much of the growing season. In contrast, in the moist meadows, more water tended to inhibit methane oxidation, probably by limiting the rate of gas diffusion in these soils (West and Schmidt 1998). Finally, in the wet meadows, soils were usually wet enough to inhibit methane oxidation to varying degrees and net methane production was most affected by soil temperature.

## Carbon Dioxide Fluxes

### Growing-Season Soil $CO_2$ Fluxes

Our data set for growing-season soil $CO_2$ fluxes is less extensive than for $CH_4$ and $N_2O$, but several intriguing patterns are apparent in these data (Fig. 13.2). The most interesting pattern is an inverse relationship between $CH_4$ and $CO_2$ fluxes in the wet meadows. The lowest $CO_2$ fluxes were observed at the Saddle site (Fig. 13.2), which conversely was the site with the highest positive $CH_4$ fluxes (Fig. 13.1). Both of these observations are what would be expected if the Saddle site was the most anaerobic of the wet meadow sites. Anaerobiosis would limit $CO_2$ production because

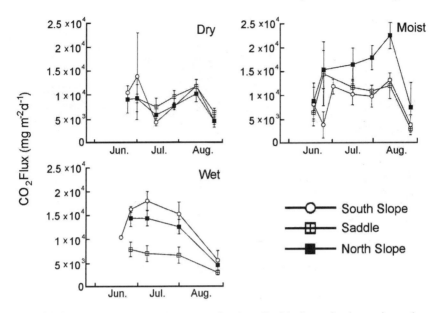

Figure 13.2. Spatial and temporal patterns of carbon dioxide fluxes for dry, moist, and wet meadows on Niwot Ridge for 1992.

fermentative microbial processes yield less $CO_2$ per substrate molecule than aerobic respiration yields. In contrast, increased anaerobiosis contributes to methane emissions by stimulating methanogenesis and inhibiting methanotrophy.

There are also interesting patterns of $CO_2$ fluxes in the dry meadow communities. The three dry meadow sites showed the least intersite variation of the community-types, and overall fluxes were lower at the dry meadows than in the moist and wet meadows. This supports the idea that microbial activity in the dry meadows is moisture limited (West and Schmidt 1998) and is consistent with results from decomposition experiments (Bryant et al. 1998). Further evidence of moisture limitation is that $CO_2$ emissions increased in the dry meadows (and moist meadows) in August following a large rainfall event (Fig. 13.2).

## Winter $CO_2$ Fluxes

Fluxes of $CO_2$ from under alpine snowpacks were greatest at sites that had the deepest and most-consistent snow cover. For example, overwinter losses of $CO_2$ ranged from 0.3 g C m$^{-2}$ for sites with inconsistent snow cover to 25.7 g C m$^{-2}$ at sites that had consistent snow cover (Brooks et al. 1997). Consistent snow cover allowed soils to thaw, which in turn led to increased microbial activity. At sites with moderate duration snowpacks, measurable $CO_2$ fluxes were observed once soil temperatures reached approximately $-5°C$, then increased throughout the remainder of the winter until approximately two weeks before the sites were snow free. While there was no significant relationship between flux and temperature, there was a signifi-

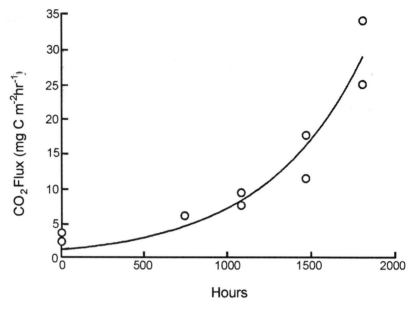

Figure 13.3. The relationship between $CO_2$ flux and number of hours since soils thawed beneath alpine snowpacks on Niwot Ridge. Data are redrawn from Figure 2a of Brooks et al. (1997).

cant exponential relationship ($r^2 = 0.93$) between $CO_2$ flux and the number of days that thawed soil was present under the snow (Fig. 13.3).

The exponential increase in $CO_2$ production (Fig. 13.3) observed by Brooks et al. (1997) is the first evidence that we know of demonstrating in situ microbial growth at temperatures below 0°C. Using the substrate-induced growth response (SIGR) method (Colores et al. 1996; Schmidt 1992), we estimated an initial cold-adapted microbial biomass of 0.78 g C m$^{-2}$ beneath these alpine snowpacks. For the 75 days that the $CO_2$ flux increased beneath the snowpack, this initial psychrophilic biomass grew (generation time = 17 days) to approximately 18 g C m$^{-2}$. The existence of large numbers of psychrophilic microbes under snowpacks in dry meadow soils has also recently been described by Lipson et al. (1999). They found a microbial biomass of 66 g C m$^{-2}$ that could grow on glutamate at 3°C. The surprising amount of microbial activity beneath alpine snowpacks needs to be explored in more detail, especially with reference to trace gas fluxes and biogeochemistry of alpine areas.

## Nitrous Oxide Fluxes

### Spatial Patterns of $N_2O$ Fluxes

On average, moist meadow communities had the highest levels of $N_2O$ production (mean flux = 44.4 µg $N_2O$ m$^{-2}$ d$^{-1}$), followed by the dry meadow (mean = 29.6

µg $N_2O$ m$^{-2}$ d$^{-1}$) and wet meadows (mean = 14.6 µg $N_2O$ m$^{-2}$ d$^{-1}$). In addition, at a south-slope site on Niwot Ridge, Neff et al. (1994) measured mean fluxes of 16.8 and 19.2 µg $N_2O$ m$^{-2}$ d$^{-1}$ for dry and wet communities, respectively.

As with methane fluxes, $N_2O$ fluxes within each plant community were spatially variable (Jaeger 1993). Dry meadows showed the least spatial variation in $N_2O$ fluxes. In the dry and moist meadows, spatial variation in $N_2O$ fluxes corresponded with spatial variation in soil moisture. Substantial $N_2O$ was emitted only from the wettest dry meadow sites and the driest moist meadow sites. This is not unexpected, as denitrification produces more $N_2O$ at intermediate soil moistures, because such intermediate soil moistures coincide with relatively high oxygen availability. The north- and south-slope moist meadow sites produced similar amounts of $N_2O$ over the season, and both showed elevated levels of $N_2O$ production following the August rainfall event. In contrast, the Saddle moist meadow site produced almost no $N_2O$ (Fig. 13.4).

The spatial pattern of $N_2O$ emission in the wet meadow sites did not correspond with soil moisture, but nevertheless appears to correspond with oxygen availability. The only wet meadow site that emitted significant amounts of $N_2O$ was the south-slope site. Compared with the other wet meadow sites, the south-slope wet meadow also had the highest rates of $CH_4$ oxidation (Fig. 13.1), $CO_2$ emission (Fig. 13.2), and nitrification (Fisk 1995), all of which indicate less anaerobic soil conditions. Excessively anaerobic conditions in the other wet meadow sites may have limited $N_2O$ production, both by increasing reduction of $N_2O$ to $N_2$ by denitrifiers and by limiting nitrification.

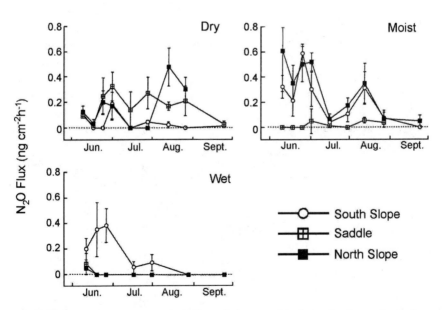

Figure 13.4. Spatial and temporal patterns of nitrous oxide fluxes for dry, moist, and wet meadows on Niwot Ridge for 1992.

## Temporal Patterns of $N_2O$ Fluxes

Large $N_2O$ fluxes occurred only early in the growing season or after a significant rainfall event late in the season (Fig. 13.4). The phenomenon of higher fluxes occurring in spring has been seen in several ecosystems, including the tallgrass prairie (Groffman et al. 1993) and the sagebrush steppe (Matson et al. 1991). The fluxes late in the growing season occurred only in the dry and moist communities. These late-season fluxes followed a wet period that had been preceded by a long dry period. Workers in other systems have also noted a significant pulse of $N_2O$ following the wetting of dry soils (Garcia-Mendez et al. 1991).

## Controls of Growing Season $N_2O$ Fluxes

In many environments, $N_2O$ fluxes are controlled by soil moisture and levels of available N. The results of Neff et al. (1994) indicate that water and available N limit $N_2O$ fluxes in the alpine. In fertilized dry meadow soils, $N_2O$ fluxes were high early in the season and tapered off as soils dried later in the summer. Fluxes increased again following the August rainy weather previously mentioned. Thus, dry meadow $N_2O$ fluxes are probably limited by a combination of water and N availability. In contrast, wet meadow $N_2O$ fluxes are limited primarily by N availability. Neff et al. (1994) showed that $N_2O$ fluxes from fertilized plots in the wet meadows continued at a high level throughout the growing season. In the nonfertilized wet meadow plots, $N_2O$ fluxes were very low or zero after mid-July presumably because of very low $NO_3^-$ levels in these soils after that date (Jaeger 1993).

## Winter $N_2O$ Production

Brooks et al. (1996, 1997) studied wintertime $N_2O$ fluxes from several moist meadow communities on Niwot Ridge. The most surprising finding of Brooks et al. (1997) was that $N_2O$ fluxes during parts of the winter are as high or higher than they are during the growing season (Table 13.2). The controls on wintertime $N_2O$ flux

Table 13.2. A Comparison of $N_2O$ Fluxes from Alpine Environments

| Site | Mean Daily Flux ($\mu g\ N\ m^{-2}\ d^{-1}$) | Range | Reference |
|---|---|---|---|
| Winter | | | |
| Moist meadow | | | |
| Shallow snowpack | 39 | 0–78 | Brooks et al. (1997) |
| Deep snowpack | 97 | 5–472 | Brooks et al. (1997) |
| Snow-free season | | | |
| Dry meadow | 16 | 0–48 | West et al. (1999) |
| Dry meadow | 17 | 0–53 | Neff et al. (1994) |
| Moist meadow | 26 | 0–94 | West et al. (1999) |
| Wet meadow | 9 | 0–42 | West et al. (1999) |
| Wet meadow | 19 | 0–170 | Neff et al. (1994) |

*Source:* Adapted from Brooks et al. (1997).

seem to be basically the same as during the growing season except that temperature is an overriding factor, with fluxes occurring only when soil temperatures were above $-2°C$. Such temperatures were only reached, however, when snow cover was deep enough (ca. 30 cm) to insulate the soil. At such times, the environment beneath the snowpack was wet, and inorganic N concentrations were high (Brooks et al. 1996). Thus, depth of snowpack is the most important environmental variable affecting wintertime $N_2O$ fluxes.

At sites with a moderate snow cover, $N_2O$ fluxes showed an exponential increase once soils had thawed beneath the snowpack. This increase began about two weeks after that seen for $CO_2$ fluxes at the same sites (Brooks et al. 1997). $N_2O$ fluxes increased from 9 $\mu g$ $N_2O$-N $m^{-2} d^{-1}$ to a maximum flux of 78 $\mu g$ N $m^2 d^{-1}$ shortly before the sites became snow free. The mean flux over the 50 days that production was observed at these sites was 40 $\mu g$ N $m^{-2} d^{-1}$.

## Summary of $N_2O$ Fluxes

Overall $N_2O$ fluxes from Niwot Ridge do not constitute a major loss of N from the system. Although the fluxes can be high at times, Brooks et al. (1997) estimated that only 15% of N inputs from the snowpack are lost as $N_2O$ during the snow-covered season. Likewise, Jaeger (1993) estimated that about 1% of N deposited on Niwot Ridge during the growing season is lost as $N_2O$.

Based on the findings of Neff et al. (1994), Jaeger (1993), and Brooks et al. (1997), we estimate that loss of N in the form of $N_2O$ from Niwot Ridge is approximately 0.01 g N $m^{-2} y^{-1}$. This value is similar to $N_2O$ losses from shortgrass prairie (0.008–0.016 g N $m^{-2} y^{-1}$) (Parton et al. 1988), a sagebrush steppe ecosystem (0.021 g N $m^{-2} y^{-1}$) (Matson et al. 1991), Northeastern spruce-fir forests (0.002 g N $m^{-2} y^{-1}$) (Castro et al. 1993), or a Rocky Mountain Douglas-fir forest (0.003–0.023 g N $m^{-2} y^{-1}$) (Matson et al. 1992). All of these values are much smaller than are $N_2O$ losses from more-productive ecosystems. For example, Groffman and Turner (1995) estimated $N_2O$ losses of 0.66 g N $m^{-2} y^{-1}$ for the Konza tallgrass prairie in eastern Kansas.

## Future Research

The most exciting finding from our studies was the dynamic behavior of trace gas producing microbes in the winter. The importance of this activity to biogeochemical cycles is only just now being explored (e.g., Brooks et al. 1997; Lipson et al. 1999). The subnivean environment is much more hospitable to microbial life than we previously imagined and is an area that deserves special attention in future alpine research. Moreover, given potential changes in the patterns and amounts of snowpack, this interest is an appropriate topic of global change research.

*Acknowledgments* This research was supported by Grant R81–9448 from the Environmental Protection Agency National Center for Environmental Research and Quality Assurance. Supplemental and logistical support was provided by the Niwot Ridge Long-Term Ecological Research project (NSF DEB 9211776).

References

Adamsen, A. P. S., and G. M. King. 1993. Methane consumption in temperate and subarctic forest soils: Rates, vertical zonation, and responses to water and nitrogen. *Applied and Environmental Microbiology* 59:485–49.

Bartlett, K. B., P. M. Crill, R. L. Sass, R. C. Harris, and N. B. Dise. 1992. Methane emissions from tundra environments in the Yukon-Kuskokwim Delta, Alaska. *Journal of Geophysical Research* 97D:16645–16660.

Brooks, P. D., S. K. Schmidt, and M. W. Williams. 1997. Winter production of $CO_2$ and $N_2O$ from alpine tundra: Environmental controls and relationship to inter-system C and N fluxes. *Oecologia* 110:403–413.

Brooks, P. D., M. W. Williams, and S. K. Schmidt. 1996. Microbial activity under alpine snowpacks, Niwot Ridge, Colorado. *Biogeochemistry* 32:93–113.

Bryant, D. M., E. A. Holland, T. R. Seastedt, and M. D. Walker. 1998. Analysis of decomposition in alpine tundra. *Canadian Journal of Botany* 76:1295–1304.

Castro, M. S., P. A. Steudler, J. M. Melillo, J. D. Aber, and S. Millham. 1993. Exchange of $N_2O$ and $CH_4$ between the atmosphere and soils in spruce-fir forests in the northeastern United States. *Biogeochemistry* 18:119–135.

Christensen, T. R. 1993. Methane emission from Arctic tundra. *Biogeochemistry* 21:117–139.

Colores, G. M., S. K. Schmidt, and M. C. Fisk. 1996. Estimating the biomass of microbial functional groups using rates of growth-related soil respiration. *Soil Biology and Biochemistry* 28:1569–1577.

Conway, T. J., P. Tans, L. Waterman, K. Thoning, D. Kitzis, K. Mesanie, and N. Zhang. 1994. Evidence for interannual variability of the carbon cycle from the NOAA air sampling network. *Journal of Geophysical Research* 99:22831–22855.

Dörr, H., L. Katruf, and I. Levin. 1993. Soil texture parameterization of the methane uptake in aerated soils. *Chemosphere* 26:697–713.

Fisk, M. C. 1995. Nitrogen Dynamics in an Alpine Landscape. Ph.D. diss., EPO Biology, University of Colorado, Boulder.

Fisk, M. C., and S. K. Schmidt. 1995. Nitrogen mineralization and microbial biomass nitrogen dynamics in three alpine tundra communities. *Soil Science Society of America Journal* 59:1036–1043.

———. 1996. Microbial responses to nitrogen additions in alpine tundra soil. *Soil Biology and Biochemistry* 28:751–755.

Fisk, M. C., S. K. Schmidt, and T. Seastedt. 1998. Topographic patterns of above-and belowground production and nitrogen cycling in alpine tundra. *Ecology* 79:2253–2266.

Garcia-Mendez, G. C., J. M. Maass, P. A. Matson, and P. M. Vitousek. 1991. Nitrogen transformations and nitrous oxide flux in a tropical deciduous forest in Mexico. *Oecologia* 88:362–366.

Groffman, P. M., C. W. Rice, and J. M. Tiedje. 1993. Denitrification in a tall grass prairie landscape. *Ecology* 74:855–862.

Groffman, P. M., and C. L. Turner. 1995. Plant productivity and nitrogen gas fluxes in a tallgrass prairie landscape. *Landscape Ecology* 10:255–266.

Hütsch, B. W., C. P. Webster, and D. S. Powlson. 1994. Methane oxidation in soil as affected by land use, soil pH and N fertilization. *Soil Biology and Biochemistry* 26:1613–1622.

Jaeger, C. H. 1993. Plant and soil nitrogen interactions in the alpine. Ph.D. diss., University of Colorado, Boulder.

Jaeger, C. H., R. K. Monson, M. C. Fisk, and S. K. Schmidt. 1999. Seasonal partitioning of nitrogen by plants and soil microorganisms in an alpine ecosystem. *Ecology* 80:1883–1891.

Lashof, D. A., and D. R. Ahuja. 1990. Relative contributions of greenhouse gas emissions to global warming. *Nature* 344:529–531.

Lipson, D. A., S. K. Schmidt, and R. K. Monson. 1999. Links between microbial population dynamics and nitrogen availability in an alpine ecosystem. *Ecology* 80:1623–1631.

Matson, P. A., S. T. Gower, C. Volkmann, C. Billow, and C. C. Grier. 1992. Soil nitrogen cycling and nitrous oxide flux in a Rocky Mountain Douglas-fir forest: Effects of fertilization, irrigation and carbon addition. *Biogeochemistry* 18:101–117.

Matson, P. A., C. Volkmann, K. Coppinger, and W. A. Reiners. 1991. Annual nitrous oxide flux and soil nitrogen characteristics in sagebrush steppe ecosystems. *Biogeochemistry* 14:1–12.

Morrissey, L. A., and G. P. Livingston. 1992. Methane emissions from Alaska Arctic tundra: An assessment of local spatial variability. *Journal of Geophysical Research* 97D:16661–16670.

Neff, J. C., W. D. Bowman, E. A. Holland, M. C. Fisk, and S. K. Schmidt. 1994. Fluxes of nitrous oxide and methane from nitrogen amended soils in a Colorado Alpine ecosystem. *Biogeochemistry* 27:23–33.

Parton, W. J., A. R. Mosier, and D. S. Schimel. 1988. Rates and pathways of nitrous oxide production in a shortgrass steppe. *Biogeochemistry* 6:45–58.

Rodhe, H. 1990. A comparison of the contribution of various gases to the greenhouse effect. *Science* 248:1217–1219.

Schmidt, S. K. 1992. A substrate-induced growth-response (SIGR) method for estimating the biomass of microbial functional groups in soil and aquatic systems. *FEMS Microbiology and Ecology* 101:197–206.

Sievering, H., D. Rusch, and L. Marquez. 1996. Nitric acid, particulate nitrate and ammonium in the continental free troposphere: Nitrogen deposition to an alpine tundra ecosystem. *Atmospheric Environment* 30:2527–2537.

Smith, L. K., and W. M. Lewis Jr. 1992. Seasonality of methane emissions from five lakes and associated wetlands of the Colorado Rockies. *Global Biogeochemical Cycles* 6:323–338.

Striegl, R. G., T. A. McConnaughey, D. C. Thorstenson, E. P. Weeks, and J. C. Woodward. 1992. Consumption of atmospheric methane by desert soils. *Nature* 357:145–147.

Torn, M. S., and J. Harte. 1996. Methane consumption by montane soils: Implications for positive and negative feedback with climatic change. *Biogeochemistry* 32:53–67.

West, A. E., P. D. Brooks, M. C. Fisk, L. K. Smith, E. A. Holland, C. H. Jaeger, S. Babcock, R. S. Lai, and S. K. Schmidt. 1999. Landscape patterns of $CH_4$ fluxes in an alpine tundra ecosystem. *Biogeochemistry* 45:243–264.

West, A. E., and S. K. Schmidt. 1998. Wetting stimulates atmospheric methane oxidation by alpine soils. *FEMS Microbiology Ecology* 25:349–353.

Whalen, S. C., and W. S. Reeburgh. 1990a. Consumption of atmospheric methane by tundra soils. *Nature* 346:160–162.

———. 1990b. A methane flux transect along the trans-Alaska pipeline haul road. *Tellus* 42B:237–249.

Whalen, S. C., and W. S. Reeburgh. 1992. Interannual variations in tundra methane emission: a 4-year time series at fixed sites. *Global Biogeochemical Cycles* 6:139–159.

# 14

# Plant-Herbivore Interactions

Denise Dearing

## Introduction

The alpine provides a tremendous opportunity for studying plant-herbivore interactions at the population, community, and ecosystem levels. For herbivores, variations in topography and microclimate result in a relatively large amount of spatial variation in plant communities within short distances (chapter 6). A large community of herbivores, from nematodes to grasshoppers to elk, occurs on Niwot Ridge. Furthermore, given the low rates of nutrient availability in alpine soils (Fisk and Schmidt 1995; chapter 12) combined with the slow-growing perennial habit of the vegetation, alpine plants should, in theory, invest heavily in defense against herbivores (Coley et al. 1985).

The goal of this chapter is to provide: (1) a summary of the feeding behaviors of the herbivores on Niwot Ridge, (2) information on the nutritional and secondary chemistry of plants on Niwot Ridge as it relates to herbivory, and (3) a review of hypotheses on community dynamics of herbivores and plants relevant to the alpine. The ultimate objective is to provide a synthesis of information that will stimulate interest in alpine tundra as a system for studying the dynamics of plant-herbivore interactions at all levels of ecological organization.

## Forage Quality and Quantity

The flora of Niwot Ridge has been divided into six communities (May and Webber 1982; chapter 6). Regardless of community association, nearly all of the plant species occurring on the ridge are perennials and several are very long lived (May and Webber 1982). Communities can change across small spatial scales (meters), and community origin and maintenance are believed to be largely determined by abiotic factors (Walker et al. 1994; chapter 6). However, several studies suggest that biotic factors such as herbivory may have a significant impact on plant community dynamics (Huntly et al. 1986; Davies 1994).

There is significant variation in the nutritional composition of plants on Niwot Ridge. Generally, and in the absence of plant secondary compounds, species that are high in nitrogen and low in fiber are presumed to be the most desirable as forage. Based solely on these nutritional variables, the clover *Trifolium parryi* is hypothesized to be one of the more-preferred forages, whereas alpine sandwort, *Minuartia obtusiloba*, should be one of the less-preferred food items (Table 14.1).

Given the low nutrient availability, high light availability, and perennial habit of the majority of plants on Niwot Ridge, theory predicts plants should produce high levels of antiherbivore defense, particularly secondary compounds that are carbon based (Coley et al. 1985). Indeed, the most common class of secondary compounds on the ridge is tannins, a carbon-based defense (Dearing 1996; Table 14.2). Of the 10 plant species studied thus far, *Acomastylis rossii* (Rosaceae), one of the most

Table 14.1. Energy, Nitrogen and Fiber Levels of Niwot Ridge Plants

| Plant Species | Energy (Kcal/g) | Nitrogen (%) | Fiber (%) |
|---|---|---|---|
| *Acomastylis rossii* | | | |
| Flowers | 4.728 | 2.49 | 25.1 |
| Leaves | 4.607 | 2.95 | 14.3 |
| *Artemisia scopulorum* | | | |
| Flowers | 4.085 | — | — |
| Leaves | 4.640 | 3.12 | 22.5 |
| *Bistorta bistortoides* | | | |
| Flowers | 4.711 | 3.00 | 25.1 |
| Leaves | 4.575 | 3.72 | 29.5 |
| *Castilleja occidentalis* | | | |
| Flowers | 4.462 | 2.46 | — |
| Leaves | 4.368 | 2.34 | 26.4 |
| *Erigeron simplex* | | | |
| Flowers | 4.574 | 2.21 | 25.6 |
| Leaves | 4.584 | 2.62 | 19.0 |
| Graminoids | 4.645 | 2.66 | 28.1 |
| *Minuartia obtusifolia* | | | |
| Leaves | 4.449 | 1.59 | 43.7 |
| *Silene acaulis* | | | |
| Flowers | — | 2.83 | 20.7 |
| Leaves | 4.778 | 3.01 | 26.2 |
| *Trifolium parryi* | | | |
| Flowers | 4.739 | 2.9 | 38.6 |
| Leaves | 4.688 | 4.1 | 24.8 |

*Notes*: Energy measured with bomb calorimetry; nitrogen was measured on a CHN analyzer; fiber measured using the acid detergent fiber technique and therefore represents the cellulose and lignin fraction.

Table 14.2. Three Measures of Phenolic Contents in Plants Occurring on Niwot Ridge

| Plant Species | Total Phenolics (mg/g TAE) | Condensed Tannins (mg/g QE) | Protein Precipitation (mg/g TAE) |
|---|---|---|---|
| *Acomastylis rossii* | | | |
| Flowers | 162.2 | 5.2 | 459.7 |
| Leaves | 204.3 | 0 | 549.5 |
| *Artemisia scopulorum* | | | |
| Flowers | 32.3 | 0 | 18.4 |
| Leaves | 31.7 | 15.0 | 40.1 |
| *Bistorta bistortoides* | | | |
| Flowers | 90.6 | 218.7 | 119.5 |
| Leaves | 53.4 | 233.4 | 98.1 |
| *Castilleja occidentalis* | | | |
| Flowers | 31.2 | — | 29.0 |
| Leaves | 31.6 | — | 33.8 |
| *Erigeron simplex* | | | |
| Flowers | 24.1 | 0 | 57.5 |
| Leaves | 11.6 | 0 | 38.5 |
| Graminoids | 31.7 | 0 | 32.0 |
| *Minuartia obtusifolia* | | | |
| Leaves | 6.0 | — | — |
| *Sibbalia procumbens* | | | |
| Leaves | 157.7 | 335.8 | — |
| *Silene acaulis* | | | |
| Flowers | 29.7 | — | 25.5 |
| Leaves | 28.9 0 | 23.9 | |
| *Trifolium parryi* | | | |
| Flowers | 63.2 | 0 | 85.8 |
| Leaves | 48.4 | 0 | 48.0 |

*Notes*: Phenolics were extracted in 85% MeOH using the methods of Torti et al. (1995). Phenolics were measured using the following assays: total phenolics, Folin-Ciocalteau; condensed tannins, butanol-HCL; protein precipitation, Hagerman and Butler (1978). The protein precipitation assay is a measure of how strongly the phenolics interact with protein. TAE is tannic acid equivalents; QE is quebracho equivalents.

abundant, produces the greatest concentration of tannins (hydrolyzable tannins) (Dearing 1997a). A few species, *Bistorta bistortoides* (Polygonaceae) and *Sibbaldia procumbens* (Rosaceae), produce very high levels of condensed tannins. Many of the other species assayed produce moderate-to-low quantities of phenolics. Although other classes of secondary chemicals are present in plants on Niwot Ridge, they are not as predominant as tannins are. For example, terpenes have been reported for only one species, *Artemisia scopulorum* (Gibbs, 1974). Thus, in addition to the challenges presented by a short growing season and relatively low nitrogen, herbivores must also contend with tannin-rich foods.

The ingestion of tannins by herbivores appears to have two different effects (Waterman and Mole 1994). Tannins are renowned for their affinity for other molecules, especially proteins. Initially, tannins were hypothesized to act on mammalian herbivores by binding proteins in the gut, thereby rendering the protein-tannin complex inaccessible for digestion (Feeny 1976). This hypothesis implied that tannins were not toxins because they were not absorbed by the gut tissue into the body, and thus they were labeled as "quantitative defenses" or "digestibility-reducers." Many studies have demonstrated that the apparent digestibility of protein decreases in herbivores consuming high-tannin diets (Lindroth and Batzli 1986; Robbins et al. 1987; Dearing 1997a). That tannins could behave as a toxin was largely ignored because tannin molecules were thought too large to be absorbed across the gut lining. However, recent work has demonstrated that some tannins are able to cross the gut lining and therefore are also toxins (Lindroth and Batzli 1983; Dearing 1997a). Whether all tannins exhibit these dual modes of action has not yet been established.

Although tannins are the most common class of secondary compounds in alpine plants, alkaloids appear to have this distinction in arctic tundra plants, as determined by a survey of 91 species growing near Meade River, Alaska (Jung et al. 1979). This result is surprising, given the similar climate and flora of the alpine and arctic as well as the extensive overlap of plant genera. Moreover, alkaloids, due to their high content of nitrogen, would not be predicted by the resource availability hypothesis to be produced by species growing in a nitrogen-limited environment (Coley et al. 1985) such as tundra. These surveys of secondary compounds of arctic and alpine plants did not utilize similar methods, however, and thus the differences may be artefactual. Simultaneous surveys of the secondary compounds produced by plants in both these habitats would be an interesting test of the resource availability hypothesis.

## The Herbivores

Across all terrestrial habitats, invertebrate herbivores, particularly nematodes, insects, and mollusks, are the most diverse and abundant herbivores and also consume the largest proportion of primary production (Batzli et al. 1981). The arctic tundra is one of the exceptions to this pattern, where low year-round temperatures ultimately limit diversity and abundance of invertebrate herbivores (Somme and Block 1991). Thus, homeothermic herbivores (i.e. mammals) have a far greater impact on the plant community than invertebrate herbivores do (Batzli et al. 1980, 1981; MacLean 1981). Because similar abiotic conditions occur in both arctic and alpine tundra, herbivory by mammals is probably more important than herbivory by invertebrates in alpine communities as well.

### Insect Fauna

Insects are the most common aboveground invertebrate herbivores across several arctic tundra habitats, with the predominant orders consisting of Lepidoptera (moths and butterflies), Hymenoptera (wasps, bees, and ants), Homoptera (scale insects, plant lice), Coleoptera (beetles), Diptera (flies), and Orthoptera (grasshoppers) (Haukioja 1981). In the Arctic, evidence suggests that many insects prefer deciduous shrubs over other available plant life forms. Plant lice (Homoptera: Psylloidea) in the Arctic occur only on a few species of willows (Hodkinson et al. 1979). Lepidopteran and dipteran larvae prefer deciduous shrubs over evergreen shrubs and graminoids least (MacLean and Jensen 1985). The preference for deciduous shrubs was attributed to their higher nutrients and lower secondary compound contents compared to other life forms (MacLean and Jensen 1985). However, these plant variables were not measured in this study.

Of the potential insect herbivores in Niwot Ridge, grasshoppers have received the most attention. The grasshopper community consists primarily of three species, all in the Acrididae; *Aeropedellus clavatus* and *Chorthippus curtipennis* (both in the subfamily Gomphocerinae) and *Melanoplus dodgei* (subfamily Catantopinae). Of the three, *A. clavataus* is the most common, reaching densities as high as 1 m$^{-2}$ (Alexander and Hillard 1964; Coxwell and Bock 1995). Both *A. clavatus* and *C. curtipennis* feed primarily on grasses and sedges, particularly *Kobresia myosuroides* and *Carex* species (Coxwell 1992). In no-choice feeding trials, *A. clavatus* readily consumed grasses and sedges (e.g., *Deschampsia caespitosa* and *Carex* species) but avoided *Acomastylis rossii* (Alexander and Hillard 1964). *Melanopus dodgei* is the least specialized and included more forbs (*Trifolium* species and *Artemesia* species) than graminoids in its diet (Coxwell 1992).

### Vertebrate Fauna

Of the mammalian species that occur on the ridge (chapter 7), approximately 12 species include some plant material in their diet. However, some of these species, for example, golden mantled ground squirrel (*Spermophilus lateralis*), least chipmunk (*Tamias minimus*), and red-backed vole (*Clethrionomys gapperi*), are more omnivorous than herbivorous (Merritt and Merritt 1978; Tevis 1952) and will not be included in this survey. The other species excluded from this discussion are transients such as porcupines (see chapter 7). Thus, the discussion focuses on those herbivores that should have the greatest impact on the plant communities.

### Gophers

The northern pocket gopher, *Thomomys talpoides* (Rodentia: Geomyidae), is common in the alpine, particularly on Niwot Ridge (Thorn 1978; Thorn 1982). Northern pocket gophers vary in weight from 78 to 130 g (Burt and Grossenheider 1976). Among moist meadow, snowbed, and dry meadow communities, aboveground evidence of gopher activity is highest in the snowbed (Davies 1994). In ridgetop areas with fellfield, dry meadow, and moist meadow, gopher activity is usually highest in moist meadows (Davies 1994). Although the effects of gopher mounds on the plant communities and differential use of plant communities has been studied (Davies 1994), the diet choices of gophers in alpine tundra remains unknown. Analyses of winter caches suggest that the nonfibrous roots of common species that produce tubers are harvested (Seastedt, personal communication). Given that belowground herbivory appears to be more important than aboveground herbivory, that is, removal of root has a 12 times greater effect on plant biomass than removal of similar quantities of leaf (Reichman and Smith 1991), a thorough study of the gopher diet is warranted.

The winter diet choices of gophers at an alpine site in Utah were determined by stomach content and cache item analysis (Stuebe and Andersen 1985). During the winter, gophers consume high-protein items (*Erythronium* corms or *Lupinus* roots), whereas low-protein items were stored (*Claytonia lanceolata*).

Gophers are typically most abundant in areas with the highest aboveground biomass (Reichman 1985), and this variable is often correlated with belowground rootmass (Reichman and Smith 1991), although this is probably not the case in the alpine. On Niwot Ridge, gopher abundance and aboveground biomass were not related. Gophers disturbed more soil and were more abundant (eight animals per hectare) in snowbed communities (Davies 1994), despite the fact that this community has the lowest aboveground biomass of the communities on the ridge (Walker et al. 1994).

Gophers appear to require early and complete cover of snow to survive the alpine winters, so that the soils they occupy do not freeze. These animals do not hibernate and continue to burrow during the winter through the soils to obtain tubers for food. Hence, gopher survival depends on a subset of the alpine landscape, an area that may vary in spatial and temporal extent from year to year. In summer, gophers are capable of exploiting almost any habitat of the tundra, but the areas of deep snow provide winter refugia.

Little is known of gopher population dynamics, therefore adequate information on burrowing, mounding, and feeding frequencies is not available. Steeper hillslopes often exhibit substantial effects of chronic gopher herbivory, including eroded topsoils, channel erosion, and surface gravel deposits. Do gophers exhibit pulses in population densities, exploiting certain regions of the alpine landscape before going extinct locally? Observations suggest that this is the case, but data in support of this contention are lacking. The Martinelli study site used by Litaor et al. (1996) was first monitored for gopher activity in 1986. The mounding intensities have exhibited large interannual variation and currently appear to be almost nonexistent. Further monitoring of this area should be very useful in connecting population and ecosystem phenomena.

## Voles

Two herbivorous voles, *Microtus longicaudus* and *M. montanus* (Rodentia: Arvocolidae), occur on Niwot Ridge. Studies on the diet composition and habitat preferences of these two species have never been conducted here. Thus, information on their feeding habits is from data gathered in other habitats. *Microtus longicaudus,* the long-tailed vole, weighs 45–50 g, and consumes primarily fruits and seeds, followed by leaves, preferably those of dicots. In the Great Basin, it has been reported to feed on leaves of sagebrush, *Artemisia tridentata,* which is notable for its high levels of terpenes (Linsdale 1938). The preferred habitat of *M. longicaudus* is shrubby areas (Smolen and Keller 1987). The most likely habitats on Niwot Ridge for *M. longicaudus* are probably willow- or forb-dominated communities. In contrast, the montane vole, *M. montanus,* is much smaller (27 g) and more of a grazer than *M. longicaudus* is (Tomasi 1985). In certain habitats, it can subsist on diets that consist exclusively of graminoids (Negus et al. 1986). Although *M. montanus* is only half the size of *M. longicaudus,* it appears to be capable of physically displacing *M. longicaudus* (Randall and Johnson 1979). Where montane and long-tailed voles occur sympatrically, both species tended to have synchronous population cycles on a 3-year term (Randall and Johnson 1979). Peak sizes of the populations were two- to fourfold greater for *M. montanus* compared with *M. longicaudus* (Randall and Johnson 1979).

## Pikas

Of all the herbivores occurring on Niwot Ridge, the North American pika, *Ochotona princeps,* has been the best studied (Conner 1983; Southwick et al. 1986; McKechnie 1990; Dearing 1995). Its diurnal habit and caching activities make it an attractive organism for foraging studies.

Pikas (Lagomorpha: Ochotonidae) are small (160–180 g), territorial, and solitary (Smith and Weston 1990). Pikas do not hibernate (Krear 1965). To survive the long winter when fresh vegetation is unavailable, individual pikas collect and store vegetation during the summer. These caches are commonly known as "haypiles" (Broadbrooks 1965). During cache construction, pikas clip fresh vegetation and store it under rocks. Although plant material may naturally desiccate in haypiles, pikas do not actively dry vegetation prior to inclusion in the haypile (Smith and Weston 1990). Pikas may live up to 7 years and haypile areas tend to be reused by the same individual in successive years. (Smith and Ivins 1984; Dearing, personal observation). The caches of pikas on Niwot Ridge are the largest caches ever measured (Dearing 1995). The average size of haypiles is 28 kg of fresh vegetation per pika. The haypile comprises the majority of the winter diet of pikas (Dearing 1997b).

Surprisingly, the composition of plants stored in the haypile ("winter diet") differs from the diet consumed during the summer ("summer diet") while pikas are caching. This differential foraging for summer and winter diets has been reported for Niwot Ridge and other locations (Huntly et al. 1986; Dearing 1996). On Niwot Ridge, pikas forage as generalists for their summer diets, consuming several plant species with no single species accounting for more than 50% of the diet. Plant species common in the summer diets (each 10–30% of fresh weight) are *Trilfolium parryi* leaves, *Silene acaulis* leaves, *Acomastylis rossii* leaves, graminoids, and *Bistorta bistortoides* leaves and flower heads. Of these, *T. parryi* comprises a larger proportion of the diet than expected from its relative abundance, *S. acaulis* and *B. bistortoides* are taken in proportion to their relative abundance, and grasses and *A. rossii* were consumed less frequently than their relative abundance.

Pikas are much more specialized with respect to selection of their winter diet. *Acomastylis rossii* leaves comprise 60–77% of the winter diet, far more than would be expected from its abundance (Fig. 14.1). This represents a significant departure from collection of *A. rossii* for the summer diet, where it was taken far below its relative abundance. No other food items comprised more than 13% of the winter diet

Figure 14.1. Pika (*Ochotona princeps*) with shoot of Alpine Avens (*Acomastylis rossii*), a preferred plant for its winter diet (photo by Denise Dearing).

and most comprised less than 10%. Species in the 5–13% category included *B. bistortoides* flowers and leaves, graminoids, and *T. parryi*.

Thus pikas simultaneously select two different diets: a summer diet that is consumed immediately and a winter diet that is stored prior to consumption. Neither differences in plant morphology nor nutritional content explained diet selection differences between summer and winter diets (Dearing 1996), but there were significant differences in levels of plant secondary compounds in the two diets. Concentrations of phenolic compounds were 2–7 times greater in the winter diet (Dearing 1996). The most abundant food type in the winter diet, *A. rossii* leaves, contained high concentrations of hydrolyzable tannins. In laboratory experiments, pikas ingesting a diet of 25% *A. rossii* leaves exhibited lower levels of fiber and protein digestion and excreted higher concentrations of detoxification metabolites in the urine compared with pikas consuming a diet without *A. rossii* (Dearing 1997a).

Why would pikas collect a diet consisting primarily of a plant high in toxins (i.e., *A. rossii*) that had demonstrable negative effects? Laboratory and field studies (Dearing 1997b) suggest that pikas gainfully manipulate the secondary chemistry of plants in their haypiles in two fashions. First, these compounds act as natural preservatives. Plant species high in phenolics preserve better, resulting in greater yield for every gram initially stored. Cached *A. rossii* yielded greater biomass and nutrient returns during the long winter storage period than other low-phenolic species consumed during the summer (Dearing 1997b). In laboratory experiments, extracts of *A. rossii* phenolics retarded bacterial growth (Dearing 1997b). However, its resistance to bacterial decay was not conferred to other plants with which it was stored. Therefore, it did not function as a natural preservative for other haypile plants. The other way pikas manipulate the toxins in plants is by preferentially caching plant species too noxious for immediate ingestion and delaying consumption of these plants until the toxins degrade to palatable levels (Dearing 1997b). This strategy allows pikas to increase utilization of plant species that it otherwise would not be able to consume in large quantities.

## Marmots

The yellow-bellied marmot, *Marmota flaviventris* (Rodentia: Sciuridae) ranges in weight from 2.8 kg for females to 3.9 kg for males (Armitage et al. 1976). They are social and typically live in colonies consisting of one male, several females, and offspring (Barash 1973). They hibernate during the winter, at which time their metabolic rate may drop as low as 0.03 mL $O_2$ $g^{-1}$ $hr^{-1}$, which is considerably lower than their summer metabolic rate of 0.54 mL $O_2$ $g^{-1}$ $hr^{-1}$ (Hock 1969). Marmots spend the vast majority (80%) of their lives in burrows (Svendsen 1976). The peripheries of burrow entrances are quite disturbed and may be devoid of plant material due to marmot burrowing and foraging activities. Soil nutrients around burrows, however, may be quite high as marmots tend to deposit feces in the same location.

The ecology and physiology of marmots has been intensively studied in subalpine meadows on the western slope of the Rockies near Gothic, CO by Armitage and others. At Gothic, marmots consume only 2.6–6.4% of the available primary production in the areas where they forage (Kilgore and Armitage 1978). Early in the summer, the majority (42–68%) of their diet is grass. As the season progresses, marmots begin incorporating more forbs into their diet, particularly *Mertensia ciliata* (Frase and Armitage 1989). Late in the season, they forage heavily on seeds (Frase and Armitage 1989). Marmots, although generalist herbivores, are selective and may avoid plants with secondary compounds (Armitage 1979). In addition, the demands of hibernation may impose foraging restrictions during the summer. Animals that hibernate seem to select foods based on both concentration and type of lipid. The latter is related to the fat's suitability for use during hibernation, with a preference for polyunsaturated over saturated fats because they function better at low temperatures (Geiser and Kenagy 1987).

Population growth of marmots is not thought to be limited by food availability, primarily because they consume such a small proportion of the available biomass (Kilgore and Armitage 1978). However, the population growth of marmots in alpine environments may be limited by the length of time that food is available (Van Vuren and Armitage 1991). Mating occurs within 2 weeks after emergence from hibernation. Females do not become reproductive until their second summer, and then only 25% of these females produce offspring (Armitage and Downhower 1970). Gestation and weaning last approximately 30 days each, and thus in the alpine, a reproductive female has little time to store fat after the young are weaned, prior to the onset of plant senescence. Not surprisingly, females in a high alpine meadow in Gothic did not produce litters in consecutive years (Johns and Armitage 1979). Moreover, snow cover duration was negatively correlated with several life-history traits, that is, frequency of reproduction, litter size, and juvenile body mass (Van Vuren and Armitage 1991). Collectively, these results suggest that the short growing season places a constraint on the female marmot's ability to store enough fat after weaning a litter to both hibernate and successfully reproduce the following spring.

## *Elk*

Elk, *Cervus elaphus* (Cervidae), are not year-round residents in the alpine. However, because of their large body size, 225–450 kg (Burt and Grossenheider 1976), their summer foraging on the tundra may have a significant impact on the vegetation. Baker and Hobbs (1982) used trained, tamed elk to study diet choices in the alpine tundra of Rocky Mountain National Park. Elk primarily consume graminoids (62%), with forbs (26%) and shrubs (11%) making up less of their diet. *Carex* species comprise the majority of the diet (39–43%), with *Deschampsia caespitosa* (4–6%) consumed less frequently than sedges are. *Acomastylis rossii* (7%), *T. parryi* (6%), and *B. bistortoides* (1%) are the forbs most commonly consumed (Baker and Hobbs 1982). This partiality to graminoids over forbs and shrubs is consistent in nonalpine habitats, and elk appear to forage for graminoids over greater distances and consume a more-diverse diet when graminoid quality is low (McCorquodale 1993). The strong preference of elk for gramnoids may be related to their avoidance of plant secondary compounds (Jakubus et al. 1994), which are typically more abundant in forbs and shrubs.

## Ecological Aspects of Herbivory in the Alpine

### *How Herbivores Contend with the Short Growing Season*

It seems paradoxical that alpine tundra with a growing season of 8–10 weeks supports as many mammalian herbivores as it does. Alpine herbivores employ various strategies to cope with the prolonged absence of fresh plant material. Marmots escape the winter by hibernating. However, successful hibernation is correlated with fat accumulation; marmots may lose up to 50% of their body mass during hibernation (Armitage et al. 1976). During the summer, marmots have a relatively low metabolic rate and are five times more efficient at tissue growth compared with other homeotherms (Kilgore and Armitage 1978). Low metabolic demands and high tissue accumulation efficiency may enable marmots to accumulate enough fat for hibernation in the short time that vegetation is available. Pikas and voles do not engage in hibernation or torpor. To survive the alpine winter, pikas depend primarily on large caches of food. Each animal caches on average 28 kg of fresh plant material (Dearing 1995). Construction of a cache this size requires approximately 14,000 collecting trips during the 8–10 weeks of the summer (Dearing 1995). Pikas invest the majority of their surface activity (up to 55%) in cache construction (Conner

1983). Voles also cache plant material (Vander Wall 1990); however, the contents and reliance on the cache is unknown in general for the two vole species occurring on Niwot Ridge. Voles are known to consume roots, and surface disturbance presumably caused by voles during the winter has been observed in the Saddle research site of Niwot Ridge (Seastedt, personal communication). Gophers are active year-round (Stuebe and Andersen 1985) and build small caches containing approximately 2 weeks of food (Vander Wall 1990).

## Degree of Dietary Specialization

Although the diets of all herbivores present have not been studied in Niwot Ridge, data complied from other localities suggest that the mammalian and insect herbivores occurring on the ridge are generalists (Alexander and Hillard 1964; Armitage 1979; Baker and Hobbs 1982; Coxwell 1992; Smolen and Keller 1987; Dearing 1996). However, the diet preferences of the different species are disparate. Such resource partitioning is found among mammalian herbivores in the Arctic (Klein and Bay 1994). Based on the diet data for pikas, marmots, and elk, the predominance of graminoids in the diet appears to increase with increasing body sizes. Graminoids typically are high in fiber and low in secondary compounds (McArthur et al. 1991). Thus, the underlying mechanisms regulating the amount of graminoids in the diet may be a function of an animal's ability to digest fiber and/or its ability to cope with plant secondary compounds. Fiber digestion efficiency has been proposed to increase with increasing body size (Parra 1978). There has been very little empirical work on species differences in detoxifying secondary compounds. The herbivore and plant communities of Niwot Ridge present an excellent system for tests of such hypotheses.

## Effects of Herbivory on Plant Community Structure

### Central Place Foragers and Vegetation Gradients

With the exception of elk, the mammalian herbivores on Niwot Ridge are central place foragers (Orians and Pearson 1979); that is, they forage out from one or a few locations. This type of foraging behavior tends to cause a gradient in the plant community composition, where the relative abundances of the more-preferred food items are lowest closest to the central place due to intensive foraging pressure. Cover and species diversity increase as distance from the central place increases (Huntly 1987). On the western slope of the Rockies, Huntly (1987) found that pikas lowered vegetation cover and plant species richness at locations close to their central place refuge and that this effect lessened with increasing distance from the refuge. The effect of central place foraging on the plant communities in the alpine is unknown. However, since the different species of mammalian herbivores tend to have different diet and habitat preferences, central place foraging may produce landscape-level effects on plant community structure.

## Succession

The effects of herbivory on succession in plant communities appear to depend on the herbivore species (Davidson 1993; Tahvanainen et al. 1991). For example, in the same habitat, moose accelerate succession by creating a nitrogen-depleted environment, whereas lemmings retard succession (Tahvanainen et al. 1991). The foraging activities of three of the herbivores that occur on Niwot Ridge (voles, pikas, and elk) produced the same effect on succession at other sites by reversing it to an earlier sere (Batzli et al. 1980; Huntly 1987; Hanley and Taber 1980). However, data on these three species were collected independently in different plant communities (i.e., voles in the Arctic, elk in coniferous forests in western Washington, pikas in subalpine meadows in western Colorado). Whether the same effect would be produced if all three herbivores were present in the same habitat is unknown.

Individual herbivores in the alpine live in a landscape composed of a mosaic of communities and potentially forage daily in a number of plant communities. Thus, an individual's foraging activities may promote succession in one community and retard it in another. Pikas are a prime candidate for this type of differential influence on succession. Pikas forage out from a central location, most intensively close to their rock refuge. The intensive foraging activities close to the talus maintains the cushion plant (fellfield) community (Huntly 1987). Farther from the talus, the cushion plant community disappears, perhaps in part because pikas forage less intensively there. Nonetheless, pikas do forage in the moist and dry meadow communities farther from the talus. In these communities, pikas tend to collect forbs, mostly *A. rossii,* for their winter diet. Higher herbivory pressures on forbs in these communities may yield an advantage to the less-grazed graminoid species. Thus, the effect of pika foraging in dry and moist meadows may accelerate succession from a forb-dominated community to a graminoid-dominated one.

The effects that herbivores have on plant communities is greater than just the removal of photosynthetic tissue. The movement of aboveground herbivores across the landscape can create well-worn paths with compacted soils that are devoid of plant material. The excavation activities of belowground herbivores may cover plants with soil (Huntly and Inouye 1988). Vegetation buried in gopher mounds decomposes more rapidly than does surface litter, and such chronic disturbance lowers the carbon storage in the top soil horizon, lowering the fertility of the site in the long term (Cortinas and Seastedt 1996; Litaor et al. 1996). In addition to negative effects on plant growth, indirect effects of herbivory may also have positive effects on plant growth of some species through nutrient transport and facilitation of microbial mineralization. Small herbivores daily consume their body weight in plant material and excrete approximately 50% of initial intake (Batzli and Cole 1979). In arctic tundra, mammalian urine and feces significantly increase the level of soil nitrogen and phosphorus (McKendrick et al. 1980). This fertilization effect may be very localized, as it is typically concentrated around burrow entrances or in small patches.

The net impact of the combined direct and indirect effects of herbivory on the plant communities in the alpine is unknown. However, theory and the results of other studies previously described suggest that both the direct and indirect effects of herbivores should promote expansion of graminoids as community dominants. Except for elk and grasshoppers, the majority of herbivores in the alpine consume forbs over grasses. Thus, direct consumption of forb leaf material should favor growth of the less-preferred species like graminoids. In addition, feces and urine produced by herbivores should serve as significant inputs of N and P to the soil. These amendments should also favor the growth of graminoids (Bowman et al. 1995).

Why is the alpine not a grassland? It is possible that the differing diet preferences of the herbivores on the Ridge counteract one another as well as competitive exclusion of the forbs by graminoids, thereby maintaining species diversity as well as a diversity of communities. These antagonistic interactions may happen on a static or dynamic basis. In the static case, different herbivores are expected to forage in the same community, with their disparate preferences nullifying the effects of one another. In this case, removal of one herbivore should alter the successional community. Also, in the static case, the boundaries of plant communities are not expected to change greatly over time. In the dynamic case, one herbivore creates a shift in the plant community either through foraging or nutrient transport. As the plant community shifts to one not preferred by the initial herbivore, other herbivores move in, which may ultimately cause reversion to the original community.

## Future Directions

The examples given in this chapter suggest that herbivores may strongly influence plant community structure and succession in the alpine. How strong is this influence? Which herbivores exert the greatest impact? Are their effects syngeristic, additive, or antagonistic? How do the effects of belowground herbivory compare with aboveground herbivory? Do mammalian herbivores have a larger impact on the plant community than insect herbivores do? Are the different diet preferences of herbivores a product of their ability to detoxify secondary compounds and utilize fiber? What regulates populations of herbivores? The plants and herbivores of Niwot Ridge are ideal for studying such questions at the individual, community, and landscape levels. Answers to such questions will advance not only our understanding of the ecology of the alpine but also our knowledge of plant-herbivore interactions in general.

### References

Alexander, G., and J. Hillard. 1964. Life history of Areopedellus clavatus (Orthoptera: Acrididae) in the alpine tundra of Colorado. *Annals of the Entomological Society of America* 57:310–317.

Armitage, K. B. 1979. Food selectivity by yellow-bellied marmots. *Journal of Mammalogy* 60:628–629.

Armitage, K. G., and J. F. Downhower. 1970. Interment behavior in the yellow-bellied marmot (*Marmota flaviventris*). *Journal of Mammalogy* 51:177–178.

Armitage, K. B., J. F. Downhower, and G. E. Svendsen. 1976. Seasonal changes in weights of marmots. *American Midland Naturalist* 96:36–51.

Baker, D. L., and N. T. Hobbs. 1982. Summer diets in Colorado. *Journal of Wildlife Management* 46:694–703.

Barash, D. P. 1973. Social variety in the yellow-bellied marmot (*Marmota flaviventris*). *Animal Behavior* 21:579–584.

Batzli, G. O., and F. R. Cole. 1979. Nutritional ecology of microtine rodents: Digestibility of forage. *Journal of Mammalogy* 60:740–750.

Batzli, G. O., R. G. White, and F. L. Bunnell. 1981. *Herbivory: A strategy of tundra consumers*. Cambridge: Cambridge University Press.

Batzli, G. O., R. G. White, J. MacLean, F. A. Pitelka, and B. D. Collier. 1980. *The herbivore-based trophic system*. Stroudsburg, PA: Dowden, Hutchinson and Ross.

Bowman, W. D., T. A. Theodose, and C. F. Fisk. 1995. Physiological and production responses of plant growth forms to increases in limiting resources in alpine tundra: Implications for differential community response to environmental change. *Oecologia* 101:217–227.

Broadbrooks, H. E. 1965. Ecology and distribution of the pikas of Washington and Alaska. *American Midland Naturalist* 73:299–335.

Burt, W. H., and R. P. Grossenheider. 1976. *A field guide to the mammals*. Boston: Houghton Mifflin.

Coley, P. D., J. P. Bryant, and S. Chapin. 1985. Resource availability and plant antiherbivore defense. *Science* 230:895–899.

Conner, D. A. 1983. Seasonal changes in activity patterns and the adaptive value of haying in pikas (*Ochotona princeps*). *Canadian Journal of Zoology* 61:411–416.

Cortinas, M. R., and T. R. Seastedt. 1996. Short- and long-term effects of gophers (*Thomomys talpoides*) on soil matter dynamics in alpine tundra. *Pedobiologia* 40:1–9.

Coxwell, C. C. 1992. Abiotic and biotic influences on the distribution and abundance of an alpine grasshopper *Aeropedellus clavatus*. Master's thesis, University of Colorado, Boulder.

Coxwell, C. C., and C. E. Bock. 1995. Spatial variation in diurnal surface temperatures and the distribution and abundance of an alpine grasshopper. *Oecologia* 104:433–439.

Davidson, D. W. 1993. The effects of herbivory and granivory on terrestrial plant succession. *Oikos* 68:23–35.

Davies, E. F. 1994. Disturbance in alpine tundra ecosystems: The effects of digging by Northern pocket gophers (*Thomomys talpoides*). Master's thesis, Utah State University.

Dearing, M. D. 1995. Factors governing diet selection in a herbivorous mammal, the North American pika, *Ochotona princeps*. Ph.D. diss., University of Utah.

———. 1996. Disparate determinants of summer and winter diet selection in a generalist herbivore, *Ochotona princeps*. *Oecologia* 108:467–478.

———. 1997a. Effects of *Acomastylis rossii* tannins on a mammalian herbivore, the North American pika, *Ochotona princeps*. *Oecologia* 109:122–131.

———. 1997b. The manipulation of plant toxins by a food-hoarding herbivore, *Ochotona princeps*. *Ecology* 78:774–781.

Feeny, P. 1976. Plant apparency and chemical defence. *Recent Advances in Phytochemistry* 10:1–40.

Fisk, M. C., and S. K. Schmidt. 1995. Nitrogen mineralization and microbial biomass N dynamics in three alpine tundra communities. *Soil Science Society of America Journal* 59:1036–1043.

Frase, B. A., and K. B. Armitage. 1989. Yellow-bellied marmots are generalist herbivores. *Ethology, Ecology and Evolution* 1:353–366.

Geiser, F., and G. J. Kenagy. 1987. Polyunsaturated lipid diet lengthens torpor and reduces body temperature in a hibernator. *American Journal of Physiology* 252:R897–R901.

Gibbs, R. D. 1974. *Chemotaxonomy of flowering plants*. Montreal: McGill-Queen's University Press.

Hagerman, A. E., and L. G. Butler. 1978. Protein precipitation method for the quantitative determination of tannins. *Journal of Agricultural and Food Chemistry* 26:809–812.

Hanley, T. A., and R. D. Taber. 1980. Selective plant species inhibition bu elk and deer in three coniferous communities in western Washington. *Forest Science* 26:978–1007.

Haukioja, E. 1981. *Invertebrate herbivory at tundra sites*. Cambridge: Cambridge University Press.

Hock, R. J. 1969. Thermoregulatory variations of high-altitude hibernators in relation to ambient temperature, season and hibernation. *Federation of American Societies for Experimental Biology* 28:1047–1052.

Hodkinson, I. D., T. S. Jensen, and S. F. MacLean. 1979. The distribution, abundance and host plant relationships of *Salix*-feeding psyllids (Homptera: Psylloidea) in arctic Alaska. *Ecological Entomology* 4:119–132.

Huntly, N., and R. Inouye. 1988. Pocket gophers in ecosystems: patterns and mechanisms. *BioScience* 38:786–793.

Huntly, N. J. 1987. Influence of refuging consumers (pikas: *Ochotona princeps*) on subalpine meadow vegetation. *Ecology* 68:274–283.

Huntly, N. J., A. T. Smith, and B. L. Ivins. 1986. Foraging behavior of the pika (*Ochotona princeps*) with comparisons of grazing versus haying. *Journal of Mammalogy* 67:139–148.

Jakubus, W. J., R. A. Garrott, P. J. White, and D. R. Mertens. 1994. Fire-induced changes in the nutritional quality of lodgepole pine bark. *Journal of Wildlife Management* 58:35–46.

Johns, D. W., and K. B. Armitage. 1979. Behavioral ecology of alpine yellow-bellied marmots. *Behavioral Ecology and Sociobiology* 5:133–157.

Jung, H. G., G. O. Batzli, and D. S. Seigler. 1979. Patterns in the phytochemistry of arctic plants. *Biochemical Systematics and Ecology* 7:203–209.

Kilgore, D. L., Jr., and K. B. Armitage. 1978. Energetics of yellow-bellied marmot populations. *Ecology* 59:78–88.

Klein, D. R., and C. Bay. 1994. Resource partitioning by mammalian herbivores in the high Arctic. *Oecologia* 97:439–450.

Lindroth, R. L., and G. O. Batzli. 1983. Detoxication of some naturally occurring phenolics by prairie voles: A rapid assay of glucuronidation metabolism. *Biochemical Systematics and Ecology* 11:405–409.

———. 1986. Lespedeza phenolics and Penstemon alkaloids: Effects on digestion efficiencies and growth of voles. *Journal of Chemical Ecology* 12:713–728.

Linsdale, J. M. 1938. Environmental responses of vertebrates in the Great Basin. *American Midland Naturalist* 19:1–206.

Litaor, M. I., R. Mancinelli, and J. C. Halfpenny. 1996. The influence of pocket gophers on the status of nutrients in alpine soils. *Geoderma* 70:37–48.

MacLean, J., and T. S. Jensen. 1985. Food plant selection by insect herbivores in the Alaskan arctic tundra: The role of plant life form. *Oikos* 44:211–221.

May, D. E., and P. J. Webber. 1982. *Spatial and temporal variation of the vegetation and its productivity on Niwot Ridge, Colorado*. Boulder, CO: Institute of Arctic and Alpine Research.

McArthur, C. A., E. Hagerman, and C. T. Robbins. 1991. *Physiological strategies of mammalian herbivores against plant defenses*. Boca Raton, FL: CRC Press.

McCorquodale, S. M. 1993. Winter foraging behavior of elk in the shrub-steppe of Washington. *Journal of Wildlife Management* 57:881–890.
McKechnie, A. 1990. The spatial and temporal use of food patches by the pika, *Ochotona princeps*, in the Rocky Mountains of Colorado. Master's thesis, Colorado State University.
McKendrick, J. D., G. O. Batzli, K. R. Everett, and J. C. Swanson. 1980. Some effects of mammalian herbivores and fertilization on tundra soils and vegetation. *Arctic and Alpine Research* 12:565–578.
Merritt, J. F., and J. M. Merritt. 1978. Population ecology and energy relationships of *Clethrionomys gapperi* in a Colorado subalpine forest. *Journal of Mammalogy* 59:576–598.
Negus, C. N., P. J. Berger, and B. W. Brown. 1986. Microtine population dynamics in a predictable environment. *Canadian Journal of Zoology* 64:785–792.
Orians, G. H., and N. E. Pearson. 1979. *On the theory of central place foraging*. Columbus, OH: Ohio State University Press.
Parra, R. 1978. *Comparison of foregut and hindgut fermentation in herbivores*. Washington: Smithsonian Institution Press.
Randall, J. A., and R. E. Johnson. 1979. Population densities and habitat occupancy by Microtus longicaudus and M. montanus. *Journal of Mammalogy* 60:217–219.
Reichman, O. J., and S. C. Smith. 1985. Impact of pocket gopher burrows on overlying vegetation. *Journal of Mammalogy* 66:720–725.
Reichman, O. J., and S. S. Smith. 1991. Responses to simulated leaf and root herbivory by a biennial, *Tragopogon dubius*. *Ecology* 72:116–124.
Robbins, C. T., T. A. Hanley, H. O. Hjeljord, D. L. Baker, C. C. Schwartz, and W. W. Mautz. 1987. Role of tannins in defending plants against ruminants: Reduction of protein availibility. *Ecology* 68:98–107.
Smith, A. T., and B. L. Ivins. 1984. Spatial relationships and social organization in a facultatively mongamous mammal. *Zeitschrift für Teirpsychologie* 66:298–308.
Smith, A. T., and M. L. Weston. 1990. Ochotona princeps. *Mammalian Species* 352:1–8.
Smolen, M. J., and B. L. Keller. 1987. Microtus longicaudus. *Mammalian species* 271:1–7.
Somme, L., and W. Block. 1991. *Adaptations of alpine and polar environments in insects and other terrestrial arthorpods*. New York: Chapman and Hall.
Southwick, C. H., S. C. Golian, M. R. Whitworth, J. C. Halfpenny, and R. Brown. 1986. Population densityad fluctuations of pikas (*Ochotona princeps*) in Colorado. *Journal of Mammalogy* 67:149–153.
Stuebe, M. M., and D. C. Andersen. 1985. Nutritional ecology of a fossorial herbivore: Protein N and energy value of winter caches made by the northern pocket gopher, *Thomomys talpoides*. *Canadian Journal of Zoology* 63:1101–1105.
Svendsen, G. E. 1976. Structure and location of burrows of yellow-bellied marmot. *Southwestern Naturalist* 55:760–771.
Tahvanainen, J., P. Niemala, and H. Henttonen.1991. *Chemical aspects of herbivory in boreal forest-feeding by small rodents, hares, ad cervids*. Boca Raton, FL: CRC Press.
Tevis, J. L. 1952. Stomach contents of chipmunks and mantled ground squirrels in northeastern California. *Journal of Mammalogy* 34:316–324.
Thorn, C. E. 1978. A preliminary assessment of the geomorphic role of pocket gophers in the alpine zone of the Colorado Front Range. *Geografiska Annaler* 60:181–187.
———. 1982. Gopher disturbance: Its variability by Braun-Blanquet vegetation units in the Niwot Ridge alpine tundra zone, Colorado Front Range, USA. *Arctic and Alpine Research* 14:45–51.
Tomasi, T. E. 1985. Basal metabolic rates and thermoregulatory abilities in four small mammals. *Canadian Journal of Zoology* 63:2534–2537.

Torti, S. D., M. D. Dearing, and T. A. Kursar. 1995. Extraction of phenolics: A comparison of methods. *Journal of Chemical Ecology* 21:117–125.

Vander Wall, S. B. 1990. *Food hoarding in animals.* Chicago: University of Chicago Press.

Van Vuren, D., and K. B. Armitage. 1991. Duration of snow cover and its influence on life-history variation in yellow-bellied marmots. *Canadian Journal of Zoology* 69:1755–1758.

Walker, M. D., P. J. Webber, E. H. Arnold, and D. Ebert-May. 1994. Effects of interannual climate variation on aboveground phytomass in alpine vegetation. *Ecology* 75:393–408.

Waterman, P. G., and S. Mole. 1994. *Analysis of phenolics plant metabolites.* London: Blackwell Scientific.

# Part IV

# Past and Future

# 15

# Paleoecology and Late Quaternary Environments of the Colorado Rockies

Scott A. Elias

## Introduction

Present-day environments cannot be completely understood without knowledge of their history since the last ice age. Paleoecological studies show that the modern ecosystems did not spring full-blown onto the Rocky Mountain region within the last few centuries. Rather, they are the product of a massive reshuffling of species that was brought about by the last ice age and indeed continues to this day.

Chronologically, this chapter covers the late Quaternary Period: the last 25,000 years. During this interval, ice sheets advanced southward, covering Canada and much of the northern tier of states in the United States. Glaciers crept down from mountaintops to fill high valleys in the Rockies and Sierras. The late Quaternary interval is important because it bridges the gap between the ice-age world and modern environments and biota. It was a time of great change, in both physical environments and biological communities.

## Late Pleistocene History of the Front Range Region

The Wisconsin Glaciation is called the Pinedale Glaciation in the Rocky Mountain region (after terminal moraines near the town of Pinedale, Wyoming; see chapter 4). The Pinedale Glaciation began after the last (Sangamon) Interglaciation, perhaps 110,000 radiocarbon years before present (yr BP), and included at least two major ice advances and retreats. These glacial events took different forms in different regions. The Laurentide Ice Sheet covered much of northeastern and north-central North America, and the Cordilleran Ice Sheet covered much of northwestern North

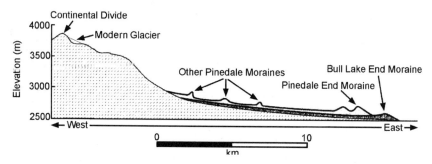

Figure 15.1. Diagram showing limits of Bull Lake and Pinedale till and moraines on the eastern slope of the Colorado Front Range (vertical scale exaggerated). Data are from Madole and Shroba (1979).

America. The two ice sheets covered more than 16 million km² and contained one third of all the ice in the world's glaciers during this period.

The history of glaciation is not as well resolved for the Colorado Front Range region as it is for regions farther north. For instance, although a chronology of three separate ice advances has been established for the Teton Range during Pinedale times, in northern Colorado we know only that there were earlier and later Pinedale ice advances. We do not know when the earlier advance (or multiple advances) took place. However, based on geologic evidence (Madole and Shroba 1979), the early Pinedale glaciation was more extensive than the late Pinedale was. Early Pinedale moraines can be seen near the town of Estes Park, whereas late Pinedale ice formed moraines several kilometers up-valley (Fig. 15.1).

During the last glaciation, Pinedale glaciers flowed out of high mountain cirques, down-valley to elevations between 2440 and 2470 m, the elevation of the lower montane forests. Pinedale ice advanced 14–15 km down slope from the Continental Divide on the eastern slope. Pinedale glaciers on the western slope were larger, extending down-slope as much as 33 km. Pinedale ice may have been as much as 450 m thick near the heads of the glaciers in the Front Range.

Most of the paleoecological data summarized in this chapter was collected from bogs and fens in the upper subalpine zone of the Colorado Rockies. This is not coincidental; it is based on the fact that organic deposits (peat bogs and organic-rich lake sediments) are scarce in the alpine zone of Colorado. Much of the alpine landscape was stripped of both vegetation and soil during the last glaciation. High-altitude landscapes that were not covered by ice remained in a permafrost zone where periglacial features were formed. These features are presently active in a patchy distribution of sites above treeline.

Good estimates are available concerning the timing of some Pinedale glacial events, based on radiocarbon ages of organic-rich sediments at several high-elevation sites. At the Mary Jane ski area, near Winter Park, Colorado (Fig. 15.2), an excavation for a ski lift tower exposed a series of alternating lake sediments and glacial tills. The oldest lake bed was dated at about 30,000 yr BP (Nelson et al. 1979). This

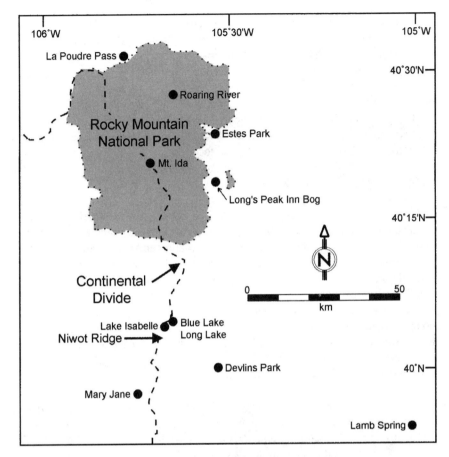

Figure 15.2. Map of northeastern Colorado, showing location of sites mentioned in text.

bed was overlain by glacial till, and the next youngest lake bed yielded a radiocarbon age at the base of about 13,750 yr BP. Based on the Mary Jane sequence, it appears that the last major ice advance of the Pinedale Glaciation took place between the time of deposition of the older and younger lake beds, between 30,000 and 13,750 yr BP.

At Devlins Park in the Indian Peaks Wilderness area (Fig. 15.2), Legg and Baker (1980) studied sediments from a lake that was dammed by late Pinedale ice. During the time that Glacial Lake Devlin existed (22,400–12,200 yr BP), ice covered the Devlins Park region. The lake drained catastrophically when the ice retreated, and sediments from the top of the lake provide a limiting age for this event. Presumably, the late Pinedale glacier that advanced down-slope from the Continental Divide west of Devlins Park area took some centuries to reach that elevation (2953 m), so the glacial advance began before 22,400 yr BP. The terminal moraine for this glacier is 2.3 km down-slope from the study site.

In a few places, it is possible to date the retreat of late Pinedale ice from montane valleys back to the alpine zone where they originated. At La Poudre Pass (Fig. 15.2), Madole (reported in Elias 1983) obtained a radiocarbon date of 10,000 yr BP from peat that formed in a bog after Pinedale ice retreated. The pass, which is located at modern tree line, was free of ice sometime prior to that time. In the Indian Peaks Wilderness, sediments began accumulating in Long Lake (3200 m), in the subalpine zone north of Niwot Ridge, by 12,000 yr BP (Short 1985). Studies in the Southern Rocky Mountains suggest that the melting of mountain glaciers began after 14,000 yr BP, so the process of deglaciation was relatively rapid, probably because the glaciers were not very thick compared with glaciers farther north, and the relatively low latitude of the Southern Rockies is associated with greater insolation than that of more northerly regions.

Alpine environments during the last glaciation were thus extremely perturbated by glacial ice and periglacial activity. The soils that currently mantle the Colorado alpine developed during the Holocene. Soil formation and revegetation of high-elevation landscapes did not occur until deglaciation. From a landscape perspective, deglaciation brings chaos. The receding glaciers leave great piles of rubble, and large quantities of finer-grained sediments come pouring out from underneath and alongside the melting ice, as torrents of sediment-choked water are liberated. Even though streamflow did not always increase much beyond what we see today, all of the water that had been locked up in the ice was liberated in a relatively short interval of time. That meltwater made its way down drainages clogged with massive loads of debris, ranging from clay and silt to large boulders, depending on local circumstances.

While there was tremendous instability in alpine landscapes undergoing deglaciation, some species of plants were already present in these disturbed habitats. In the subalpine zone, prostrate junipers were some of the first woody plants to colonize the recently deglaciated landscapes of the Rockies (Whitlock 1993). When these were shaded-out by other conifers, they became much less common. Likewise, it appears that sage (*Artemisia*) was an early colonizer in the alpine. Sage pollen is relatively important in alpine and subalpine fossil pollen assemblages from the end of the last glaciation, then it decreased as other alpine plant taxa became established (Short 1985).

## Late Pinedale Environments of the Front Range

Paleoclimatic reconstructions for the Rocky Mountain region indicate that the Colorado Front Range received less moisture than did ranges to the north (in the Yellowstone region) and the south (the San Juan Mountains). Major glacial advances also occurred in the San Juan Mountains of Colorado. For example, the Animas Glacier filled the Animas River valley with ice to the town site of Durango (Carrara et al. 1984). In contrast, the mountain glaciers of the Front Range region were small, and glaciers from most drainages did not coalesce to form larger glaciers or ice sheets.

Paleoclimatic reconstructions for the Front Range region (Elias 1986, 1996a) indicate that mean July temperatures were as much as 10–11°C colder than are modern parameters as late as 14,500 yr BP, and mean January temperatures were depressed by 26–29°C compared with modern climate. The temperature regime was therefore cold enough to foster the growth of glacial ice but lacked sufficient winter precipitation to develop the necessary snowpack for glacial development.

At the Mary Jane site (Fig. 15.2), pollen in lake sediments laid down during an interstadial interval before the last major Pinedale ice advance (circa 30,000 yr BP) records a sequence of vegetation beginning with open spruce-fir forest with herbs and shrubs adjacent to the lake. This was followed by a colder phase, in which alpine tundra replaced the subalpine forest. The alpine vegetation is indicated by pollen of such taxa as *Bistorta vivipara* and Caryophyllaceae (cushion plants) (Short and Elias 1987). The youngest (uppermost) sediments in this lake bed reflect the return of spruce forest to the vicinity before the advance of late Pinedale ice. The Mary Jane site is at an elevation of 2882 m, in the lower part of the modern subalpine forest. The existence of alpine tundra at the site in mid-Pinedale times translates into a depression of treeline by more than 500 m. This, in turn, corresponds to a climatic cooling of at least 3°C from average modern summer temperatures.

The next indication of late Pinedale environments in the region comes from the oldest lake sediments from Glacial Lake Devlin (22,400 yr BP). Legg and Baker (1980) found pollen evidence for a treeless landscape at the site. The pollen diagram is dominated by sagebrush. Scattered pollen grains of tundra plants (which are not abundant pollen producers) and low numbers of conifer pollen grains suggest that the site was above treeline in late Pinedale times. The tundra pollen types include *Bistorta*, *Pedicularis*, *Koenigia islandica*, Caryophyllaceae (possibly *Silene acaulis*, *Paronychia sessiliflora*, *Sagina saginoides*, *Minuartia* spp., and *Melandrium* spp.), *Phlox*, *Polemonium*, *Saxifraga*, and *Ranunculus*. Some of these genera also contain species found at lower elevations, but the overall composition of the pollen flora is suggestive of alpine tundra vegetation. The pollen assemblages from this site are in many ways similar to those described by Baker (1976) and Whitlock (1993) from the earliest postglacial environments in Grand Teton and Yellowstone National Parks. In both instances, mixtures of sagebrush and alpine tundra plants suggest steppe-tundra. This vegetation probably developed in the sort of cold, dry climate seen today in parts of northern Siberia (Elias 1995).

Recently, the author has begun reconstructing seasonal temperatures based on mutual climatic range (MCR) analyses of fossil insect assemblages in the Rocky Mountain region. The MCR technique is based on the assumption that the present climatic tolerances of a species can be applied to the Quaternary fossil record, and thus fossil occurrences of a given species imply a paleoclimate that was within the same tolerance range. MCR studies focus on predators and scavengers, as these groups show the most rapid response to climate change. The predators are nearly all generalists that prey on a wide variety of small arthropods. Plant-feeding groups are not considered because these species cannot become established in new regions until their host plants arrive. In contrast to this, predators and scavengers have been shown to be able to shift distributions on a continental scale in a few tens or hun-

Table 15.1. Summary of Age Ranges and References for Sites Discussed in Text

| Site | Age Range | References |
| --- | --- | --- |
| La Poudre Pass | 10,000 yr BP to recent | 1,4 |
| Mount Ida Bog | 900 yr BP | 2,4 |
| Roaring River | 2400 yr BP | 5,4 |
| Longs Peak Inn | 3000 yr BP to recent | 8,4 |
| Lake Isabelle delta | 9000–8000 yr BP | 2,4 |
| Lake Isabelle fen | 7000 yr BP to recent | 2,4 |
| Mary Jane | 13,800–ca. 12,3000 yr BP | 4,9 |
| Lamb Spring | 17.850 and 14,500 yr BP | 3,4,6,7,10 |

1. Elias (1983).
2. Elias (1985).
3. Elias (1986).
4. Elias (1996).
5. Elias et al. (1986).
6. Elias and Nelson (1989).
7. Elias and Toolin )1990).
8. Richmond (1960).
9. Short and Elias (1987).
10. Stanford et al. (1982).

dreds of years (Coope 1977; Elias 1991, 1994). The fossil assemblages considered in the Rocky Mountain regional study include 74 species in the families Carabidae (ground beetles), Dytiscidae (predaceous diving beetles), Hydrophilidae (water scavenger beetles), Staphylinidae (rove beetles), Scarabaeidae (dung beetles), and Coccinellidae (ladybird beetles).

MCR analysis was applied to 17 beetle assemblages in Colorado, spanning the interval 14,000–400 yr BP (Tables 15.1 and 15.2). The fossil sites range from La Poudre Pass near the Continental Divide in northern Colorado to Lamb Spring, a site in the piedmont zone south of Denver (Fig. 15.2). While there are many gaps in the transect (both spatially and temporally), the available sites provide sufficient information to allow an initial paleoenvironmental scenario.

The earliest indications of climatic amelioration were found at the Mary Jane site, where peat layers were deposited after the retreat of late Pinedale ice. Short and Elias (1987) reported on pollen and insect remains from peat layers ranging in age from 13,740 to 12,350 yr BP. Fossil evidence from layers dated 13,740–12,700 yr BP indicate the existence of alpine tundra based on the flora and insect fauna. Elias (1996a) performed an MCR reconstruction of mean July and January temperatures from a fossil beetle assemblage dated 13,200 yr BP and at 12,800 yr BP. These assemblages showed that mean July temperatures had risen quite dramatically from previous full glacial conditions. Mean July temperatures reconstructed for these Mary Jane assemblages were only 3.2–3.6°C cooler than at present, although mean January temperatures remained 19–20°C cooler than present.

After 12,600 yr BP, montane and subalpine insects increased as alpine species decreased, and willow shrubs expanded across the bog, suggesting a climatic warm-

Table 15.2. Site Data and Summary of Modern and Paleoclimatic Data

| | | | | Late Quaternary Mean | | Temperature (°C) Modern Mean | | Change ΔT | |
|---|---|---|---|---|---|---|---|---|---|
| Site | Elevation (m) | Sample Age[a] | Calibrated Age[b] | July | Jan. | July | Jan. | July | Jan |
| Lamb Spring | 1731 | 14,500 ± 500 | 17,370 | 10–11 | −31 to −27 | 21.4 | −1.3 | −11.4 to −10.4 | −29.7 to −25.7 |
| Mary Jane | 2882 | 13,200 | N/A | 9.8–10.2 | −29.3 to −27.6 | 13.4 | −8.6 | −3.6 to −3.2 | −20.7 to −19 |
| Mary Jane | 2882 | 12,800 | N/A | 10–10.2 | −29.1 to −27.6 | 13.4 | −8.6 | −3.4 to −3.2 | −20.5 to −19 |
| La Poudre Pass | 3100 | 9850 ± 300 | 11,000 | 15–18 | −17.5 to −7 | 11.3 | −7.7 | 3.7 to 6.7 | −9.8 to 0.7 |
| Lake Isabelle Delta | 3323 | 9000 ± 285 | 9980 | 11.75–14.5 | −31.25 to −15 | 10.8 | −8.2 | 1 to 3.7 | −23 to −6.8 |
| La Poudre Pass | 3323 | 8800 ± 90 | 9860 | 13.5–16.5 | −19.5 to −9 | 11.3 | −7.7 | 2.2 to 5.2 | −13.6 to −1.3 |
| Lake Isabelle Delta | 3323 | 8500 | N/A | 10.5–13 | −23.5 to −16 | 10.8 | −8.2 | −0.3 to 2.2 | −15.3 to −7.8 |
| Lake Isabelle Delta | 3323 | 7800 ± 255 | 8530 | 11–13 | −14 to −9 | 10.8 | −8.2 | 0.2 to 2.2 | −5.8 to −0.8 |
| Lake Isabelle Fen | 3325 | 7080 ± 90 | 7850 | 10.25–13 | −14.75 to −7.5 | 10.8 | −8.2 | −0.6 to 2.2 | −6.6 to 0.7 |
| La Poudre Pass | 3100 | 5360 ± 90 | 6180 | 12.5–13.5 | −12.5 to −11.5 | 10.8 | −8.2 | 1.7 to 2.7 | −4.3 to −3.3 |
| La Poudre Pass | 3100 | 3485 ± 180 | 3760 | 11.75–15 | −21.25 to −14.5 | 11.3 | −7.7 | 0.4 to 3.7 | −13.6 to −6.8 |
| Lake Isabelle Fen | 3325 | 3000 | N/A | 10.25–13 | −14.5 to −7.5 | 10.8 | −8.2 | −0.6 to 2.2 | −6.6 to 0.7 |
| Longs Peak Inn | 2732 | 2965 ± 75 | 3110 | 12–15.5 | −26.5 to −14 | 14.5 | −5 | −2.5 to 1 | −21.5 to −9 |
| Longs Peak Inn | 2732 | 2680 ± 80 | 2770 | 13.5–15.5 | −24 to −15 | 14.5 | −5 | −1 to +1 | −19 to −10 |
| Roaring River | 2800 | 2400 ± 130 | 2360 | 14.25–14.75 | −18.25 to −16.5 | 14.3 | −5.5 | 0 to 0.5 | −12.75 to −11 |
| Mount Ida Bog | 3520 | 900 ± 150 | 790 | 10.25–12 | −14.75 to −9 | 9.3 | −11.5 | 1 to 2.7 | −3.25 to 2.5 |
| Longs Peak Inn | 2732 | 395 ± 100 | 470 | 13.5–15.5 | −24.5 to −15 | 14.5 | −5 | −1 to +1 | −19.5 to −10 |

[a] $^{14}$C yr BP.
[b] Calendar year BP; calibration of radiocarbon ages from Stuiver and Reimer (1993).

ing. By 12,300 yr BP, spruce trees were probably growing adjacent to the site, after an absence of about 18,000 years. In the Indian Peaks Wilderness area, pollen spectra from Long Lake indicate tundra vegetation from the earliest sediments laid down after deglaciation (12,000 yr BP) to 10,500 yr BP (Short 1985). This vegetation was dominated by sagebrush and grasses with some birch and willow. Rock selaginella (*Selaginella densa*) was interpreted as a tundra indicator species.

## Early Holocene Environments

During the Holocene (10,000 yr BP to present), the Colorado Front Range experienced a series of climatic fluctuations. Insect assemblages from several sites are indicative of warmer-than-present summer temperatures and colder-than-present winter temperatures. The earliest Holocene record in the insect fossil study transect comes from La Poudre Pass. Here, an assemblage dated 9850 yr BP yielded MCR estimates showing mean July temperatures from 3 to 6°C warmer than modern. In fact, this assemblage represents the greatest degree of summer warming of the entire 14,000-year record in the Rocky Mountain region. Winter temperatures were as much as 10°C colder than modern, however, so the degree of continentality also reached a peak at this time. These predictions based on fossil beetle data agree well with Berger's (1978) reconstruction of incoming solar radiation (insolation), based on the Milankovitch model of variations in Earth's orbit around the sun (Fig. 15.3). The Milankovitch model predicts a summer insolation maximum and winter insolation minimum in the mid-latitudes of the Northern Hemisphere from about 9000 to 12,200 $^{14}$C yr BP (10,000–14,200 calendar yr BP). This peak in summer insolation coincides with the MCR estimates of the postglacial warming in the Colorado Rockies. The fossil insect record is the only fossil data source from the Rocky Mountain region to register this degree of warming in early postglacial times, consistent with glaciological data that suggest rapid melting of regional glaciers before 12,000 yr BP (Madole and Shroba 1979). Evidence from the San Juan Mountains indicates that the major glaciers in that region had melted as early as 15,000 yr BP (Carrara et al. 1984).

The extreme seasonality of climates indicated by the fossil insect data may have affected plant colonization of the region. Species adapted to extremes of summer warmth and winter cold were the only ones suited to this climatic regime. This perhaps explains the abundance of *Artemisia* pollen in regional fossil assemblages from the end of the last glaciation. Though *Artemisia* pollen cannot readily be identified at the species level, it seems likely that those species best adapted to very continental climate would have had a competitive advantage over other plant species at that time. Such species of *Artemisia* are found today in the Great Basin desert regions, immediately west of the Rockies.

By 9000 yr BP, the fossil insect data indicate that summer temperatures were already declining from an early Holocene peak, though still above modern values (Fig. 15.3). A fossil insect assemblage from Lake Isabelle yielded MCR reconstructions indicating mean July temperatures 1–3.7°C warmer than modern and mean January temperatures well below modern levels.

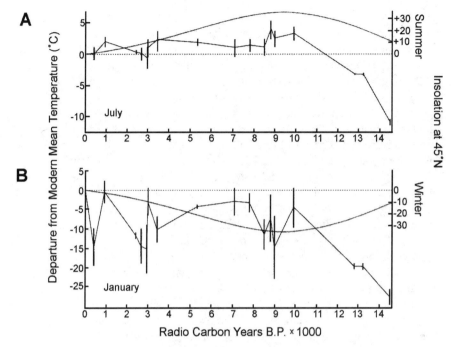

Figure 15.3. MCR reconstruction of late-Quaternary temperatures as indicated by fossil beetle assemblages from Colorado, overlain by summer and winter insolation curves for 45° N latitude, predicted by the Milankovitch model of orbital forcing (data from Berger 1978). The vertical bars represent the mutual climatic range of the species in the fossil assemblages. Trends in the TMAX and TMIN reconstructions taken by connecting the midpoints of each bar in the data sets. (a) mean July temperatures (TMAX). (b) mean January temperatures (TMIN).

The forest vegetation of the Front Range apparently lagged behind the changes in climate, especially during the early postglacial interval. The insolation curve and insect data both indicate that summer temperatures following deglaciation were warm enough to allow spruce-fir forest to grow up to its modern limit and beyond. Regional pollen records (Short 1985) indicate that subalpine forests did not reach modern treeline until 9500–9000 yr BP. The fossil insect record suggests that summer temperatures reached modern values by about 11,000 yr BP. Thus it appears that the trees lagged behind the postglacial climatic amelioration on the order of 1500–2000 years. The pollen record from the Indian Peaks Wilderness region indicates that spruce and fir trees were present in the modern upper subalpine zone as early as 10,500 yr BP, but that the density of trees on the landscape remained low for many centuries afterwards, perhaps reflecting slow soil development following deglaciation. Additionally, greater seasonality of regional climates may have made subalpine tree establishment more difficult. Whatever the reason(s), the broad belt of alpine tundra that developed in Pinedale times at elevations down to about 2900 m (9500 ft) retreated upslope to the mountaintops. The tundra peninsula became an archipelago of tundra islands. We do not know very much about the composition of alpine tundra vegetation during the Pinedale Glaciation for two reasons. First, most tundra plants do not produce much pollen, so they are poorly represented in Pinedale-age pollen assemblages, even those that come from high-elevation sites. Second, nearly all pollen grains are identified only to the generic or family level, so we do not know the species of plants identified in pollen assemblages. Based on identifications of spruce and fir pollen, Short (1985) found that treeline remained below its modern elevation as late as 6800 yr BP at La Poudre Pass. The regional vegetation history is not consistent, however. At Long Lake in the Indian Peaks Wilderness, the early Holocene pollen spectra reflect well-established spruce-fir forest by 9500 yr BP.

In contrast to the Front Range scenario, spruce and fir macrofossils from sites above modern treeline in the San Juan Mountains indicate that the trees advanced up-slope during three Holocene intervals. The oldest evidence of higher-than-modern treelines (hence, warmer-than-modern climate) comes from wood dated from 9600 to 7800 yr BP. The second warm interval has been dated between 6700 and 5600 yr BP, and the third has been dated at 3100 yr BP. The plant macrofossil data, combined with regional pollen studies, suggest that treeline was above its modern elevation from 9600 to 3000 yr BP (Carrara et al. 1984). This proxy data signal for warmer-than-modern climates in the San Juan Mountains agrees closely with the fossil insect reconstruction for the Front Range. Fewer snags of ancient trees have been found above treeline in the Mosquito Range, and those that have been found are all late Holocene in age (Brunstein and Yamaguchi 1992).

## Early Peoples of the Rockies

The chronology of archaeological sites in the Rocky Mountains begins with Paleoindian occupation, shortly after 12,000 yr BP. There are several Paleoindian sites in mountain regions north of the Front Range, including Mammoth Meadows (in southwestern Montana), Mummy Cave, and Indian Creek in west-central Montana. The shores of Boulder Lake, near Pinedale, Wyoming, yielded Paleoindian artifacts made from obsidian that came from Obsidian Cliffs in Yellowstone Park (Cannon and Hughes 1993). The earliest documented Paleoindian culture in North America is the Clovis culture, named after a site in New Mexico where the characteristic fluted projectile points that typify this culture were first discovered. Clovis projectile points have been found at several localities on the plains east of the Front Range, including the Dent site, near the South Platte river in northeastern Colorado. Paleoindian archaeology of this region is well summarized in Frison (1991).

Evidence for high-country occupation by Paleoindians has been more difficult to obtain. Benedict (1991) concluded that evidence for human occupation in the Front Range prior to 10,000 yr BP remains inconclusive. Single projectile points of Clovis and Folsom age have been found there, but these points may have been carried into the mountains as amulets by later people. From 9500 until after 8500 yr BP, at least two groups of people made use of high-mountain terrain. Based on the type of projectile points and the origin of stones used to fashion these tools, one group probably lived mostly on the plains east of the mountains and came to the high country for hunting during the summer and fall. The other group apparently lived in the mountains, rarely venturing east beyond the foothills. This dichotomy of mountain and plains cultures was seen as late as the nineteenth century, in the relationship between the Ute tribe (mountain people) and the Arapaho tribe (plains dwellers who made use of the mountains on a seasonal basis). The hunting strategies and occupational patterns of regional peoples are detailed in Benedict (1992).

## Mid-Holocene Environments

From 7800 to 3000 yr BP, insect fossil assemblages from La Poudre Pass and Lake Isabelle show a gradual summer cooling trend. The 7800 yr BP assemblage from Lake Isabelle yielded an MCR reconstruction of mean July temperature of 0.2–2.2°C warmer than modern. The 7080 yr BP assemblage from Lake Isabelle yielded a mean July temperature estimate that ranged from 0.6°C cooler than modern to 2.2°C warmer than modern. This is the oldest Holocene assemblage indicating a mean July temperature that dipped below the modern baseline level. However, assemblages from La Poudre Pass and Lake Isabelle that date to 5360 and 3485 yr BP both provided MCR reconstructions of mean July temperatures that were warmer than modern by 2–3°C. According to the MCR reconstructions of mean January temperature values, winter temperatures remained below modern levels throughout the mid-Holocene. Winter temperatures below modern mean values persisted in the study region until the last 1000 years. Again, this is in agreement with the insolation curve for mid-latitudes in the Northern Hemisphere (Fig. 15.3).

The fossil insect MCR reconstructions shed new light on the question of a mid-Holocene warming event. The concept of a hot, dry, "altithermal" climatic regime from 7500 to 4000 yr BP was first invoked by Antevs in 1948, based on archaeological evidence from the northern Great Basin. Benedict (1979) called for a mid-Holocene altithermal period from 7500 to 5000 yr BP, based on shifting land-use patterns in the Archaic cultural period. However, his hypothesis dealt more with increasing aridity rather than with increasing temperatures. Benedict's interpretation called for two severe droughts, at 7000–6500 yr BP and again at 6000–5000 yr BP. Based on radiocarbon-dated charcoal, Benedict inferred that the Indian Peaks region of the Front Range was an important refuge for Archaic populations during these drought intervals. At least three cultural complexes made frequent use of sites near the upper treeline in this region during the mid-Holocene. The fossil insect record of the Rockies does not support the theory of a mid-Holocene thermal maximum. In terms of summer temperatures, the thermal maximum for the postglacial period in the Rocky Mountains of Colorado, based on MCR reconstructions, took place between 11,000 and 9000 yr BP.

Pollen evidence from the Front Range suggests that treeline shifted up-slope to elevations beyond its modern limit during the interval from 6500 to 3500 yr BP. During this interval, pine also achieved its maximum representation in Front Range subalpine forests. The pollen spectra from Blue Lake in the Indian Peaks Wilderness (Fig. 15.2) indicate that pine trees (possibly lodgepole pine) first arrived at the lake (3450 m) around 6200 yr BP (Short 1985).

Pollen studies from sites in the central Colorado Rockies (Fall 1985) indicate that from 7000 to 4000 yr BP, subalpine forest covered a broader elevational range there than it does today. Upper treeline moved up-slope beyond its modern limit, and the lower boundary shifted down-slope into what is now the upper montane forest zone. Fall (1985) interpreted the upward shift in treeline as a climatic warming and the downward extension of subalpine forest species as an indication of increased available moisture. The conflicting interpretations of insect, pollen, and archaeological data during the mid-Holocene interval are puzzling. If the archaeological inferences are valid, then there were two major drought cycles in the mid-Holocene. Taken in combination with the insect data, these drought cycles took place in a time of cooler-than-modern temperatures. If these two sets of data can be relied on, then it is very difficult to explain the expansion of subalpine forests during this interval because this type of expansion would presumably take place during an interval of optimal climate, not during an interval of cool climate with persistent droughts. Additional studies need to be conducted in this region to clarify the climatic reconstruction of the mid-Holocene.

## Late Holocene Environments

Late Holocene insect records from the Colorado Front Range show a progression from warmer-than-modern to cooler-than-modern summers, and back to warm again. At 3000 yr BP, mean July temperatures reconstructed from an assemblage at Lake Isabelle were just slightly above modern values. An assemblage just a few decades younger (and in fact, overlapping in radiocarbon age) from Longs Peak Inn yielded mean July temperatures as much as 2.5°C cooler than modern. However, this cooling of summer temperatures was neither long lived or extreme because by 2680 yr BP, the Longs Peak Inn fauna was reflecting mean July temperatures that were essentially modern. MCR analysis of the fossil insect record from the Front Range indicates that summer temperatures remained within 1–2°C of modern parameters from 2900 yr BP until the last millennium. A brief warming pulse was inferred from a 900-yr-BP assemblage from Mount Ida Bog, then temperatures returned to near-modern levels by 400 yr BP. Winter temperatures finally warmed to near modern levels at 900 yr BP, then cooled again by 400 yr BP. The 900-yr-BP warming may correspond to what historians refer to as the "Medieval warm period." This was a time when Norse colonists established outposts on Iceland and then Greenland. The subsequent cooling, or "Little Ice Age," is suggested by the cooling in summer temperatures, but more strongly indicated by a cooling of mean January temperatures by perhaps 15°C below modern levels. Additional well-dated late Holocene insect assemblages are needed to clarify the timing and intensity of climatic change during the last few thousand years.

Pollen spectra from La Poudre Pass, Long Lake, and Lake Isabelle show substantial changes after 3500 yr BP. Herb pollen increased, at the expense of tree pollen in regional subalpine pollen assemblages. The herb pollen was dominated by sagebrush and grasses. This pollen change has been interpreted as either a lowering of treeline or a thinning of subalpine forests or both (Short 1985). Pine pollen also became less abundant in subalpine pollen spectra at this time. These changes were inferred to be a response by the vegetation to climatic cooling. This cooling may have intensified during the last few hundred years, based on regional pollen spectra.

## Neoglacial Ice Advances

There is unequivocal evidence for a series of glacial advances in the Rocky Mountains during the Holocene, but the timing of these events has been the subject of considerable controversy (Davis 1988). The best evidence for the timing of post-Pleistocene, or Neoglacial, ice advances in Colorado begins with moraines that have been dated between 2000 and 1000 yr BP. Moraines are notoriously difficult to date, however, so about all that can presently be said is that the glacial ice that formed these moraines advanced sometime between 2400 and 1200 yr BP (Davis 1988). Moraines dating to the Little Ice Age are the most common Holocene glacial deposits in the Front Range. These glacial advances reached their maximum extent by the mid-nineteenth century, and have been retreating ever since.

## Important Research Questions in Front Range Paleoecology

Beginning with Nichols's (1982) review of the history of regional vegetation, several important questions have arisen from paleoecological research in the Front Range region. These questions have served as catalysts for additional paleoenvironmental research, but they also have important ramifications for modern ecological research, as the current biological communities are the product of thousands of years of biotic responses to environmental change. I list these questions here, then attempt to provide the most up-to-date answers based on a variety of sources.

### *Is Front Range Vegetation in Equilibrium with Climate?*

As we have seen in this chapter, the orbital forcing model of climate indicates that the peak of summer insolation came well before any noticeable response in regional vegetation. The fossil insect record corroborates the orbital model assertion that summer temperatures were warm enough to support spruce-fir forests up to at least the modern tree limit by 11,000 yr BP. This is not to say that vegetation does not respond to climate change; rather it points to the complexity of vegetational response to changing environments. Unlike predatory and scavenging insects that appear to respond rapidly to climate change, the response of vegetation is linked with other biotic and abiotic factors. This is an area of research that needs more input from ecologists. The role of soil development in controlling the up-slope advance of spruce and fir has been referred to in this chapter, but other factors, such as the seasonality of precipitation, snow depth, and length of growing season, may well have played key roles in controlling the rate and direction of plant communities. Regional pollen studies have not as yet benefited from some of the more-elegant statistical treatments that have been given to the analysis of pollen spectra from low-elevation regions in the eastern United States and Canada (e.g., see Overpeck et al. 1985; Webb 1988). There are some difficult problems in the Rockies that have caused palynologists to shy away from such statistical treatments as pollen response surfaces. The topographic complexity of even relatively small regions makes such modeling efforts extremely difficult, if not impossible.

### Was There a Mid-Holocene Altithermal in the Alpine?

Both the orbital forcing model and the fossil insect record indicate that the peak of summer warming took place well before the mid-Holocene, certainly well before the 7000-yr-BP boundary set by Antevs (1948) for an altithermal interval. However, changes in precipitation patterns, especially changes that affected the plains regions more than they did the mountains, would not have been detected by montane proxy records. Benedict (1979) argued that human use of the high country actually increased substantially during the mid-Holocene, probably in response to severe drought conditions on the plains. Modern climatological studies (Greenland et al. 1985) from the Front Range indicate that precipitation patterns are not consistent across an elevational transect, so a drought on the plains did not necessarily affect the high country.

### How Has the Size of the Alpine Tundra "Habitat Island" Archipelago Changed?

Lapse rates vary from site to site in the Front Range, but based on an average lapse rate of 0.6°C per 100 m of elevation, a 10°C cooling of mean July temperature translates into a lowering of mountain vegetation zones by about 600 m during the last glaciation. This estimate is borne out by the vegetational history of the Mary Jane site, where alpine tundra occurred more than 500 m down-slope from its present elevation at 30,000 yr BP and again at 13,700 yr BP. A treeline depression of 600 m was also inferred from a study of late Pinedale sites in the San Juan Mountains (Carrara et al. 1984). The downslope expansion of alpine tundra during the last glaciation would have connected many high mountain regions into a more-or-less continuous band of tundra, lapping from one mountainside to another. This expansion of the tundra biome has significant biogeographic implications. Island biogeograhic theory (MacArthur and Wilson 1967) suggests that large islands support greater numbers of species than do small islands. The same holds true for habitat "islands," that is, small, isolated patches of habitat surrounded by other types of habitats (see Shafer 1990). Alpine tundra currently represents a habitat island surrounded by a "sea" of coniferous forest. During the last glaciation, that habitat was not only larger, but much less fragmented than it is today. A comparison of modern and predicted late Pleistocene alpine tundra zones in Colorado (Fig. 15.4) shows that tundra was much more extensive in the late Pleistocene. More importantly, tundra habitats were connected in a much more continuous fashion at that time. The tundra archipelago of today was a tundra peninsula in the late Pleistocene, with some outlying tundra islands in the southeastern quadrant of the Colorado Rockies. High-altitude climates were much colder than today, and the areal extent of habitable land-

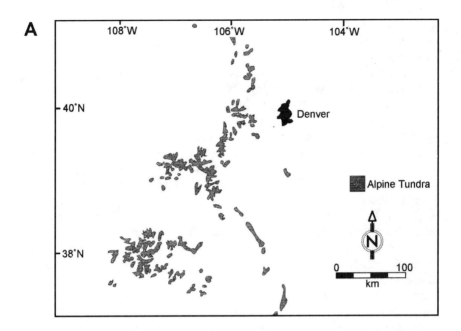

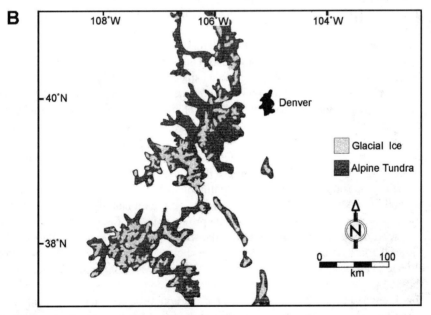

Figure 15.4. (a) Map of Colorado, showing approximate modern extent of alpine tundra (shaded regions). (b) Map of Colorado, showing approximate late Pinedale extent of alpine tundra and glacial ice.

scape was probably not as great as is shown in Fig. 15.4b, but the alpine tundra zone that existed in Colorado during the last glaciation probably served as a refuge for many species of alpine biota.

During the late Pleistocene, many species that today are associated with alpine environments expanded their ranges to encompass suitable habitats at lower elevations. This is an aspect of the Rocky Mountains that makes this region valuable for reconstructing the history of biological communities. It sets this region apart of adjacent lowland regions, where species' options were more limited. For example, species that shifted southward in front of glacial ice advances in the interior of North America had only two options for survival when the climate began warming and the ice receded at the end of the last glaciation. They had to migrate either northward or up the slopes of mountains, in search of suitably cold climate. Thus, cold-adapted beetles that lived on the plains south of Denver toward the end of the last glaciation have retreated to the alpine tundra in the Colorado Front Range (Elias 1986). In contrast to the Rocky Mountain fauna, cold-adapted species living in the central Great Plains region were not able to migrate north at the end of the last glaciation. The southern edge of the Laurentide ice sheet blocked northward migration as regional climates began to ameliorate. Between ca. 16,700 and 15,300 yr BP, climatic warming extirpated much of the cold-adapted beetle fauna in this region, with small populations of arctic and subarctic species surviving in alpine habitats of the Appalachians or in specialized habitats associated with stagnant ice (Schwert and Ashworth 1988).

The evidence from the late Quaternary fossil record of the Rocky Mountains has several implications for modern biologists. Current communities and ecosystems of the Rocky Mountain region developed only since the end of the last glaciation. They are the product of their Quaternary environments. Changes in distribution have taken place in a relatively short time. These changes show that some groups of organisms shift distributions more rapidly than do others, but a wide variety of taxa were able to move substantial distances to find suitable habitats in the face of changing climates. These changes also suggest that the current biological communities of the Rocky Mountain region are simply the latest reshuffling of species, and that species composition is probably in a continuous state of flux. The transitory nature of community composition is scarcely visible during human lifetimes and only becomes obvious over the longer time scale provided by the fossil record.

According to the fossil insect record, the warming of summer temperatures at the end of the last glaciation in the Rockies was both rapid and large scale. This rapid climate change, and the other types of concomitant environmental changes (e.g., shifts in vegetation patterns, changes in precipitation patterns, seasonality, and soil conditions) were the driving forces in the disruption of previous biological communities and led to episodes of species extirpation. Many species of cold-adapted beetles that today are found in arctic and subarctic regions were extirpated from Colorado since the last glaciation. These extirpations mostly took place between 12,500 and 10,000 yr BP, the interval in which regional summer temperatures rose rapidly, finally attaining levels above modern parameters. However, many other cold-adapted species simply retreated up-slope to the modern alpine tundra zone, where they persist today. It is important to keep in mind that glacial periods account for approximately 90% of the last two million years. Glacial climates have therefore been the norm, not the exception, over the time frame when we presume that most of the alpine flora and faunas has evolved. The current interglacial conditions constitute a brief interruption from the normal (glacial) conditions. This truly longterm perspective must be remembered if we are to develop a good understanding of regional ecosystems, because the forces that have shaped the various biological communities act over long time scales (Elias 1996b).

*Acknowledgments* Support for the analysis of insect fossil assemblages from the Rocky Mountains was provided by NSF grants for Long-Term Ecological Research at Niwot Ridge, Colorado, DEB-9211776. Support for the MCR reconstructions was provided by a grant to the author from the National Science Foundation, ATM-9219040. I thank Kathy Anderson, INSTAAR, for her help in the preparation of climate envelopes for the species used in the MCR analyses. Eric Leonard, Geology Department, Colorado College, Colorado Springs, provided a useful discussion on paleotemperature reconstructions based on ELAs of regional glaciers.

References

Antevs, E. 1948. Climatic changes and pre-White man. *University of Utah Bulletin* 38:168–191.
Baker, R. G. 1976. Late Quaternary vegetation history of the Yellowstone Lake Basin, Wyoming. *U.S. Geological Survey Professional Paper* 729-E:1–48.
Benedict, J. B. 1979. Getting away from it all: A study of man, mountains and the two-drought altithermal. *Southwestern Lore* 45:1–12.
———. 1991. Along the Great Divide: Paleoindian archaeology of the high Colorado Front Range. In *Ice Age hunters of the Rockies,* edited by D. J. Stanford and J. S. Day. Denver, CO: Denver Museum of Natural History.
———. 1992. Footprints in the snow: High-altitude cultural ecology of the Colorado Front Range, U.S.A. *Arctic and Alpine Research* 24:1–16.
Berger, A. L. 1978. Long-term variations in caloric insolation resulting from the Earth's orbital elements. *Quaternary Research* 9:139–167.
Brunstein, F. C., and D. K. Yamaguchi. 1992. The oldest known Rocky Mountain bristlecone pines (*Pinus aristata* Egelm.). *Arctic and Alpine Research* 24:253–256.
Cannon, K. P., and R. E. Hughes. 1993. Obsidian source characterization on Paleoindian projectile points from Yellowstone National Park. *Current Research in the Pleistocene* 10:54–56.
Carrara, P. E., W. N. Mode, M. Rubin, and S. W. Robinson. 1984. Deglaciation and postglacial timberline in the San Juan Mountains, Colorado. *Quaternary Research* 21:42–55.
Coope, G. R. 1977. Fossil Coleopteran assemblages as sensitive indicators of climatic changes during the Devensian (Last) cold stage. *Philosophical Transactions of the Royal Society of London* Ser. B 280:313–340.
Davis, P. T. 1988. Holocene glacier fluctuations in the American Cordillera. *Quaternary Science Reviews* 7:129–157.
Elias, S. A. 1983. Paleoenvironmental interpretations of Holocene insect fossil assemblages from the La Poudre Pass site, northern Colorado Front Range. *Palaeogeography, Palaeoclimatology, Palaeoecology* 41:87–102.
Elias, S. A. 1985. Paleoenvironmental interpretations of Holocene insect fossil assemblages

from four high altitude sites in the Front Range, Colorado, U.S.A. *Arctic and Alpine Research* 17:31–48.

———. 1986. Fossil insect evidence for Late Pleistocene paleoenvironments of the Lamb Spring site, Colorado. *Geoarchaeology* 1:381–386.

———. 1991. Insects and climate change: Fossil evidence from the Rocky Mountains. *BioScience* 41:552–559.

———. 1994. *Quaternary insects and their environments.* Washington: Smithsonian Institution Press.

———. 1995. *Ice age history of Alaskan National Parks.* Washington: Smithsonian Institution Press.

———. 1996a. Ice-age environments of National Parks in the Rocky Mountains. Washington: Smithsonian Institution Press.

———. 1996b. Late Pleistocene and Holocene seasonal temperatures reconstructed from fossil beetle assemblages in the Rocky Mountains. *Quaternary Research* 46:311–318.

Elias, S. A., and A. R. Nelson. 1989. Fossil invertebrate evidence for Late Wisconsin environments at the Lamb Spring site, Colorado. *Plains Anthropologist* 34:309–326.

Elias, S. A., S. K. Short, and P. U. Clark. 1986. Paleoenvironmental interpretations of the Late Holocene, Rocky Mountain National Park, Colorado, U.S.A. *Revue de Paléobiologie* 5:127–142.

Elias, S. A., and L. J. Toolin. 1990. Accelerator dating of a mixed assemblage of late Pleistocene insect fossils from the Lamb Spring site, Colorado. *Quaternary Research* 33:122–126.

Fall, P. L. 1985. Holocene dynamics of the subalpine forest in central Colorado. *American Association of Stratigraphic Palynologists Contribution Series* 16:31–46.

Frison, G. C. 1991. *Prehistoric hunters of the high plains.* 2d ed. San Diego, CA: Academic Press.

Greenland, D., J. Burbank, J. Key, L. Klinger, J. Moorhouse, S. Oaks, and D. Shankman. 1985. The bioclimates of the Colorado Front Range. *Mountain Research and Development* 5:251–262.

Legg, T. E., and R. G. Baker. 1980. Palynology of Pinedale sediments, Devlins Park, Boulder County, Colorado. *Arctic and Alpine Research* 12:319–333.

MacArthur, R. H., and E. O. Wilson. 1967. *The theory of island biogeography.* Princeton, NJ: Princeton University Press.

Madole, R. F., and R. R. Shroba. 1979. Till sequence and soil development in the North St. Vrain drainage basin, East Slope, Front Range, Colorado. In Guidebook for postmeeting field trips held in conjunction with the 32nd annual meeting of the Rocky Mountain section of the Geological Society of America, May 26–27, 1979, Colorado State University, edited by F. G. Ethridge. Fort Collins, CO: Geological Society of America.

Nelson, A. R., A. C. Millington, J. T. Andrews, and H. Nichols. 1979. Radiocarbon-dated upper Pleistocene glacial sequence, Fraser Valley, Colorado Front Range. *Geology* 7:410–414.

Nichols, H. 1982. Review of late Quaternary history of vegetation and climate in the mountains of Colorado. In *Ecological studies of the Colorado alpine: A festschrift for John W. Marr,* edited by J. C. Halfpenny. Occasional Paper 37, Institute of Arctic and Alpine Research, University of Colorado, Boulder.

Overpeck, J. T., T. Webb III, and I. C. Prentice. 1985. Quantitative interpretation of fossil pollen spectra: dissimilarity coefficients and the method of modern analogs. *Quaternary Research* 23:87–108.

Richmond, G. M. 1960. Glaciation of the east slope of Rocky Mountain National Park, Colorado. *Geological Society of America Bulletin* 71:1371–1382.

Schwert, D. P., and A. C. Ashworth. 1988. Late Quaternary history of the northern beetle fauna of North America: A synthesis of fossil distributional evidence. *Memoirs of the Entomological Society of Canada* 144:93–107.

Shafer, C. L. 1990. *Nature reserves: Island theory and conservation practice.* Washington: Smithsonian Institution Press.

Short, S. K. 1985. Palynology of Holocene sediments Colorado Front Range: Vegetation and treeline changes in the subalpine forest. *American Association of Stratigraphic Palynologists Contribution Series* 16:7–30.

Short, S. K., and S. A. Elias. 1987. New pollen and beetle analysis at the Mary Jane site, Colorado: Evidence for Late-Glacial tundra conditions. *Geological Society of America Bulletin* 98:540–548.

Stanford, D., W. R. Wedel, and G. R. Scott. 1982. Archaeological investigations of the Lamb Spring site. *Southwestern Lore* 47:14–27.

Stuiver, M., and P. J. Reimer. 1993. Extended $^{14}$C data base and revised CALIB 3.0 $^{14}$C age calibration. *Radiocarbon* 35:215–230.

Webb, T., III. 1988. Eastern North America. In *Vegetation history,* edited by B. Huntley and T. Webb III. Dordrecht, The Netherlands: Kluwer Academic.

Whitlock, C. 1993. Postglacial vegetation and climate of Grand Teton and southern Yellowstone National Parks. *Ecological Monographs* 63:173–198.

# 16

## Environmental Change and Future Directions in Alpine Research

Jeffrey M. Welker
William D. Bowman
Timothy R. Seastedt

### Introduction

Alpine tundra is an important indicator system of environmental change (Grabherr et al. 1996; Beniston and Fox 1996). This ecosystem occurs at all latitudes, with its lower altitudinal limit at timberline inversely related to latitude (Woodward 1993). Thus, the global distribution of alpine tundra and the fact that this system exists near the limits of vascular plant tolerance to temperature, moisture, and growing season duration makes it an excellent system for monitoring environmental change across the globe (e.g., Körner and Larcher 1988). In addition, the functional integrity of this system is critical to lower elevation ecosystems because substantial amounts of water and elements are intercepted by the alpine, filtered, and transported to lower elevations (Williams et al. 1996; chapter 4).

Alpine tundra has been spared most of the large-scale disturbances associated with human development and resource extraction that have occurred in lower altitudinal ecosystems. This is probably due to its climatic severity and lack of renewable resources that can be exploited (e.g., trees, fast-growing forage for grazing). Some high-altitude sites have been impacted by recreational development (e.g., ski areas, trails, roads), mining, and grazing. In addition, changes in native herbivore populations, particularly elk and deer, due to extirpation of predators by humans (chapter 12), may have significantly influenced tundra vegetation (chapter 14). However, the indirect effects of human activities associated with the burning of fossil fuels are of greater current and future concern as the dominant anthropogenic influences on alpine tundra ecosystems. At a global scale, climate change is of concern for all ecosystems, whereas at a more-regional scale, there is concern for the impact of N and acid deposition near centers of industrial and urban growth (Galloway et al. 1995).

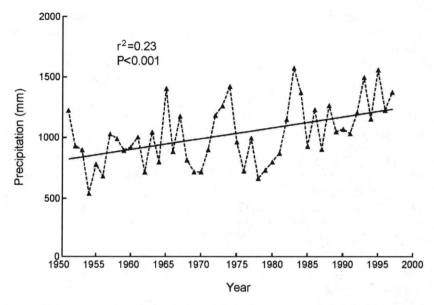

Figure 16.1. Temporal trend of annual precipitation at the D1 weather station, Niwot Ridge (3743 m; see chapter 2 for additional site information). Data were collected as part of the Mountain Climate Program of the Mountain Research Station. Line fit using a linear regression. Seasonal evaluation of the trend indicates most of the precipitation increase is occurring in the autumn.

Niwot Ridge, and possibly much of the Colorado Front Range, has experienced significant increases in the rate of N deposition (Sievering et al. 1996; chapter 3) and precipitation (Fig. 16.1; chapter 2) in the past several decades. Local sources of $NO_x$ compounds that potentially contribute to the elevated rates of N deposition in the Front Range include power plants in western Colorado and automobile and industrial emissions in the urban corridor at the base of the mountains (Fort Collins-Boulder-Denver-Colorado Springs megalopolis), which is home to around four million people. The continued rate of population growth in this Front Range urban corridor, one of the highest in the United States (Riebsame 1997), will result in continued increases in N deposition. The increase in precipitation, which is most pronounced in the autumn, may be related to regional warming and subsequently increased evapotranspiration in low elevational areas (Manabe and Broccoli 1990).

The focus for much of our research efforts has been on monitoring and experimentation to detect and predict alpine ecosystem responses to increased N deposition, snow cover, and warming growing season temperature. These factors are not mutually exclusive, and wherever possible their interactions are being investigated. For example, snow is an important determinant of the spatial variation in N deposition inputs through its redistribution by winds across the alpine landscape (Bowman 1992) as well as an important control on wintertime microbial activity in the soil (Brooks et al. 1996). This chapter focuses on N deposition and climate change, although we recognize that other anthropogenically enhanced factors (e.g., increased UVB irradiation; Caldwell et al. 1995; invasive species) could be significant to the future of the alpine.

## Nitrogen Deposition

### Terrestrial Ecosystem Impacts

Consideration of the N cycle for the Niwot Ridge alpine ecosystem (Fig. 16.2) indicates several unusual features relative to other herbaceous and forest ecosystems, including low rates of internal cycling and biotic uptake (chapter 12) and relatively high rates of symbiotic $N_2$-fixation (Bowman et al. 1996). Even though the current rate of N deposition (0.6 g N m$^{-2}$ yr$^{-1}$) (chapter 3) is low relative to ecosystems where associated ecosystem changes have been reported (e.g., central Europe, northeastern United States), the rate is high relative to estimated rates of net N mineralization for alpine tundra on Niwot Ridge (spatially integrated: 1.2 g N m$^{-2}$ yr$^{-1}$) (Fisk and Schmidt 1995). This, coupled with relatively low rates of primary production, suggest that N deposition could have impacts at levels lower than those considered critical limits in other ecosystems (Baron et al. 1994).

An increase in N supply that results from an increase in N deposition would impact primary production and plant and microbial species composition. Biomass production has been shown to be N limited in several alpine tundra communities

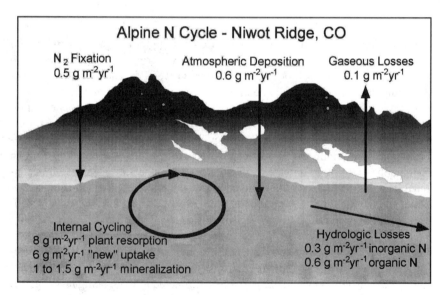

Figure 16.2. Fluxes of N for the Niwot Ridge alpine ecosystem. N deposition value is from chapter 3, $N_2$-fixation from Bowman et al. 1996, internal cycling from chapter 12, gaseous losses from chapter 13, and groundwater and stream losses from chapter 5 and Caine (unpublished data).

(Bowman et al. 1993, 1995), and microbial biomass also is enhanced toward the end of the alpine growing season in N fertilized soils (Fisk and Schmidt 1996). The changes in N sequestration that accompany increased N deposition is of critical importance to determining N loss to adjacent aquatic ecosystems, where accompanying eutrophication and associated chemical changes can have large impacts on ecosystem health and water quality. Unfortunately, we know little regarding the capacity of alpine soils to act as a sink for inorganic N, although for some forested systems it can be relatively large (Johnson 1992).

The changes in plant biomass production that accompanies N fertilization experiments in the alpine are primarily associated with changes in species composition (Bowman et al. 1993, 1995; Theodose and Bowman 1997). The increase in production is relatively small and the timing of the response delayed as a result of strong developmental constraints on growth for the dominant species (Walker et al. 1994; Diggle 1997; chapter 10). Thus, a N-limited biotic system can exist even though the initial stages of N-saturation may be occurring (Williams et al. 1996; chapter 5).

Plant community structure is related to N supply, as competitive interactions are influenced by N availability, and some species are more capable of establishing viable populations when N availability is increased (Theodose and Bowman 1997). Microbial functional groups are affected by higher N supply, as indicated by greater $N_2O$ production and changes in the balance between $CH_4$ production and consumption (Neff et al. 1994; Fig. 16.3). While large-scale changes in the vegetation have not been detected over the past 40 years, there have been significant increases in the cover of species that show large increases in cover under N fertilization (Korb 1997; Table 16.1).

The changes in species composition that could accompany increased N deposition would have profound feedbacks to N cycling, enhancing the potential for N loss from alpine tundra. These biotic changes may have proportionally greater effects on alpine ecosystem function relative to the increase in production associated with greater N deposition (Bowman and Steltzer 1998). Plant species exert control on N mineralization rates through variation in biomass turnover rates, differences in litter quality, and production of secondary compounds such as phenolics (Steltzer and Bowman 1998). The alpine tundra community most susceptible to biotic change that results from N deposition is the moist meadow, which receives greater N input from melting snow (mean of 1.2 g $m^{-2}$ $yr^{-1}$), an important reservoir of wintertime N deposition (Bowman 1992). Moreover, the moist meadow contains plant species that are both N limited (*Deschampsia*) and N insensitive (*Acomastylis*). Net N mineralization rates are tenfold higher in soils under *Deschampsia* than under adjacent *Acomastylis* patches, independent of changes in soil microclimate (Steltzer and Bowman 1998). The replacement of *Acomastylis* as a dominant in the moist meadow by *Deschampsia,* along with similar changes in other communities (e.g., increase in *Poa* and *Festuca* cover at the expense of *Kobresia* in dry meadows) may induce an important positive feedback to increased N deposition, potentially increasing $NO_3^-$ leaching from alpine soils (Fig. 16.4), leading to greater N export from alpine catchments.

As a result of the biotic feedbacks to increasing N inputs, we would expect a nonlinear response of the alpine system to increases in N deposition (Fig. 16.5). Export

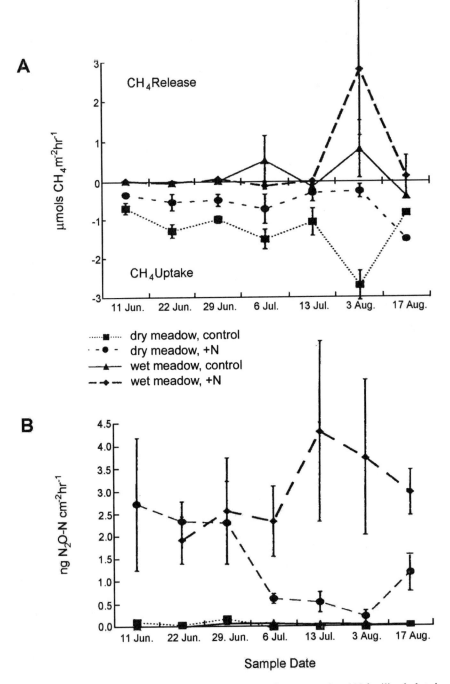

Figure 16.3. (a) Methane and (b) nitrous oxide fluxes from control and N fertilized plots in dry and wet meadow communities on Niwot Ridge during the growing season in 1992. Error bars are standard errors ($n = 5$) (from Neff et al. 1994).

Table 16.1. Alpine Plant Species That Have Increased over a 4-Decade Period and Those Which Increase with Greater N Availability

|  | Korb 1997[1] | Theodose & Bowman[2] | Thomas & Bowman[3] |
|---|---|---|---|
| *Allium geyeri* |  |  | √ |
| *Arenaria fendleri* | √ |  |  |
| *Artemisia spp.* |  | √ |  |
| *Bistorta bistortoides* | √ | √ | √ |
| *Campanula rotundifolia* |  | √ |  |
| *Carex albonigra* | √ |  |  |
| *Carex rupestris* | √ | √ |  |
| *Erysimum capitatum* | √ |  |  |
| *Mertensia lanceolatum* | √ | √ |  |
| *Oreoxis alpina* |  |  | √ |
| *Poa alpina* |  | √ |  |
| *Poa arctica* |  | √ |  |
| *Poa glauca* | √ | √ | √ |
| *Potentilla spp.* | √ | √ | √ |
| *Sedum lanceolatum* |  |  | √ |
| *Taraxacum ceratophorum* | √ |  |  |

[1]Incresed in frequency in dry meadows on Niwot Ridge during the period between 1953 and 1996.
[2]Increased in abundance in N fertilized plots relative to control plots over a 4-year period.
[3]Increases in abundance in the presence of $N_2$-fixing *Trifolium dasyphyllum*.

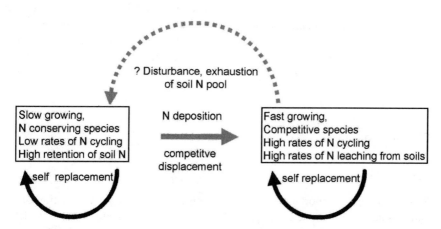

Figure 16.4. Conceptual model showing the generalized pattern of plant species self-replacement in communities with low or high N availability (from Bowman and Steltzer 1998). In the face of higher N deposition, N conserving species of infertile environments would be replaced by more-competitive species, with biotic characteristics promoting higher N cycling in soils. The new state would be preserved, even in the absence of additional N inputs. A return to the initial state would require some disturbance or exhaustion of soil N reserves, allowing species adapted to low N conditions to reestablish.

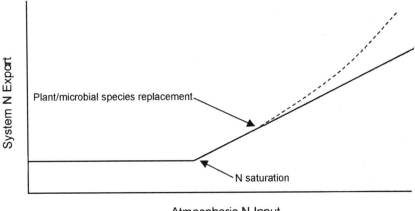

Figure 16.5. Hypothetical response of system N exports to increasing N deposition with and without changes in plant and microbial species replacements. As N deposition increases, a point is reached where system N export begins to increase during the growing season (N saturation). As N availability continues to increase, species replacements result in greater N export as a result of greater rates of N cycling and lower retention.

of inorganic N from the system would accelerate as species replacements occur, enhancing N cycling in the soil. In this scenario, it is possible that exports could exceed imports if biomass production cannot keep up with the increased production of inorganic N in the soil.

Impacts of N deposition on higher trophic levels would be indirect and mediated through changes in plant production and species composition. Grazers such as grasshoppers may benefit from higher biomass production and higher quality (higher tissue N) food (Coxwell and Bock 1995; chapter 14), but plant species replacements may result in lower forage quality, as grasses replace forbs (Bowman et al. 1995; chapter 14).

## Climate Change

### Increase in Winter Snow

Snow is the most important factor controlling the distribution and function of organisms across the alpine landscape. Plant community structure is well correlated with maximum winter snow depth (May and Webber 1982; Walker et al. 1993; chapter 6) as is the distribution of most perennial endothermic animal species (chapter 7), invertebrates (Addington and Seastedt 1999), and microbial activity (Brooks et al. 1996). Snow depth is an important determinant of the growing season length, soil moisture temperature, and subsequently the multitude of physiological activities that are dependent on them. Therefore, determination of the response of the alpine to altered snow regime is critical to understanding climate change impacts.

Snow augmentation and removal experiments have demonstrated the importance of the growing season length to reproductive output of alpine plants. In a study in the San Juan Mountains, Bock (1976) found that 70% of the species studied produced fewer viable seeds in snowfence treatments than under natural snow cover. Seed number and mass are correlated with natural variation in growing season length in *Ranunculus adoneus,* a common snowbed species (Galen and Stanton 1993), and augmentation or removal of snow will decrease or increase seed mass relative to the ambient condition (Galen and Stanton 1995). These results suggest that increased snow cover may favor those species that are able to reproduce vegetatively over those that rely on sexual reproduction, although the general success of sexual reproduction in Rocky Mountain alpine tundra is largely unknown.

In order to assess the impacts of increased snow cover on alpine community and ecosystem dynamics, a long-term snowfence experiment was initiated in the fall of 1993 on Niwot Ridge (Fig. 16.6). The fence was established in a transition zone between shrub and moist meadow tundra and dry meadow tundra. Approximately 4800 m$^2$ of tundra receives up to 200% greater snow, which is deposited earlier in the fall than would normally occur. Where snow depth is at a maximum, the growing season is 40–50% shorter (Walker et al. 1999).

Initial results from the snowfence experiment indicate that although initiation of growth is delayed by the late melt out, some species are able to accelerate their development, reaching the same phenological stage as control plants by the end of the

Figure 16.6. The Niwot Ridge snowfence experiment, initiated in 1993. The fence is 60 m long and 2.6 m high, influencing approximately 4800 m$^2$ of dry meadow tundra (photo by William D. Bowman).

growing season (P. L. Turner and M. D. Walker, unpublished data). However, the percentage of the growing season that some species spend in flower is greater under augmented snow relative to controls, and other phenological stages, especially vegetative development, are greatly accelerated. *Acomastylis rossii* will actually produce more and larger seeds with snow augmentation. Additionally, some species have decreased in abundance in response to the additional snow cover, including *Kobresia,* which is very sensitive to the depth of winter snow cover (Bell and Bliss 1979; chapter 6), while other species have increased in abundance (e.g., *Trifolium parryi*). Increased snow cover could lead to substantial changes in the composition of plant growth forms in some communities. For example, the encroachment of shrubs, primarily willows, which are presently found almost exclusively in moist and wet meadow communities, would occur into areas dominated by graminoids and forbs. These changes in species composition will in turn modify the soil microbial and faunal composition (O'Lear and Seastedt 1994; Fisk et al. 1998) and could dramatically alter the allocation of biomass, biogeochemistry, and microenvironment of these sites.

The Niwot snowfence experiment has contributed to our understanding of the importance of snow cover depth and timing for regulating elemental cycles during the winter (Brooks et al. 1997). Significant microbial activity can occur during the winter and is related to the depth and timing of snow cover. Rates of wintertime N cycling under deep, early-accumulating snowpacks are relatively high, contributing to a significant pool of early season inorganic N (Brooks et al. 1996; chapter 12). $N_2O$ emission from tundra soils are several orders of magnitude greater under deep snowpack relative to emissions from tundra covered by shallow snow cover (Brooks et al. 1997).

Significant C fluxes may also occur under deep snow. A low level of heterotropic respiration occurs throughout the winter (Brooks et al. 1997). Although the rates of respiration are not as high as they are in the summer, the long snow-covered period results in significant amounts of carbon being effluxed from alpine tundra under snowpack to the atmosphere in winter (Sommerfeld et al. 1993, 1996). The extent to which these short-term changes in carbon inputs and losses under increased snow cover will alter the long-term carbon balance is unclear. However, the long-term consequences of higher snow pack can be inferred from natural gradients and long-term snow manipulation studies. Soil C in the alpine is highest in anaerobic wet meadows but exhibits a secondary peak in aerobic soils of intermediate snowpacks (e.g., dry to mesic meadows) (Burns 1980; chapter 8). The aerobic sites have this C storage due to lower rates of winter and growing season decomposition. Inferences about the long-term consequences of increases in snow cover can also be derived from field studies in the subalpine meadows of Montana. Welker and Weaver (unpublished data) have found that after 20 years of increased snowpack, soil carbon content is reduced by 50%, and soil nitrogen content is 25% lower. The reduction in carbon content is likely the result of higher rates of $CO_2$ efflux in winter under deeper snow, since primary production in ambient and deep snow conditions are the same. This interpretation is also consistent with Niwot Ridge decomposition studies (Bryant et al. 1998).

## Increases in Growing Season Temperatures

Low temperature is one of the primary determinants of the alpine ecosystem and thus increases in growing season temperature could have profound impacts on community composition and ecosystem processes. Temperature optima for the existing flora are close to existing ambient conditions, and developmental controls of plant growth (Körner and Palaez Menendez-Riedl 1989; Diggle 1997; chapters 9, 10) may be strong enough to overide any positive impacts of increasing temperatures on growth (Körner and Larcher 1988). Thus, the major effects of increased temperature on the alpine are likely to be the result of indirect effects on nutrient availability, growing season length, and species replacements, including possibly the establishment of subalpine species in alpine tundra.

Two experiments have investigated the influence of increased growing season temperatures on Niwot Ridge tundra. Both have employed open-top chambers to warm the tundra (Marion et al. 1997; Welker et al. 1997), and both are associated with the International Tundra Experiment (Henry and Molau 1997). One of these has investigated the interactive effects of warming with increased snow cover, whereas the other has stressed the impact of warming on a single, circumpolar community dominated by *Dryas octopetala* (Fig. 6.5; Welker et al. 1997). On average, the chambers increased air temperatures at the *Dryas* site by 1.8°C, while at the snowfence site, the chambers resulted in a 2.5°C increase in air temperaturer and a 1.3°C increases in soil temperatures (Fig. 16.7).

The growing season under the warming treatment is extended relative to the ambient controls, due to a delay in senescence rather than to more-rapid early season development (Turner, P. L. and M. D. Walker, unpublished data). Vegetative growth of *Acomastylis rossii* and *Bistorta bistortoides* is greater under a combined treatment

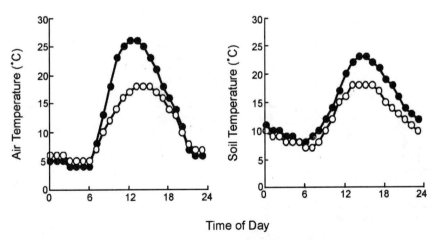

Figure 16.7. Representative diurnal trend in air and soil temperatures for plots with ITEX chambers (closed symbols) and ambient plots (open symbols) during the growing season for *Dryas* tundra on Niwot Ridge.

of warming and snow augmentation, probably as a result of greater soil resource supply (water, nutrients). These results are preliminary, and additional measurements are needed to determine the long-term trends in plant growth and development under a warmer temperature regime.

Results to date indicate *Dryas octopetala* produces greater numbers of leaves and total leaf mass per ramet, although this response was not consistent in all years (Table 16.2). In addition, the individual mass of *Dryas* seeds produced in the chamber treatment were approximately 30% greater than that of control plots, although this was observed only in the second year of field manipulations (Table 16.2). The time lag in these responses is probably associated with preformation of buds, which is common in alpine species (Diggle 1997). The changes in organismic traits may lead to greater ecosystem productivity, although senescing leaves of warmed *Dryas* plants had lower leaf N content and subsequently higher C/N ratios, which may retard decomposition, lowering nutrient availability, and possibly lowering soil and plant N levels in subsequent years (Fig. 16.8; Welker et al. 1997).

In addition to comparative measures of soil respiration, the net flux of $CO_2$ from *Dryas* dominated tundra under ambient and warmed conditions has been measured (Brown 1997; Welker et al. 1999). Under ambient conditions, this community is a carbon sink of about 8 g $CO_2$-C m$^{-2}$ per growing season (Fig 16.9). However, after 4 years of experimental warming, this community switches to a C source of almost 10 g $CO_2$-C m$^{-2}$ per summer (Fig 16.9). This sink-source switch is due primarily to increases in the rate of ecosystem respiration under warmer conditions, without a corresponding increase in photosynthesis, a trend also reported for *Dryas* communities in the Arctic (Jones et al. 1998; Welker et al. 1999). A switch in carbon dioxide sink-source activity of arctic tundra was also reported to have occurred in the Arctic over the past 30 years due to warmer and drier conditions (Oechel et al. 1993).

Table 16.2. Morphological Traits and Seed Mass of *Dryas Octopetala* Under Four Treatments

| Treatment | Leaves/Ramet | | Leaf Mass (mg) | | Leaf Mass/ Ramet (mg) | | Seed Mass (μg) | |
|---|---|---|---|---|---|---|---|---|
| | 1993 | 1994 | 1993 | 1994 | 1993 | 1994 | 1993 | 1994 |
| −T−W[1] | 3.2[a] | 4.8[a] | 4.0[a] | 4.1[a] | 12.5[a] | 19.8[a] | 230[a] | 310[a] |
| −T+W[2] | 3.7[b] | 3.7[b] | 4.4[a] | 4.7[b] | 16.5[b] | 17.8[a] | 170[a] | 370[a] |
| +T−W[3] | 3.9[b] | 4.8[a] | 4.8[b] | 4.8[b] | 18.9[b] | 19.8[a] | 190[a] | 400[b] |
| +T+W[4] | 3.8[b] | 4.1[a] | 4.9[b] | 4.4[a] | 19.2[b] | 18.3[a] | 390[b] | 410[b] |

*Notes:* Portions of clones were harvested in late August and leaves per ramet were counted (leaves/ramet), individual leaf mass determined after drying (leaf mass), and average leaf mass per ramet calculated as the result of the number of leaves times their individual weights (leaf mass/ramet). Individual seed mass was determined from culms collected in September following drying in an oven.

[1]Ambient temperatures and ambient rainfall.
[2]Ambient temperatures and 50% increase in rainfall.
[3]Warmed conditions and ambient rainfall.
[4]Warmed and 50% increase in rainfall.

Superscripts depict significant treatment effects within each year as tested by a factorial analysis of variance and means separation ($p < 0.05$).

## Effects of Changing Climates on Dryas Proccesses

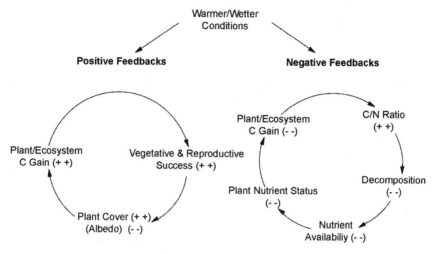

Figure 16.8. Conceptual model of the direct impacts of climate change on *Dryas octopetala* plants and ecosystems and the positive and negative feedbacks on functional attributes of these habitats (from Welker et al. 1997).

These carbon flux measurements reflect the inherent variability in soil carbon dynamics across the Niwot Ridge landscape (Burns 1980; Brooks et al. 1997). In addition, these findings emphasize the importance of using similar manipulations and measurements across community-types to estimate whole-ecosystem response to climate change. However other communities may be more unpredictable and may be influenced by other parameters, such as soil water or nutient availability, and may not exhibit the degree of temperature sensitivity as seen in the *Dryas* tundra.

### Increases in Growing Season Precipitation

Soil moisture has been implicated as an important control on plant and microbial distrubution and productivity on Niwot Ridge (chapters 6, 9, 12, 13). Thus any change in growing season soil moisture, whether resulting from increased snow pack or increased growing season precipitation, could have a significant impact on alpine ecosystem structure and function.

Watering experiments in dry and moist meadow tundra (50% greater than average growing season precipitation) for two growing seasons did not influence primary production (Bowman et al. 1995). A water limitation of dry meadow tundra has been inferred from long-term measurements of production (Walker et al. 1994) and water balance estimation (Isard 1986; Greenland et al. 1984). However, the conservative growth strategy of the dominant meadow species, *Kobresia* (Bell and Bliss 1979) may have resulted in only modest increases in production after augmentation. Individual species may have responded to the watering treatment, but their response

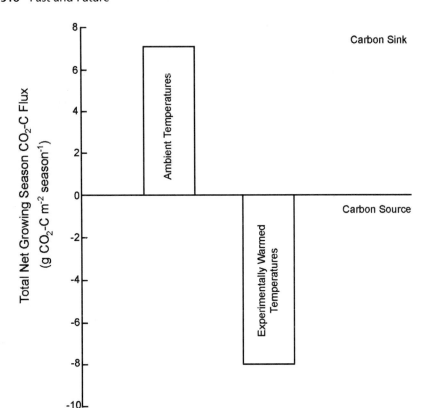

Figure 16.9. Net flux of carbon dioxide from dry tundra on Niwot Ridge under conditions of ambient temperatures and under conditions where dry tundra has been experimentally warmed each summer for 4 years. Net flux measures are growing season averages based on diurnal measures of $CO_2$ flux in June, July, August, and September (from Welker et al. 1999).

was not detected in the measurement of biomass. There were increases in the rate of net N mineralization in watered plots relative to control plots, consistent with the suggestion that soil microbes may be water limited for part of the growing season (Fisk and Schmidt 1995; chapter 12). Thus the long-term impact of increased water availability may be mediated primarily by increased soil nutrient availability (Bowman et al. 1995).

Welker et al. (1997) conducted a summer watering experiment in conjunction with field warming studies mentioned earlier at the *Dryas* tundra site. Summer rainfall was increased by approximately 50%. *Dryas* that received additional summer water produced significantly greater numbers of leaves and leaf mass per ramet (Table 16.2) and produced seeds that were 30% heavier than plants receiving ambient summer rainfall. These responses were, however, not consistent across time. Additional water also resulted in higher rates of in situ soil respiration. In general, soil respiration in watered plots was 20% higher than those in unwatered plots. Whether these increases in respiration were the result of greater root respiration, increases in microbial respiration, or increases in both of these activities remains unknown.

The Century Model (Parton et al. 1987) has been used to predict responses of the tundra to changes in nitrogen deposition (Baron et al. 1994) and climate change (Conley et al. 2000). These simulations have been fairly consistent with predictions based solely on empirical evidence. The most consistent finding is that the tundra is likely to respond to increased N additions with increased N exports in ground and surface waters. In general, the simulated response of the tundra to climate change scenarios is highly variable and strongly dependent on shifts in the functional characteristics of the tundra biota.

## Summary

The primary response of the alpine biota to environmental change will be a change in the composition and abundance of species. The conservative growth responses of many of the dominant plant species to environmental change will result in increases in abundance of more rare species, which could serve as indicator species (Lesica and Steele 1996). Experiments can and have been used to determine which species and communities are particularly sensitive and appropriate for indicating that the alpine is responding to specific environmental change. In conjunction with long-term monitoring of plant species' composition across the tundra landscape, these experiments provide a means of determining indirect anthropogenic impacts on the regional environment.

A potentially more-sensitive and rapid biotic indicator of environmental change is the soil microbial community. Changes in the fluxes of trace gases occur in response to increases in N and water availability (Neff et al. 1994; chapter 13), indicating changes in the functional group composition of soil microbes. These microbial responses may occur at levels of environmental change below the detection limit of other biotic indicators. Work is currently underway to directly measure changes in microbial functional groups in response to environmental change (Colores et al. 1996).

In addition to indicating a response of the alpine system to environmental change, these shifts in the composition of plants and microbes may also exacerbate the impacts (e.g., N deposition, Bowman and Steltzer 1998; warming, Welker et al. 1999). The critical role of alpine biota in controlling production and biogeochemistry of the tundra will result in important, potentially nonlinear feedbacks to the systems' response to anthropogenically driven environmental change. Predicting and detecting these responses remains one of the greatest challenges of alpine tundra ecology.

## Future Directions in Alpine Research

Each of the preceding chapters has identified lines of inquiry that will benefit our understanding of the current and future structure and functioning of the alpine ecosystem. Here, we highlight and attempt to integrate some of these into broader research questions.

Elias (chapter 15) has emphasized that the climatic conditions that have dominated over the last two million years, which have influenced the current composition of the alpine flora and fauna, were quite different than those experienced today. This lack of climatic stability is also observed when patterns are examined at decade-to-century time scales (chapter 2). We therefore might expect the alpine to be in some state of transition due to the absence of any sort of climatic equilibrium. To this natural variability, we must add the anthropogenic atmospheric inputs that affect acidity, acid neutralizing capacities, and nitrogen loadings leading to biogeochemical changes at both the local and watershed scales (chapters 3 and 5). Certainly, change is to be expected, but the mechanisms responsible for these changes, the responses of the biota, and the consequences that these responses have on the biogeochemical properties of the alpine remain exciting and relevant research questions. While we have attempted to address some of these with our snowfence and nutrient enrichment experiments, a large list of manipulations to address additional effects imposed by biotic and abiotic variables remains to be initiated.

We have argued that the alpine tundra is a potential bellwether of environmental change. This is due, in part, to the relatively low level of biotic processing of energy and materials relative to the magnitude of these inputs and to the fact that on an aerial basis only a portion of the alpine landscape contains a plant and soil canopy. Thus, changes in outputs, in terms of the water and energy budgets or material export, may be detected more readily than in other ecosystems. However, the alpine also offers us some rather unique opportunities to study the role that the biota may play in amplifying or attenuating change. This has been discussed here in terms of plant species shifts or changes in microbial composition and subsequent changes in N biogeochemistry as well as in chapter 8 on how tree colonization can alter tundra soils.

Niwot Ridge also possesses some traits that will allow us to understand how soil development can impact atmospheric changes. The age of a soil (the time that the surface horizons have had to weather) is an important control on ecosystem structure and function (e.g., Vitousek et al. 1997). Since Niwot Ridge includes soils of at least three different ages, as a result of different extents and ages of glaciation (chapter 4), paired-plot studies of older and younger soils should demonstrate the relative importance of soil age and nutrient availability as mediators of changes due to natural or anthropogenic sources. This work is just beginning, and at this time only speculations exist. For example, a phosphorus limitation is likely to express itself more on the older soils than in the intermediate-age soils. This limitation may mediate plant composition and the positive feedbacks discussed here regarding nitrogen additions and system responses.

In similar fashion, understanding how nutrient dynamics and storage in the soil relates to stream water chemistry and its affects on aquatic biota in the alpine and subalpine catchments remains a major research challenge. The nitrogen budget, while tentatively balanced at both the local (chapter 12) and watershed scales (chapter 5; this chapter), still contains major unknowns. The fate of soil organic N and the relationship with this material and stream organic materials is a topic of future research that will integrate soil microbiology and terrestrial and aquatic biogeochemistry. Controls and consequences of alpine and subalpine water quality are of substantial societal concern. Further, the relationship between the sequestration of soil N and soil C has significance in understanding current regional patterns in atmospheric C fluxes.

Finally, we return to a theme addressed at the beginning of this chapter. The alpine functions as an island (of tundra) in a sky (of trees). Understanding the extent to which the alpine and subalpine ecosystems are interactive components of a larger functional unit is a current challenge for the Long Term Ecological Research (LTER) program. We know that the majority of the vertebrate species that use the alpine as seasonal habitat or breeding habitat require the existence of at least the subalpine forest to persist (chapter 7). We know that the water and nutrient subsidies from the alpine that are redeposited in the subalpine are at least partially responsible for the structure and functioning of that ecosystem. Finally, we know that the biotic border—treeline—that composes the boundary of these two ecosystem, while constrained by local climate, also controls local climate downwind of these borders for potentially hundreds of years. Understanding the wealth of biotic-abiotic interactions that create the alpine ecosystem and those interactions that define its borders provides a wealth of science questions and opportunities for all interested in high-elevation research.

References

Addington, R. N., and T. R. Seastedt. 1999. Activity of soil microarthropods beneath snowpack in alpine tundra and subalpine forest. *Pedobiologia* 43:47–53.

Baron J. S., D. S. Ojima, E. A. Holland, and W. J. Parton. 1994. Analysis of nitrogen saturation potential in Rocky Mountain tundra and forest: Implications for aquatic systems. *Biogeochemistry* 27:61–82.

Bell K.L. and L.C. Bliss. 1979. Autecology of *Kobresia belardii:* Why winter snow accumulation limits local distribution. *Ecological Monographs* 49: 377–402.

Beniston, M., and D. G. Fox. 1996. Impacts of climate change on mountain regions. In *Climate change 1995, impacts, adaptations and mitigations of climate change: Scientific-technical analysis. Intergovernmental Panel of Climate Change (IPCC)*, edited by R. T. Watson, M. C. Zinyowera, and R. H. Moss. Cambridge University Press, Cambridge.

Bock, J. H. 1976. The effects of increased snowpack on the phenology and seed germinability of selected alpine species. In H. W. Steinhoff and J. D. Ives. *Ecological impacts of snowpack augmentation in the San Juan Mountains of Colorado.* Final report to the Division of Atmospheric Water Resources Management, Bureau of Reclamation, Denver, Colorado.

Bowman, W. D. 1992. Inputs and storage of nitrogen in winter snowpack in an alpine ecosystem. *Arctic and Alpine Research* 24:211–215.

Bowman, W. D., J. C. Schardt, and S. K. Schmidt. 1996. Symbiotic $N_2$-fixation in alpine tundra: Ecosystem input and variation in fixation rates among communities. *Oecologia* 108:345–350.

Bowman, W. D., and H. Steltzer. 1998. Positive feedbacks to anthropogenic nitrogen deposition in Rocky Mountain alpine tundra. *Ambio* 27:514–517.

Bowman, W. D., T. A. Theodose, and M. C. Fisk. 1995. Physiological and production responses of plant growth forms to increases in limiting resources in alpine tundra: Implications for differential community response to environmental change. *Oecologia* 101:217–227.

Bowman, W. D., T. A. Theodose, J. C. Schardt, and R. T. Conant. 1993. Constraints of nutrient availability on primary production in two alpine tundra communities. *Ecology* 74:2085–2097.

Brooks, P. D., S. K. Schmidt, and M. W. Williams. 1997. Winter production of $CO_2$ and $N_2O$ from alpine tundra; environmental controls and relationship to inter-system C and N fluxes. *Oecologia* 110:403–413.

Brooks P. D., M. W. Williams, and S. K. Schmidt. 1996. Microbial activity under alpine snow packs, Niwot Ridge, CO. *Biogeochemistry* 32:93–113.

Brown, K. A. 1997. Organismic and ecosystem responses of dry alpine tundra to simulated and natural climate variation. Master's thesis, Department of Rangeland Ecosystem Science, Colorado State University, Fort Collins.

Bryant, D. M., E. A. Holland, T. R. Seastedt, and M. D. Walker. 1998. Analysis of litter decomposition in alpine tundra. *Canadian Journal of Botany* 76:1295–1304.

Burns, S. F. 1980. Alpine soil distribution and development, Indian Peaks, Colorado Front Range. Ph.D. diss., University of Colorado.

Caldwell, M. M., A. H. Teramura, M. Tevini, J. F. Bornman, L. O. Bjorn, and G. Kulandaivelu. 1995. Effects of increased solar ultraviolet-radiation on terrestrial plants. *Ambio* 24: 166–173.

Colores, G. M., S. K. Schmidt, and M. C. Fisk. 1996. Estimating the biomass of microbial functional groups using rates of growth-related soil respiration. *Soil Biology and Biochemistry* 28:1569–1577.

Conley, A. H., E. A. Holland, T. R. Seastedt, and W. J. Parton. 2000. Simulation of carbon and nitrogen cycling in an alpine tundra. *Arctic, Antarctic, and Alpine Research* 32:147–154.

Coxwell, C. C., and C. E. Bock. 1995. Spatial variation in diurnal surface temperatures and abundance of an alpine grasshopper. *Oecologia* 104:433–439.

Diggle, P. K. 1997. Extreme preformation in alpine *Polygonum viviparum:* An architectural and developmental analysis. *American Journal of Botany* 84:154–169.

Fisk, M. C., and S. K. Schmidt. 1995. Nitrogen mineralization and microbial biomass N dynamics in three alpine tundra communities. *Soil Science Society of America Journal* 59:1036–1043.

———. 1996. Microbial responses to excess nitrogen in alpine tundra soils. *Soil Biology and Biochemistry* 28:751–755.

Fisk, M. C., S. K. Schmidt, and T. R. Seastedt. 1998. Topographic patterns of above- and belowground production and N cycling in alpine tundra. *Ecology* 79:2253–2266.

Galen, C., and M. L. Stanton. 1993. Short-term responses of alpine buttercups to experimental manipulations of growing season length. *Ecology* 74:1052–1058.

———. 1995. Responses of snowbed plant species to changes in growing-season length. *Ecology* 76:1546–1557.

Galloway, J. N., W. H. Schlesinger, H. Levy II, A. Michaels, and J. L. Schnoor. 1995. Nitrogen fixation: Anthropogenic enhancement-environmental response. *Global Biogeochemical Cycles* 9:235–252.

Grabherr, G., M. Gottfried, A. Gruber, and H. Pauli. 1996. Patterns and current changes in alpine plant diversity. In *Arctic and alpine biodiversity,* edited by F. S. Chapin and Ch. Körner. New York: Springer-Verlag.

Greenland, D., N. Caine, and O. Pollak. 1984. The summer water budget and its importance in the alpine tundra of Colorado. *Physical Geography* 5:221–239.

Henry, G. H. R., and U. Molau. 1997. Tundra plants and climate change: The International Tundra Experiment (ITEX). *Global Change Biology* 3 (suppl. 1):1–9.

Isard, S. A. 1986. Factors influencing soil moisture and plant community distribution on Niwot Ridge, Front Range, Colorado, USA. *Arctic and Alpine Research* 18:83–96.

Jones, M. H., J. T. Fahnestock, D. A. Walker, M. D. Walker, and J. M. Welker. 1998. $CO_2$ flux in moist and dry arctic tundra: Responses to increases in summer temperature and winter snow accumulation. *Arctic and Alpine Research* 30:373–380.

Johnson, D. W. 1992. Nitrogen retention in forest soils. *Journal of Environmental Quality* 21:1–12.

Korb, J. 1997. Vegetation dynamics in four ecosystems along an elevational gradient in the Front Range, Colorado. Master's thesis, University of Colorado, Boulder.

Körner Ch., and W. Larcher. 1988. Plant life in cold climates. In *Plants and temperature,* S. P. Long and F. I. Woodward. Cambridge, England: The Company of Biologists.

Körner Ch., and S. Pelaez Menendez-Riedl. 1989. The significance of developmental aspects in plant growth analysis. In *Variation in growth rate and productivity of higher plant.* H. Lambers, M. L. Cambridge, H. Konings, and T. L. Pons. SPB Academic Publishing, The Hague.

Lesica, P., and B. M. Steele. 1996. A method for monitoring long-term population trends: An example using rare arctic-alpine plants. *Ecological Applications* 6:879–887.

Manabe, S., and A. J. Broccoli. 1990. Mountains and arid climate of middle latitudes. *Science* 247:192–195.

Marion, G., G. Henry, D. Freckman, J. Johnstone, G. Jones, M. Jones, E. Levesque, U. Molau, P. Molgaard, A. Parsons, J. Svoboda, and R. Virginia. 1997. Open-top designs for manipulating field temperature in high-latitude ecosystems. *Global Change Biology* 3 (suppl. 1):20–32.

May, D. E., and P. J. Webber. 1982. Spatial and temporal variation of vegetation and its productivity on Niwot Ridge, Colorado. In *Ecological studies in the Colorado Alpine: A festschrift for John Marr,* edited by J. Halfpenny. Occasional paper 37, University of Colorado, Institute of Arctic and Alpine Research.

Neff, J. C., W. D. Bowman, E. A. Holland, and S. K. Schmidt. 1994. Fluxes of nitrous oxide and methane from nitrogen amended soils in the Colorado alpine. *Biogeochemistry* 27:23–33.

O'Lear, H. A., and T. R. Seastedt. 1994. Landscape patterns of litter decomposition in alpine tundra. *Oecologia* 99:95–101.

Oechel, W. C., S. J. Hastings, G. L. Vourlitis, M. A. Jenkins, G. Riechers, and N. Grulke. 1993. Recent changes of arctic tundra ecosystems from a carbon sink to a source. *Nature* 361:520–523.

Parton, W. J., D. S. Schimel, C. V. Cole, and D. S. Ojima. 1987. Analysis of factors controlling the soil organic matter levels in Great Plains grasslands. *Soil Science Society of America Journal* 51:1173–1179.

Riebsame, W. (ed.) 1997. *Atlas of the new West: Portrait of a changing region.* New York: W. W. Norton.

Sievering, H., L. Marquez, and D. Rusch. 1996. Nitric acid, particulate nitrate and ammonium in the continental free troposphere: Nitrogen deposition to an alpine tundra ecosystem. *Atmospheric Environment* 30:2527–2537.

Sommerfeld, R. A., W. J. Massman, and R. C. Musselman. 1996. Diffusional flux of $CO_2$ through snow: Spatial and temporal variability among alpine-subalpine sites. *Global Biogeochemical Cycles* 10:473–482.

Sommerfeld, R. A., A. R. Mosier, and R. C. Musselman. 1993. $CO_2$, $CH_4$, and $N_2O$ flux through a Wyoming snowpack. *Nature* 361:140–143.

Steltzer, H., and W. D. Bowman. 1998. Differential influence of plant species on soil N transformations within the alpine tundra. *Ecosystems* 1:464–474.

Theodose, T. A., and W. D. Bowman. 1997. Nutrient avialability, plant abundance, and species diversity in two alpine tundra communities. *Ecology* 78:1861–1872.

Thomas, B. D., and W. D. Bowman. 1998. Influence of a N2-fixing Trifolium on plant species composition and biomass production in alpine tundra. *Oecologia* 115:26–31.

Vitousek, P. M., J. D. Aber, R. W. Howarth, G. E. Likens, P. A. Matson, D. W. Schindler, W. H. Schlesinger, and D. G. Tilman. 1997. Human alteration of the global nitrogen cycle: Sources and consequences. *Ecological Applications* 7:737–750.

Walker, D. A., J. C. Halfpenny, M. D. Walker, and C. A. Wessman. 1993. Long-term studies of snow-vegetation interactions. *Bioscience* 43:287–301.

Walker, M. D., P. J. Webber, E. A. Arnold, and D. Ebert-May. 1994. Effects of interannual climate variation on aboveground phytomass in alpine vegetation. *Ecology* 75:393–408.

Walker, M. D., D. A. Walker, J. M. Welker, A. M. Aarft, T. Bardsley, P. D. Brooks, J. T. Fahenstock, M. H. Jones, M. Losleben, A. N. Parsons, T. R. Seastedt, and P. L. Turner. 1999. Long-term manipulation of winter snow regime and temperature in arctic and alpine tundra. *Hydrological Processes* 13:2315–2330.

Welker, J. M., K. Brown, and J. T. Fahnestock. 1999. $CO_2$ flux in arctic and alpine dry tundra: Comparative field responses under ambient and experimentally warmed conditions. *Arctic, Antarctic and Alpine Research* 31:272–277.

Welker, J. M., U. Molau, A. N. Parsons, C. H. Robinson, and P. A. Wookey. 1997. Responses of *Dryas octopetala* to ITEX environmental manipulations: A synthesis with circumpolar comparisons. *Global Change Biology* 3:61–73.

Williams, M. W., J. S. Baron, N. Caine, R. Sommerfeld, and R. Sanford. 1996. Nitrogen saturation in the Rocky Mountains. *Environmental Science and Technology* 30:640–646.

Woodword, F. I. 1993. The lowland-to-upland transition-modeling plant responses to environmental change. *Ecological Applications* 3:404–408.

# Index

*Note:* Page numbers in *italic type* indicate figures or tables. Fauna are indexed by common name. Flora are indexed by scientific name.

δ-acetylornithine, in *B. bistortoides* rhizomes, 209
*Acomastylis rossii*, 191, *206,* 255
   dominance in moist meadows, 110, 178, 205–6
   in elk diet, 275
   foliage decay, 224
   gopher mounding and, 169
   grasshoppers' avoidance of, 270
   growth pattern, 205–6
   growth under snow, 117
   linkage to other regions, 111
   litter decay rates, 227, *228,* 229
   nitrogen mineralization under, 307
   phenolics' preservative function, 274
   in pika diets, 272, 273–4, *273*
   soil acidity under, *168*
   soil carbon content under, 165, *168*
   soil phosphorus content under, 165, 168, *168*
   soil temperature effect on, 114
   tannin production, 267–8
Advection, 25–6
Aerosol transport, wind patterns and, 19–20
Air chemistry, free tropospheric processes and, 34
Air flow. *See* Wind

Air temperature, 17, 18, *18, 19,* 118
   early Holocene, 292, *293*
   growing season increases, 313–15, *315*
   increase after last glaciation, 300–1
   long-term trends, 21
   paleoclimate and modern compared, 290, *291*
   paleoclimate reconstruction, Milankovich model, 292, *293*
   paleoclimate reconstruction, mutual climatic range technique, 289–90, *293*
   photosynthesis and, 183–85
   plant community site variations, 118–20, *119*
   primary production and, 183–85
Albedo, 22–4
Albion catchment area
   annual peak flows, *79*
   diurnal flow pattern, *85*
Alkaloids, 269
Alpine, defined, 4
Alpine avens. *See Acomastylis rossii*
Alpine plants. *See* Plants
Alpine tundra
   arctic compared, 5–6, 29
   arctic decomposition patterns compared, 232–3

**324** Index

Alpine tundra (continued)
 arctic methane fluxes compared, 256–7, *257*
 characteristics, 4–6
 climate variation, 4, *6*
 as environmental change indicator, 4, 304, 318
 extent, Pinedale and modern compared, 298–300, *299*
 geomorphic provinces, 158
 habitat island size change, 298–301
 research needs, 317–19
 response to fertilization, 188, *189*
 snow cover variation, 4, *5*
 vascular plants in, 4–5
Amino acids
 in *B. bistortoides* rhizomes, 209, *211*
 in dry meadows, 216
 glycine uptake, carbon and, 215–16
 glycine uptake, rates, 215–16, *216*
 resorption of, 223
 in soils, 216–17
Ammonia (atmospheric), 34–5
 atmosphere-biosphere nitrogen exchange and, 35, 39
 bidirectional fluxes, 40
 in Front Range, 41
 growing season concentrations, 39, 40
 snow season concentrations, 39
 transport to plants, 35
Ammonium (atmospheric), 34–5
Ammonium
 in snowmelt, 83, *84*
 in soil under snowpack, 94
 uptake inability of *K. myosuroides*, 215
Ammonium bisulfate (atmospheric), 35
Ammonium sulfate (atmospheric), 35
Amphibians, 130
Animals. *See* Fauna
Arapaho Indians, Chief Niwot, 7, *9*
Arctic tundra
 alpine compared, 5–6, 29
 alpine decomposition patterns compared, 232–3
 alpine methane fluxes compared, 256–7, *257*
 plants, alkaloids in, 269
Arginine, in *B. bistortoides* rhizomes, 209
Arikaree Glacier, 57, 58, *58,* 66
*Artemesia scopulorum,* terpene in, 268

*Artemisia tridentata,* vole diet, 271
Arthropods, 231, 232
Atmospheric deposition
 acidity, 77, 91–2
 anthropogenic increases in, 32
 hydrochemical responses, 90–4, *90*
 solutes and, 77
 *See also* Nitrogen deposition
Avalanches, 61
Avifauna. *See* Birds

Bacteria, 231, 232
Badger, food preferences, 143
Bear, black, 142
Bear, grizzly, 130
Beetles, fossil assemblages, 290
Bioclimatic zones, 20–1
Biogeochemistry, 81–3
Biomass (microbial), 246, *246,* 250, 307
Biomass (plant), 177–8
 allocation patterns, nutrient acquisition and, 199–201, *200*
 alteration ability of plants, 192, *192, 193*
 belowground, 182, *184,* 201, *203*
 belowground:aboveground ratios, 181–3, *184,* 199
 humus, 201, *203*
 nitrogen allocation patterns and, 191–3, *192,* 199–201, *200,* 238–9, *238,* 241
 production, fertilization and, 188, 307
 production phenology, 178–9
 standing crops, 177
 *See also* Primary production
Biomass (vertebrate), *140, 141,* 145
Birds, 130
 breeding population, 134, 136
 on Niwot Ridge, listed, *131, 135*
 *See also common names of specific birds*
Bison, 130, 135–6, 145
*Bistorta,* Pinedale pollen, 289
*Bistorta bistortoides,* 191
 amino acids in, 209, *211*
 in elk diet, 275
 growth response to fertilization, 204
 growth under snow, 117
 leaf water potential vs stomatal conductance in, 187, *187*
 nitrogen allocation, seasonal, 207–9, *208, 209*

nutrient storage dynamics, 209, *211*
  in pika diets, 272, 273
  rhizome nitrogen dynamics, 209, *210*
  stomatal conductance, 186
  tannin production, 268
*Bistorta vivipara,* 289
*Bistorta viviparum,* fungus association with, 213
Blockfield zone, 66
Bluebird, mountain, 130
Boreo-Cordilleran species, 136
Bryophytes, 103
*Bryum algovicum,* 115
Bull Lake glaciation, 55–7, *56*

Calcium, in surface waters, 81–3, *82*
*Caltha leptosepala,* 113, 115
  growth lag, 192–3
  growth under snow, 117
  wet meadow dominance, *113,* 178, 181
Carbon
  costs of nitrogen storage, 209
  glycine and, 215–16
  in soils, 158, 164, 169, 170, 233, 277, 312
Carbon dioxide, 32, 254
  ambient air concentrations, 33–4, *33*
  Mt. Pinatubo eruption and, 33–4
  plant growth and, 34
Carbon dioxide fluxes (soil)
  in growing season, 258–9, 314, *316*
  plant community types and, 258–9, *259*
  under snowpacks, *160,* 259–60
  temporal patterns, *259*
  in winter, 259–60
*Cardamine cordifolia,* 115
*Carex* spp., *113*
  in elk diet, 275
  wet meadow dominance, *113*
*Carex aquatilis,* 113
*Carex brevipes,* 111
*Carex firma,* fungal infection, 213
*Carex leptosepala,* 113, *113*
*Carex rupestris,* 103, 109
*Carex scopulorum,* 113, 255
Carnivores, winter activity, 143
Caryophyllaceae (cushion plants), Pinedale pollen, 289
Catena concept, 161–2
Cattle, 136

*Cenococcum geophilum,* 213
Century ecosystem model, 233, 317
Chipmunk, least (*Tamias minimus*), 137, 142, 270
  food preferences, 143
Chipmunk (*Tamias*), 137–40
Chipmunk (*Tamias umbrinus*), 137
Clays, 160, 164
*Claytonia lanceolata,* in pocket gopher diet, 271
*Claytonia megarhiza,* 115
Cliffs, 50
Climate, 337
  characteristics, 16–17, 100–1
  classification, of Niwot Ridge, 18–19
  ecosystem comparisons, 6
  global-scale influences, 17–18, 304
  Holocene, 57–8
  macroclimate, 17–19, 29
  mesoscale climate, 19–20, 29
  microclimate, 22–8, 29
  paleoclimate, vegetation equilibrium, 297–8
  paleoclimate reconstructions, 288–92, *291*
  paleoclimate reconstructions, age ranges and sites, *290*
  soil formation and, 160–1
  *See also specific aspects of climate or weather*
Climate change, snow increase, 310–12
Coarse debris system, 59–61, *59,* 65
Collembolans, 232
Colorado. *See names of specific places or features in Colorado*
Conifers
  in early Holocene, 293
  fir, early Holocene presence, 293
  pine, mid-Holocene arrival, 295
  spruce, early Holocene presence, 293
Cordilleran species, 136–7
Couloirs, 50, *53*
Cushion plants
  in fellfields, *107,* 117
  in late Pinedale, 289
Cysteine, 217
*Cystopteris fragilis,* 115

*Danthonia intermedia,* 111
Debris flows, 61, *61, 62*

Decomposition
  biotic processes, 231–2
  controlling variables, 223, 233
  defined, 233
  fungi and, 231
  gopher mounds and, 277
  microbial activity and, 226
  moisture, decay rate and, 227–8, *227*
  nitrogen and, 229
  plant productivity and, 224–6
  process, 223–6
  snowpack correlation with, 222, 226–8, *228*, 232
  soil invertebrates and, 226, 231–2
  soil moisture and, 226, *227*
  soil temperature and, 226, *227*, 232–3
  substrate quality and, 228–30, 232
  within soil, 224, 228
Deer, mule, 145
*Deschampsia caespitosa*
  biomass allocation alteration, site fertility and, 192, *192*
  competition with *K. myosuroides*, 121
  dominance in moist meadows, 110, 178
  in elk diet, 275
  growth pattern, 205–6
  growth response to nitrogen, 241
  linkage to other regions, 111
  nitrogen mineralization under, 307
  root:shoot ratio, 192, *192*
Devlin Park, Pinedale glaciation, 287
*Draba fladnizensis*, 115
*Draba lonchocarpa*, 115
*Dryas octopetala*, 108, *108*
  air temperature effects on, 313–15, *314, 315*
  carbon dioxide flux, 314
  climate change effects on, *315*
  fungus association with, 213
  leaf mass, 314, *314*, 316
  precipitation effects on, 316–17
  seed mass, 314, *314*, 316
  soil temperature effects on, 313–15, *314, 315*
Dry meadows, 24, 109–10, *109*, 178
  amino acids in, 216
  biomass production phenology, 178
  carbon dioxide flux patterns, 258–9, *259*

*Kobresia myosuroides* dominance, 109, *109*, 110, 178
  methane flux correlates, 257–8
  methane flux patterns, 255–6, *256*
  microbial activity, 259
  nitrogen immobilization, 244, *245*
  nitrogen in, 238
  nitrogen mineralization, 244, *245*
  nitrous oxide fluxes, 260–1, *261*, 262, *262*
  plant associations, 109–10
  plant nitrogen-use efficiencies, 202
  snow cover, 178
  soil acidity, 255
  winter drought stress in, 120

Ecosystems
  alpine, subalpine and, 319
  catena concept of, 161–2
  Century model, 233
  dynamics, atmospheric deposition and, 75–6
Elk (*Cervus elaphus*), 145, 149, 275
  diet, 278
  foraging effect on plant succession, 277
  in wet meadows, 114
El Niño-Southern Oscillation (ENSO), Niwot Ridge and, 21
Energy budgets, 22
  long-term changes in, 28
  in snowmelt season, 24–5, *24*
  in summer, 22–4, *23*
Entisols, 158
Environmental change
  alpine biota response to, 317
  alpine tundra as indicator, 4, 304, 318
  soil microbial community as indicator, 317
*Epilobium angallidifolium*, 115
*Eriophorum* spp., 113
*Eriophorum vaginatum*, 215
*Eritrichium aretioides*, 103
Erosion, 63, 68, 161
*Erythronium* spp., corms, in pocket gopher diet, 271
Evapotranspiration, 26, *26*, 27
Experimental Ecological Reserves, Niwot Ridge as, 7

Fauna (vertebrate), 128–9, 151, 270
  areographic analysis, 136
  biomass, *140, 141,* 145
  body masses, 144–5, *146–7*
  Boreo-Cordilleran species, 136
  carnivores, winter activity, 143
  in Colorado, early studies, 129
  composition of, 130–6, *131*
  Cordilleran species, 136–7
  development, 136–7
  ecological distribution, 137–42, *138–9*
  fiber digestion efficiency, body size and, 276
  food specialization, 143–4
  as geomorphic influences, 145–9
  habitats, 137, *138–9*
  hibernal activity, *138–9,* 142
  hibernation, diet and, 274
  home ranges, *146–7*
  on Niwot Ridge, listed, *131*
  occurrence criteria, 130
  population density, *146–7*
  similarity in alpine and tundra, 137
  trophic guilds, *138–9,* 143–5
  wind factors in, 117
  winter activity, 142–3
  *See also common names of specific species; specific types of fauna e.g., Birds, Herbivores*
Fellfields, 24, 103, *107,* 108, *108*
  biomass production phenology, 178
  cushion plant growth form in, *107,* 117
  *Dryas octopetala* dominance, 108, *108*
  energy budget, 22, *23*
  plant associations, 108
  pocket gophers and, 145, 148, *148*
Fertilization
  aboveground production and, 188, *189*
  from animal excreta, 277
  competitive exclusion and, 188
  growth response to, 188, 205, 307
  leaf gas exchange and, 191
  methane fluxes and, 307, *308*
  nitrogen cycling and, 241
  nitrous oxide fluxes and, 307, *308*
  photosynthesis and, 191
Fiber, digestion of, 276
Fine sediment system, *60,* 62–4, 65–6

Flora. *See* Plant communities; Plants; Vegetation
Forbs, 109, 121
  fertilization effects, 188, *190*
  graminioid competition, 188, *190*
  nitrogen-use efficiency, 202, 205
  nutrient-use efficiency, 191
Forest, subalpine, mid-Holocene vs modern, 296
Freeze-thaw events
  microbes and, 226
  soil formation and, 166–7
Frog, striped chorus (*Pseudacris triseriata*), 130
Front Range
  ammonia in atmosphere, 41
  geologic history, 54–5, 166
  geomorphic activity, 68
  Indian Peaks, 46, *46,* 55, 58
  inversion layer of atmosphere in, 34
  late Pleistocene history, 285–8
  nitrogen deposition increase, 305
  paleoclimate and vegetation equilibrium, 297–8
  paleoecology, 297
  paleoindian occupation, 149, 294–5
  Pleistocene history, 285–8
Front Range Ecology Project, 7
Fungi
  arbuscular mycorrhizal, 212–13
  dark septate, 212, 213
  decomposition and, 231
  ectomycorrhizal, 212, 213
  infection levels in roots, 213–14, *214*
  mycorrhizal relationship with plants, 198, 213, 217
  phosphorus uptake and, 213
  plant nutrient status and, 212–15

Geochemical denudation, 64–5, *65*
  surface-water solutes from, 81–3
Geochemical system, 64–5, *65,* 66
Geologic history
  Holocene, 57–9
  Pleistocene, 55–7, *56,* 285–88
  pre-Quaternary, 54–5
Geomorphic impacts
  by humans, 149, 166
  by nonhuman vertebrates, 145–9

**328** Index

Geomorphology, 45–6, *60*
Glacial zone, 66
Glaciation. *See* Bull Lake glaciation; Pinedale glaciation; Pleistocene glaciation
Glaciers, 61, 285, 286
Glutamate, 217
Glycine, 216
 carbon and, 215–16
 uptake rates by plants, 215–16, *216*
Goat, mountain (*Oreamnos americanus*), 135
Gopher, northern pocket (*Thomomys talpoides*), 140, 141, *141,* 163, 274–5
 abundance, aboveground biomass and, 271
 as badger prey, 143
 diet choices, 121, 270–1
 effect on plant species composition, 121
 geomorphic impacts, 145, 151
 mounding activity, 167, *167,* 169, 271, 277
 nitrogen mineralization and, 121
 population density, 148, *148*
 snow-cover needed by, 271
 soil disturbance by, 145, 148, *148,* 166, 167–8, 240, 271
Graminoids, 121, 164
 in diet, body size and, 276
 in elk diet, 275
 fertilization effects, 188, *190*
 forb competition, 188, *190*
 nitrogen-use efficiency, 203
 nutrient-use efficiency, 191
 in pika diets, 272, 273
Grasshoppers (Acrididae), 270, 310
 food preferences, 270
Green Lake 4, 5, *53*
 annual peak flows, *79*
 diurnal flow pattern, *85*
 meltwater impoundment in, 86–7
Green Lakes Valley, 46, *46, 50*
 basin areas, *48, 49*
 basin morphometry, *47,* 49
 geochemical denudation in, 64–5, *65*
 geologic history, 54–9
 geomorphologic studies, 45–6
 Holocene landform development, 57–9
 hydroclimate, *88*
 landscape characteristics, 46–54
 location, *11, 46*
 neoglaciation, 57, 58
 relief, 49–50, *49*
 *See also specific features or processes, e.g.,* Ridges; Streamflow
Ground heat flux, 28
Growing season length
 herbivores' coping strategies, 275–6
 nitrogen cycling and, 237, 249
 phenology and, 116–17
 primary production and, 116–17, 188–91
 snow quantity and, 116, 311
 soil moisture effect on, 186

Hare, 142
Herbivores, 121, 269, 278
 in alpine vs arctic tundra, 128
 central place foragers, vegetation gradients and, 276
 diet preferences, 278
 effects on plant communities, 277–8
 growing season length, coping strategies, 275–6
Holocene
 altithermal interval, 298
 early, air temperatures, 292, *293*
 early, environment characteristics, 292–4
 landform development, 57–9
 late, environment characteristics, 296–7
 mid-Holocene, environment characteristics, 295–6
 soil formation, 288
Humans
 ecosystem impacts, 149, 166, 306
 impact on Niwot Ridge, 149
 occupation of Front Range, 149
 soil modifications by, 166, 169
Humus
 belowground biomass patterns, 201, *203*
 formation, 224
Hydrologic budget. *See* Water budget

Inceptisols, 158
Indian Peaks, 46, *46,* 55, 58
 Pinedale glaciation, 287, 288
Insects, 269–70
 arctic, decidous shrub preference, 269–70

dietary specialization, 276
grasshoppers, 270
*See also* common names of specific insects
International Biological Programme, Niwot Ridge participation, 7, 15
Invertebrates (soil)
densities, 231–2
winter activity, 226
Isabelle, Lake, 57

Jackrabbit, white-tailed, 136

Kansas, Konza tallgrass prairie, nitrous oxide losses, 263
Kiowa Peak, *50*
talus slope, *52*
*Kobresia myosuroides,* 102, 255
ammonium uptake inability, 215
biomass allocation alteration, site fertility and, 192, *192*
characteristics, 110
competition with *Deschampsia caespitosa,* 121
dominance in dry meadows, 109, *109,* 110, 178
fertilization effects, 188, *189*
foliage decay, 224
fungus association with, 213
glycine uptake, 215–16, *216*
habitat, 118, 121
nitrogen use, 241
root:shoot ratio, 192, *192*
soil acidity under, *168*
soil carbon content under, 165, *168*
soil moisture effects, 120
soil phosphorus content under, 168, *168*
wind sensitivity, 117–18
*Koenigia islandica,* 114
Pinedale pollen, 289
Krummholz
defined, 129
gopher mounding and, 169

Lakes
morphometry, 54, *54*
sediments, 63–4
*See also* Green Lakes Valley
Landforms, 46–54, 68
Holocene development of, 57–9, *60, 68*

morphoclimatic, 66–8, *67*
in mountain areas, *63*
periglacial, 66–8, *67*
Lark
Horned, 130, 133, 134, 136
winter migration, 142
Leaf water potential, stomatal conductance and, *183,* 186
Lemmings, 277
Lichens, 103, 109, 115
Light, primary production and, 183
Litter, decomposition rates, 224, *225*
Loess, 160, 169, 233
Long Term Ecological Research (LTER) program
on Niwot Ridge, 7–10, 15, 319
site climate comparisons, *6*
Long-term Intersite Decomposition Experiment Team (LIDET) study, 223, 224, 229
*Lupinus* spp., roots, in pocket gopher diet, 271
Lynx (*Lynx lynx*), 135

Macroclimate, 17–19, 20
Mammals, 130, 270
small, biomass, *140*
small, density in vegetational community-types, *140*
small, survivorship, 141
winter activity, 142–3
*See also* common names of specific mammals
Marmot, yellow-bellied (*Marmota favientris*), 133, *134,* 140, 274
biomass, *141*
density, *141*
diet, 274
growing season length, coping strategies for, 275
hibernation, 142, 274, 275
population growth, 274–5
snow-cover duration effect on, 275
survivorship, 141
Martinelli catchment area
annual peak flows, *53*
daily streamflow pattern, *85*
seasonal hydrograph, *80*
Meadows *See* Dry meadows; Moist meadows; Wet meadows

*Melandrium* spp., Pinedale pollen, 289
Meltwater. *See* Snowmelt
Meristematic potential, nutrient use and, 204–5
*Mertensia ciliata,* in marmot diet, 274
Mesoscale climate, 19–20, 29
Mesotopographic gradient, *107,* 116, 178
Methane, 254
 oxidation, 256, 257, 258
Methane fluxes (soil)
 controls, 257–8
 fertilization and, 307, *308*
 plant community types and, 255–7, *256*
 precipitation and, 257–8
 soil moisture and, 258
 soil temperature and, 258
 temporal patterns, *256*
Microbes
 as environmental change indicators, 317
 freeze-thaw events and, 226
 nitrogen mineralization, 244, 248
 nitrogen pools, 239, 244
 nitrogen transformations, snow-covered season, 248–9
 nitrogen transformations, snow-free season, 244–8, *244, 246*
 soil moisture and, 245, 259
 soil trace gas and, 263, 317
Microclimate, 22–8, 29
 spatial heterogeneity, 22, 28
 *See also specific aspects of climate or weather*
Milankovich model, for paleoclimate air temperature reconstruction, 292, *293*
*Minuartia* spp., Pinedale pollen, 289
*Minuartia obtusiloba,* 103, 267
Mites, 232
 oribatid, 231
 prostigmatid, 231
Moist meadows, 24, 178
 aboveground production in, 181
 *Acomastylis rossii* dominance, 110, 178, 205–6
 carbon dioxide flux patterns, 258–9, *259*
 *Deschampsia caespitosa* dominance, 110, 178
 early-melting snowbank, 110–12, *111*
 late-melting snowbank, 112
 methane flux patterns, 255–6, *256*
 nitrogen immobilization in, 244, *245*
 nitrogen mineralization in, 244, *245*
 nitrous oxide fluxes, 260–1, *261*
 plant associations, 110 12, 113 14
 pocket gopher activity in, 148, *148*
 soil acidity, 255
 susceptibility to biotic change, 307
Moist shrub tundra community, 24
 energy budget, 22–4, *23*
Mollisols, 164
Montana, nitrogen pools at alpine sites in, 239
Moose, 277
Moraines, 55, *56,* 57, 297
 Bull Lake, *286*
 Pinedale, 286, *286*
Morphoclimatic zones, 68
Mosses, 115
Mountain Research Station, 7, *11*
Mount Pinatubo eruption, atmospheric carbon dioxide and, 33–4
Mouse, deer (*Peromyscus maniculatus*), 137
 population fluctuations, 142
 survivorship, 141
Mouse, western jumping, 142
Mutual climatic range (MCR) technique, for paleoclimate air temperature reconstruction, 289–90, *293*

National Oceanic and Atmospheric Administration (NOAA), Climate Montoring and Diagnostics Laboratory, 33
Navajo catchment area, daily streamflow pattern, *85*
Nematodes, 232
Neoglaciation, 57, 58, 297
Nitrate
 in atmosphere, 34, 35
 atmospheric deposition of, 77
 in Green Lake 4 outlet stream, annual concentrations, *90,* 91
 nitric acid concentration ratio in free troposphere, 35
 in snowmelt, 83, *84*
 surface-water concentrations, basin area and, 91, *93*
Nitric acid
 in atmosphere, 34, 35

Index  331

nitrate concentration ratio in free troposphere, 35
  seasonal concentrations, 39–40
Nitrite, in atmosphere, 34
Nitrogen, 32, 34
  allocation in alpine plants, 199–200, *200*
  atmosphere-biosphere exchange, 39, 41–2
  availability, species self-replacement and, 307, *309,* 310
  in biomass, quantities, 238–9, *238*
  decomposition and, 229
  in dry meadows, 238
  exports, atmospheric deposition and, 307, 310, *310*
  in free troposphere, 34
  fungi-plant mycorrhizal relationship and, 213–14, 217
  immobilization, 244–5, *245*
  leaf concentrations, 202, 205
  microbial, 239
  microbial transformations, snow-covered season, 248–9
  microbial transformations, snow-free season, 244–8, *244, 246*
  mineralization, 121, 244–5, *245,* 248, 307
  in organic pools, 238–40, *238*
  primary production and, 187
  resorption of, 223
  rhizome N dynamics, in *Bistorta bistortoides,* 209–11, *210*
  seasonal allocation, in *Bistorta bistortoides,* 207–9, *208, 209*
  in soils, 169–70, 238
  source regions of, 38–9, *39,* 41
  species distribution over Niwot Ridge, 34–5
  storage, carbon costs of, 209
  storage, soil disturbance and, 240
  in streamwater, 161
  uptake, fungal infection and, 214–15
  uptake, inorganic vs organic, 215–17
  uptake, in snow-free season, 240–3, *240, 242*
  use efficiency, in wet meadow plants, 203
  in wet meadows, 238
  *See also* Nutrients
Nitrogen cycling, 237–8, 306, *306*
  microbial activity in soils and, 94
  primary production and, 187
  in snow-covered season, 243–4

  in snow-free season, 240–3, *240, 242*
  topographic position and, 187
Nitrogen deposition, 32–3, 91, 93–4, 248
  ammonia transport to plants, 35
  anthropogenic activities and, 35–6
  dry deposition fluxes, 36–7, *37*
  ecosystem impacts, 306–10, *306*
  exports and, 307, 310, *310*
  in growing season, 36, 37, *37*
  increase, 305
  mechanisms, 36–7
  primary production and, 306–7
  rates of, 35–6
  in snow season, 37–8, *37*
  in subalpine areas, 41–2
  wet deposition fluxes, 37, *37*
  *See also* Atmospheric deposition
Nitrogen immobilization, water availability and, 187
Nitrous oxide, 254
Nitrous oxide fluxes (soil), 263
  available nitrogen limit and, 262
  fertilization and, 307, *308*
  in growing season, 262, *262*
  oxygen availability and, 261
  plant community types and, 260–1, *261, 262*
  under snowpack, 263
  temporal patterns, *261,* 262
  water availability and, 262
  in winter, 262–3, *262*
Niwot Ridge, *50,* 318
  described, 6–7, *8–9*
  history of research on, 7–10, 15–16
  location, 7, *11, 287*
  patterned ground, 50, *51*
Niwot Ridge LTER program, 7–10, 15, 162
  web site, 10
Niwot Ridge snowfence experiment, 311–12, *311*
Niwot Saddle research site. *See* Saddle research site
North Boulder Creek basin, Pleistocene glaciation, 55, *56*
Nutrients
  aquisition, biomass allocation and, 199–201, *200,* 217
  availability, soil moisture and, 120
  availability, topographic position and, 187

Nutrients (continued)
   immobilization as decay buffering mechanism, 234
   limitation, internal growth controls and, 203–6
   in soils, 158, 164
   storage, 206–12
   use, fertilization and, 205
   use, meristematic potential and, 204–5
   use, species' variation, 202–3
   *See also* Nitrogen; Phosphorus; Plants
Nutrient use efficiency, 181
   components, 201–2
   defined, 201
   of forbs, 181
   of graminoids, 181
   photosynthesis and, 202
   resorption, 202
   species' variation, 202–3

*Orthotrichum indet,* 115

Paleoclimate, vegetation equilibrium, 297–8
Paleoindians
   in Front Range, 149, 294–5
   game-drive systems, 149
*Paronychia pulvinata,* 103, *107*
*Paronychia sessiliflora,* Pinedale pollen, 289
Peat, formation, post-Pinedale glaciation, 288
*Pedicularis,* pollen, 289
*Pedicularis groenlandica,* 113
*Pedicularis parryi,* 111
*Pelitgera malacea,* 115
*Penstemon whippleanus,* 112
*Pentaphylloides floribunda,* 179
Phenolics, 268, *268*
   preservative function, 274
Phenology
   of biomass production, 178
   growing season length effect on, 116–17
   snowbank effects on, 114, 116
*Phialocephala fortinii,* 213
*Phlox,* Pinedale pollen, 289
Phosphorus
   leaf concentrations, 202, 205
   plant use of, 168
   primary production and, 187
   resorption of, 223

   in soils, 165, 168, *168*
   uptake, fungi and, 213
   use efficiency, in dry meadow plants, 203
   utilization by *Ranunculus adoneus,* 211, *212*
   *See also* Nutrients
Photosynthesis, 183, 191
   air temperature and, 183–85
   biomass allocation and, 198–9
   fertilization and, 191
   herbivores' effect on, 277
   nutrient-use efficiency and, 202
   water availability and, 185–6
*Physcia caesia,* 115
Pika (*Ochotona princeps*), 133, *133,* 141, 151, 272, *273*
   biomass, *141*
   density, *141*
   diets, summer and winter compared, 272–3
   diets, toxins in, 272–3
   distribution, 140
   food caches, 272, 275
   foraging effect on plant succession, 277
   growing season length, coping strategy for, 272, 275
   survivorship, 141
   winter activity, 143
Pinedale glaciation, *56,* 58, 285–6
   late, environment characteristics, 288–92
Pine trees, mid-Holocene arrival, 295
Pipit, American, 134, 136
   winter migration, 142
Plant communities
   air temperature variations, 118–20, *119*
   biotic controls, 120–2
   carbon dioxide fluxes and, 258–9, *259*
   composition
      snow cover and, 312
      species interactions and, 121–2
   heat budgets for, 22, *23*
   herbivores' effects on, 277–8
   methane flux patterns and, 255–7, *256*
   mutualisms in, 122
   nitrous oxide flux patterns and, 260–1, *261*
   physical controls, 99, *100,* 116
   rock crevices, 115
   scree slopes, 115
   site factor variations, *119*
   spatial scale hierarchy, 99, *101*

species self-replacement, nitrogen availability and, 307, *309*, 310
springs, 115
structure, nitrogen availability and, 307
*See also* Dry meadows; Fellfields; Moist meadows; Shrub tundra; Snowbeds; Wet meadows
Plant growth
atmospheric carbon dioxide concentrations and, 34
internal controls, nutrient limitations and, 203–6
Plant production, growing season length effect on, 116–17
Plants
belowground:aboveground biomass ratios, 199
biomass alteration ability, 192, *192*, 193
developmental control in, 191–3
dominant species, 177
dominant species, growth patterns of, 205–6
increase with greater nitrogen availability, 307, *309*
internal growth controls, 203–6
mycorrhizal relationships, 122, 198, 213, 217
nitrogen insensitive, 307
nitrogen limited, 307
nutritional composition, 267–8, *267, 268*
phenolic contents, *268*
physiological control in, 191–3
preformation, 192–3
productivity, decomposition and, 224–6
rooting depths, 201, *202*
root system functions, 201
species increases, *309*
succession, pocket gophers and, 145
*See also* Biomass; Nutrients; Vegetation; *botanical names of specific plants*
Pleistocene glaciation, 49, 55–7, *56*, 68, 159–60, 166
Bull Lake glaciation, 55–7, *56*
Pinedale glaciation, *56*, 58, 285–6
*Pohlia cruda*, 115
*Polemonium*, Pinedale pollen, 289
Pollen
change in late Holocene, 296–7
late Pinedale types, 289
mid-Holocene, 295

*Polygonum (Bistorta) vivipara*, 192
Precipitation, 17, 18, *19, 20, 26,* 76, 161
effects on methane fluxes, 257–8
growing season increases, 315–17
long-term trends, 305, *305*
rainfall, *88,* 89–90, *87*
temporal distribution, 186
*See also* Snow
Precipitation chemistry, 77
Pressure patterns, 17–18
Primary production, 193–4
aboveground, 179, *180,* 181–2
aboveground, soil moisture and, 185
air temperature and, 183–85
belowground, 179, 179–81, *180*
environmental controls, 183–93
growing season length and, 188–91
light and, 183
nitrogen cycling and, 187–8
nitrogen deposition and, 306–7
nutrient availability, 187–8
plant disturbance and, 193
rates, 177
snowmelt and, 116
soil disturbance and, 193
soil moisture and, 185–6
topographic patterns, water availability and, 186
topographic position and, 187–8
water availability and, 185–7
*Primula parryi,* 115
Ptarmigan, white-tailed (*Lagopus leucurus*), *132,* 133, 134, 136
food preferences, 143
winter habitats, 142

Rainfall, *88,* 89–90, *87*
effects on methane fluxes, 257–8
*Ranunculus,* Pinedale pollen, 289
*Ranunculus adoneus*
fungal infection in, 121, 213–14, *214*
growth under snow, 117
nitrogen uptake in, 214
phosphorus utilization, 211, *212*
seed mass, snow and, 311
*Ranunculus escholtzii,* 115
Remote sensing, 28, 29
*Rhodiola integrifolia,* 113
Ridges, 50, *50, 51*

Robin, American, 130
Rock crevices, plant communities in, 115
Rockfall, 59, 65
Rocky Mountains
  lakes, water quality, 75–6
  physiographic regions, *10*
  treeline limit, 4
  water supply for semiarid regions, 75
Rodents
  density in vegetational community-types, *140*
  winter activity, 143

Saddle research site
  location, *11*
  soil trace gas flux studies, 255
  winds at, 19
*Sagina saginoides,* Pinedale pollen, 289
Salamander, tiger (*Ambystoma tigrinum*), 130
*Salix* spp., 179
*Salix glauca,* 24, 115
  growth response to fertilization, 204
*Salix planifolia,* 114, 115
*Salix villosa,* 114, 115
San Juan Mountains, early Holocene treeline, 294
*Saxifraga,* Pinedale pollen, 289
*Saxifraga broncialiis* spp. *austromontana,* 115
*Saxifraga caespitosa,* 115
*Saxifraga odontoloma,* 115
*Saxifraga rhomboidea,* 191
Scree slopes, plant communities on, 115
Sediment budgets, *59,* 65–6
  coarse debris system and, 59–61, *59*
  fine sediment system and, *59,* 62–4
  geochemical system and, *59,* 64–5
Sediment transport, 65–6
  avalanches and, 61
*Selaginella densa,* 109
Sheep, bighorn (*Ovis canadensis*), 136, 145, 149
Sheep, domestic (*Ovis aries*), 136, 149
Shortgrass Steppe LTER site, temperature trends, 21
Shrews
  density in vegetational community-types, *140*
  food preferences, 143
  winter activity, 143

Shrub tundra, 114–15
  belowground biomass, 182
  plant associations, 114–15
*Sibbaldia procumbens*
  dominance in snowbeds, 178
  tannin production, 268
*Silene acaulis,* 103
  in pika diets, 273
  Pinedale pollen, 289
Snow
  depths, 116
  early plant growth under, 117
  effect on soils, 162–3, 169
  growing season length and, 116
  increase in, 310–12
  relocation by wind, 76
  *See also* Precipitation
Snowbanks, effect on phenology, 114, 116
Snowbeds
  aboveground production in, 181
  belowground biomass, 182
  biomass production phenology, 178
  energy budget, 22–4, *23*
  pocket gopher abundance in, 271
  soils in, 163
  *Sibbaldia procumbens* dominance, 178
Snow cover
  in arctic tundra, 232
  microclimate spatial heterogeneity and, 22
  plant community composition and, 312
  production responses to, 312
  variation in, 4, *5*
Snow dams, effect on streamflow, 86–7, *89*
Snowfence experiment, 311–12, *311*
Snowmelt, 94
  ammonium in, 83, *84*
  chemistry, daily changes in, 83
  chemistry, seasonal changes in, 83
  energy budget, 24–5, *24*
  microbial biomass changes, and, 248, 249
  nitrate in, 83, *84*
  primary production and, 116
  slush flows, 86–7
  solute removal and, 64, 83
  streamflow and, 83–6
  sulfate in, 83, *84*
Snowpack
  ammonium in soil under, 94

carbon dioxide flux under, 259–60, *260*
decomposition correlation with, 222, 226–8, *227,* 232
microbial biomass changes under, 248–9
nitrogen cycling and, 243–4
nitrous oxide flux under, 263
temperature, 76
Sodium, in surface waters, 81–3, *82*
Soil disturbance
nitrogen storage and, 240
by pocket gophers, 148–9, *148,* 167–8, 240
primary production and, 193
Soil erosion, 63
Soil formation
climate role in, 160–1
disturbance role in, 166–9
freeze-thaw processes, 166–7
in Holocene, 288
Soil moisture, 27–8, 29, 177, 255
aboveground production and, 185–6
decomposition and, 226, *227*
deficits, 26–7
growing season length, 186
methane flux and, 258
microbial activity and, 245, 259
nutrient availability and, 120
primary production and, 185–6
from semipermanent snow banks, 22
*See also* Water availability
Soils
acidity, 158, 162, 255
age, ecosystem structure and, 318
amino acids in, 216–17
carbon storage, 158, 164, 169, 170, 233, 277, 312
cation exchange capacity (CEC), 158–9, 162, *163*
classification of Niwot Ridge, 158–9, *159*
clays, 160, 164
decomposition within, 224, 228
development, 157, 169
development, biota role in, 163–6
Entisols, 158
fertility, forest and meadow compared, 164
gophers' effects on, 167–8
human modifications, 169
Inceptisols, 158

litter and, 159, 164
loess, 160, 169
mineralization, 198, 217
Mollisols, 164
nitrogen in, 164, 169–70, 238
nutrient storage, 158, 164
organic matter in, 159, 162, *163,* 164
parent materials, 159–60
peat, 288
phosphorus in, 168, *168*
snow effect on, 162–3, 169
Spodosols, 164
surface horizons, 159, 168, 170, 318
topographic effects on, 161–3
trace gas fluxes, 254–5, 263, 317
water movement through, 161
Soil temperature, 162, 163, 255
decomposition rates and, 226, *227,* 232–3
growing season increases, 313–15, *315*
methane flux and, 258
Soil water, 120, 158, 162, 233
nitrogen immobilization and, 244, *245*
nitrogen mineralization and, 244, *245*
*See also* Water availability
*Solidago multiradiata,* 178
*Solidago spathulata,* 111
Solifluction, 62
Solifluction zone, 66
Solutes, 64–5, *65*
from geochemical denudation, 64–5, 81–3
release from snowpack, 64, 83, *84*
seasonal variations, 64, 83
snowmelt and, 77
South St. Vrain Creek valley, *50*
Sparrow, Brewer's, 130
Sparrow, Vesper, 130
Sparrow, White-crowned, 130, 133, 134, 136
winter migration, 142
Springs, 115
Squirrel, golden-mantled ground (*Spermophilus lateralis*), 140, 142, 270
food preferences, 143
Stomatal conductance
leaf water potential and, *187,* 187
water availability and, 185

Streamflow, 76
  annual hydrograph, 77–81
  annual peak flows, 77, 79
  annual variations, 77–8, *78*
  diurnal hydrographs, 83–6, *85*
  drainage area vs, 77, *78*
  nitrate concentrations and, 91, *92*
  recession coefficients, *81*
  seasonal hydrograph, 80–1, *80, 81*
  seasonal patterns, 78–80, *80*
  snow dam effects on, 86–7, *89*
  snowmelt and, 83–6
  water yield, 77–8
Streams
  acidification, 91–2, *93*
  characteristics, *48,* 53, 63
Subalpine forest
  atmospheric ammonia in, 40
  nitrogen exchange in, 40–1, 42
Substrate
  lignin:nitrogen ratios, 229, *230*
  quality, decomposition and, 228–30, 232
Sulfate, in snowmelt, 83, *84*

Talus slopes, 50–2, *52, 53,* 60, 65
Tannins, 267–9, *268*
*Taraxacum officinale,* 143
Temperature. *See* Air temperature; Soil temperature
Terpenes, 268
*Thlaspi arvense,* 109
Toad, boreal (*Bufo boreas*), 130
*Toninia* spp., 112
Topographic position
  nutrient availability and, 187–8
  primary production and, 187–8
Topography
  effect on soils, 161–3
  nitrogen quantities in biomass and, 238
  wind and, 19
*Tortula norvegica,* 115
Tree islands, 159, 164, *165*
  movement, 165, 240
  soils under, 164
Treeline
  defined, 129
  early Holocene San Juan Mountains, 294
  mid-Holocene Front Range, 295
  in southern Rocky Mountains, 4
Trees. *See* Conifers

*Trifolium* spp., 122, 177
*Trifolium dasphyllum,* 103, 122
*Trifolium parryi,* 267
  in elk diet, 273
  in pika diets, 272, 273
Troposphere, inversion layer and, 34
T-van site, 33

*Vaccinium caespitosum,* 112
*Vaccinium scoparium,* 112
Vascular plants, in alpine tundra, 4–5
Vegetation, 99, 122–3
  classification, 103, *104–6*
  elevation gradient and, 100
  flora origins, 101–2
  habitats, 103, *104–6*
  paleoclimate equilibrium, 297–8
  physical controls, 99, *100,* 116
  treeline boundary, 100
  *See also* Plants
Vegetation communities. *See* Plant communities; *names of specific communities*
Vertebrates. *See* Fauna
Vole (*Microtus*), 137, 271–2
  food caches, 276
  foraging effect on plant succession, 277
  short growing season, coping strategy for, 276
Vole, long-tailed (*Microtus longicaudus*), 271–2
Vole, montane (*Microtus montanus*), 271–2
Vole, heather (*Phenacomys intermedius*), 137
Vole, southern red-backed (*Clethrionomys gapperi*), 142, 270

Water
  chemistry, 318
  loss by sublimation, 76
  movement through soils, 161
  nitrogen in, 161, 318
Water availability
  nitrogen immobilization and, 187
  photosynthesis and, 186
  primary production and, 185–6
  *See also* Soil moisture; Soil water
Water budget, 22, 26–8, *26, 27*
  long-term changes in, 28
Water quality
  atmospheric deposition and, 75–6
  of lakes, 75–6

Watersheds. *See names of specific catchment areas*
Weasel, short-tailed (*Mustela erminea*), *144*
Wet meadows, 24, 112–14, 162, 178
  aboveground production in, 181
  biomass production phenology, 178
  carbon dioxide flux patterns, 258–9, *259*
  *Carex* spp. dominance, 113, *113*
  *C. leptosepala* dominance, *113*, 178, 181
  elk grazing in, 114
  methane flux patterns, 255–6, *256*
  nitrogen immobilization in, 244, *245*
  nitrogen in, 238
  nitrogen mineralization in, 244, *245*
  nitrous oxide fluxes, 260–1, *261*
  plant phosphorus-use efficiencies, 203
  soil acidity, 255
  soil erosion, 181
Wind, 160–1
  aerosol transport and, 19–20
  in snow-free places, 116
  snow relocation by, 76
  topographic influences on, 19
  up-slope, 19–20
Wisconsin glaciation. *See* Pinedale glaciation
Wolf, gray, 130
Wolverine (*Gulo gulo*), 135
Worms, enchytraeid, 232
Wren, Rock, 134
Wyoming, nitrogen pools at alpine sites in, 239